Advances in
IMMUNOLOGY

VOLUME **112**

Regulatory T-Cells

ASSOCIATE EDITORS

Advances in IMMUNOLOGY

VOLUME 112

Regulatory T-Cells

Edited by

ALEXANDER RUDENSKY

Memorial Sloan-Kettering Cancer Center, New York, USA

SHIMON SAKAGUCHI

Osaka University, Osaka, Japan

ELSEVIER

AMSTERDAM • BOSTON • HEIDELBERG • LONDON
NEW YORK • OXFORD • PARIS • SAN DIEGO
SAN FRANCISCO • SINGAPORE • SYDNEY • TOKYO
Academic Press is an imprint of Elsevier

Academic Press is an imprint of Elsevier
525 B Street, Suite 1900, San Diego, CA 92101-4495, USA
225 Wyman Street, Waltham, MA 02451, USA
32 Jamestown Road, London, NW1 7BY, UK
Radarweg 29, PO Box 211, 1000 AE Amsterdam, The Netherlands

First edition 2011

ISBN: 978-0-12-387827-4
ISSN: 0065-2776

CONTENTS

CONTRIBUTORS

Numbers in parentheses indicate the pages on which the authors' contributions begin.

Jhoanne Lynne Bautista
Department of Internal Medicine, Division of Rheumatology, Washington University in St. Louis, St. Louis, Missouri, USA (*25*)

Carolin Daniel
Department of Cancer Immunology and AIDS, Laboratory of Lymphocyte Biology, Dana Farber Cancer Institute; Department of Pathology, Harvard Medical School, Boston, Massachusetts, USA (*177*)

Shohei Hori
Research Unit for Immune Homeostasis, RIKEN Research Center for Allergy and Immunology, Yokohama, Kanagawa, Japan (*1*)

Chyi-Song Hsieh
Department of Internal Medicine, Division of Rheumatology, Washington University in St. Louis, St. Louis, Missouri, USA (*25*)

Hyang-Mi Lee
Department of Internal Medicine, Division of Rheumatology, Washington University in St. Louis, St. Louis, Missouri, USA (*25*)

Rick M. Maizels
Institute of Immunology and Infection Research, University of Edinburgh, Edinburgh, United Kingdom (*73*)

Ethan M. Shevach
Laboratory of Immunology, National Institutes of Allergy and Infectious Diseases, National Institutes of Health, Bethesda, Maryland, USA (*137*)

Katherine A. Smith
Institute of Immunology and Infection Research, University of Edinburgh, Edinburgh, United Kingdom (*73*)

Harald von Boehmer
Department of Cancer Immunology and AIDS, Laboratory of Lymphocyte Biology, Dana Farber Cancer Institute, Boston; Harvard Faculty of Arts and Sciences, Harvard University, Cambridge, Massachusetts, USA (*177*)

CHAPTER 1

Stability of Regulatory T-cell Lineage

Shohei Hori

Contents

Abstract

Regulatory T (Treg) cells expressing the transcription factor Foxp3 constitute a unique T-cell lineage committed to suppressive functions and play a central role in maintaining self-tolerance and

Research Unit for Immune Homeostasis, RIKEN Research Center for Allergy and Immunology, Yokohama, Kanagawa, Japan

Advances in Immunology, Volume 112
ISSN 0065-2776, DOI: 10.1016/B978-0-12-387827-4.00001-2

immune homeostasis. While their differentiation state is remarkably stable in the face of various perturbations from the extracellular environment, recent studies have also revealed their adaptability to the changing environment; in response to extrinsic cues, Treg cells differentiate further into distinct substates to regulate different classes of immune responses effectively. In contrast, some other recent studies have challenged this notion of a committed Treg cell lineage and suggested that Treg cells might lose their identity and be reprogrammed to various effector helper T cells under certain circumstances, although this issue of environment-induced Treg cell reprogramming remains highly controversial. This review will focus on recent advances in our understanding of how the stability and adaptability of Treg cell lineage is regulated and how it might be perturbed in a changing environment.

1. INTRODUCTION

One of the central questions in immunology has been to understand how the adaptive immune system naturally acquires and maintains immunological tolerance to molecular components of "self." Since the proposition of the clonal selection theory by Burnet (1957) and its subsequent modification by Lederberg (1959), recessive tolerance, based on purging of autoreactive lymphocyte repertoires by clonal deletion and anergy, has been considered to be the central mechanism of self-tolerance. Although numerous studies have demonstrated operation of these cell-intrinsic mechanisms in the immune system (Kappler *et al.*, 1987; Kisielow *et al.*, 1988; Ramsdell *et al.*, 1989; Roberts *et al.*, 1990), however, their significance in naturally acquired self-tolerance still remains elusive (von Boehmer, 2009). Over the last two decades, evidence has accumulated that natural self-tolerance may be rather dominant, based on cell-extrinsic regulation of autoaggressive lymphocytes by other functional classes of lymphocytes (Coutinho *et al.*, 2005; Sakaguchi, 2004). While several T-cell subsets have been shown to be involved in such cell-extrinsic regulation and accordingly called regulatory T (Treg) cells, recent studies have established that a small subpopulation of $CD4^+$ T cells characterized by expression of the transcription factor Foxp3 represents essential mediators of dominant self-tolerance (Kim *et al.*, 2007; Lahl *et al.*, 2007).

While $Foxp3^+$ Treg cells were initially studied in models of self-tolerance and autoimmunity (Sakaguchi, 2004), it has become apparent that they are engaged in negative regulation of any forms of immune responses and probably nonimmune-mediated inflammation as well. They are capable of preventing immunopathology triggered by pathogens or allergens (Belkaid and Tarbell, 2009; Curotto de Lafaille *et al.*, 2010; Demengeot *et al.*, 2006), maintaining homeostasis with microbial flora (Belkaid and Tarbell, 2009; Izcue *et al.*, 2009), promoting therapeutic tolerance toward transplanted organs (Kendal and Waldmann, 2010), and sometimes helping tumor cells or certain pathogens escape from immune surveillance (Belkaid and Tarbell, 2009; Yamaguchi and Sakaguchi, 2006). Moreover, Treg cells are able to suppress different classes of immune responses driven by distinct effector helper T (Th) cell subsets including Th1, Th2, Th17, and follicular Th (Tfh) cells (Chaudhry *et al.*, 2009; Chung *et al.*, 2011; Fontenot *et al.*, 2003; Linterman *et al.*, 2011). Thus, the differentiated state of Treg cells is remarkably stable yet adaptable in the face of various unpredictable perturbations from the environment. These observations led to the notion that Treg cells represent a unique cell lineage committed to the function of immune suppression, which is thus distinct from $CD4^+$ helper T (Th) cell as well as $CD8^+$ cytotoxic T-cell lineages (Coutinho *et al.*, 2005). The stability and adaptability of the Treg cell lineage ensures the robustness of self-tolerance and immune homeostasis in a changing environment, and has provided the rationale for the ongoing and future clinical trials of human Treg cell therapy for various immune-mediated diseases (Edinger and Hoffmann, 2011; Riley *et al.*, 2009).

In recent years, however, increasing numbers of studies have challenged this view of a committed Treg cell lineage, and instead suggested that Treg cells might retain developmental plasticity to be reprogrammed into various effector Th cell subsets under some circumstances (Hori, 2010; Zhou *et al.*, 2009a). Although this emerging recognition of Treg cell reprogramming has raised fundamental questions as to the robustness of self-tolerance and immune homeostasis as well as the validity and safety of human Treg cell therapy, however, this issue remains highly controversial. Here, I will review recent advances in our understanding of how the stable differentiated state of Treg cells is maintained, how it is modified, and how it might be disrupted, in the face of diverse perturbations from the environment.

2. THE TREG CELL LINEAGE

2.1. Thymus-derived $CD25^+CD4^+$ Treg cells

The seminal work by Le Douarin and colleagues and that by Nishizuka and colleagues provided initial evidence for the existence of thymus-derived Treg cells mediating dominant self-tolerance. Using chick-quail embryonic chimeras and athymic mice, Le Douarin and colleagues showed that transplantation of pure xeno- or allogeneic thymic epithelium rudiment induces transplantation tolerance to grafted tissues of the donor-type (Le Douarin *et al.*, 1996; Ohki *et al.*, 1987; Salaun *et al.*, 1990).

This tolerance was independent of clonal deletion and anergy but mediated by donor-specific $CD4^+$ Treg cells capable of suppressing graft rejection (Modigliani *et al.*, 1995; Salaun *et al.*, 1990). Importantly, adoptive transfer experiments revealed that they are able to maintain their regulatory function for a long period of time even in the complete absence of donor antigens (Modigliani *et al.*, 1995). These findings led to the hypothesis that high-affinity/avidity TCR interactions with self-antigens presented by thymic epithelial cells "imprint" autoreactive T cells with regulatory functions, which are maintained even in environments where the original inductive signals are absent (Modigliani *et al.*, 1996).

Nishizuka and colleagues, on the other hand, studied organ-specific autoimmune diseases provoked in mice thymectomized around day 3 after birth, and found that adoptive transfer of thymocytes or splenocytes from adult euthymic mice into the thymectomized neonates prevents the disease development (Kojima *et al.*, 1976; Nishizuka and Sakakura, 1969). These observations suggested that a population of T cells capable of preventing autoimmunity is produced in the thymus after 3 days of life.

Efforts to distinguish these self-protective T cells from pathogenic effector T cells led to two landmark papers by Sakaguchi and colleagues, who identified a population of $CD4^+$ T cells expressing the IL-2 receptor α-chain (CD25) capable of preventing the autoimmune diseases induced in neonatally thymectomized mice and nude mice transferred with $CD25^-CD4^+$ T cells (Asano *et al.*, 1996; Sakaguchi *et al.*, 1995). They also identified a $CD25^+CD4^+CD8^-$ population in the thymus, which also displayed similar regulatory activity (Itoh *et al.*, 1999). By using TCR x antigen double transgenic systems, later studies have demonstrated that $CD25^+CD4^+$ Treg cells can differentiate in the thymus when the congnate antigen is presented by thymic stromal cells, providing compelling evidence for the prediction that high-affinity/avidity TCR interactions with self-antigens drive intrathymic commitment of the Treg cell lineage (Apostolou *et al.*, 2002; Jordan *et al.*, 2001; Kawahata *et al.*, 2002). Although initial *in vitro* studies suggested that $CD25^+CD4^+$ Treg cells are "anergic," that is, unable to proliferate and produce IL-2 upon TCR stimulation *in vitro* (Takahashi *et al.*, 1998; Thornton and Shevach, 1998), they proliferate robustly under steady state (Fisson *et al.*, 2003; Hori *et al.*, 2002), when challenged with specific antigen (Klein *et al.*, 2003), or when adoptively transferred into lymphopenic mice (Annacker *et al.*, 2000; Gavin *et al.*, 2002). Importantly, suppressive functions of $CD25^+CD4^+$ Treg cells are stably inherited to their progeny even after many rounds of cell division, suggesting that some epigenetic mechanisms keep their differentiated state stable (Gavin *et al.*, 2002; Klein *et al.*, 2003). These findings collectively enforced the notion that $CD25^+CD4^+$ Treg cells represent a stable cell lineage committed to suppressive functions.

2.2. Foxp3 and its essential role in Treg cells

Because CD25 is an activation marker, however, it was not entirely clear whether $CD25^+CD4^+$ Treg cells represent a unique T-cell lineage distinct from conventional $CD4^+$ T cells. The direct evidence that they adopt a unique differentiation pathway was provided when the *Foxp3* gene encoding a forkhead/winged-helix transcription factor was identified as the causative gene for the fatal multiorgan autoimmune and inflammatory disorder that develops in *scurfy* mutant mice (Brunkow *et al.*, 2001) and human patients with immune dysregulation, polyendocrinopathy, enteropathy, X-linked (IPEX) syndrome (Bennett *et al.*, 2001; Chatila *et al.*, 2000; Wildin *et al.*, 2001). Foxp3 mRNA and protein were soon found to be expressed predominantly in thymic as well as peripheral $CD25^+CD4^+$ Treg cells (Fontenot *et al.*, 2003, 2005c; Hori *et al.*, 2003; Khattri *et al.*, 2003; Wan and Flavell, 2005). Importantly, enforced expression of Foxp3 in $CD25^-CD4^+$ T cells using retroviral vectors conferred a Treg cell phenotype and function similar to, if not identical to, $CD25^+CD4^+$ Treg cells (Fontenot *et al.*, 2003; Hori *et al.*, 2003). Moreover, loss-of-function mutations of the *Foxp3* gene resulted in defective development of functional $CD25^+CD4^+$ Treg cells (Fontenot *et al.*, 2003; Khattri *et al.*, 2003), which was found to be the primary cause of the autoimmune disease that develops in *Foxp3* mutant mice (Fontenot *et al.*, 2003, 2005c; Kim *et al.*, 2009; Komatsu and Hori, 2007; Liston *et al.*, 2007).

These findings collectively led to the prevailing notion that Foxp3 represents a "specific" molecular marker of Treg cells and acts as their "lineage-specification factor" or "master regulator" (Fontenot and Rudensky, 2005; Hori and Sakaguchi, 2004). However, accumulating evidence has suggested that Foxp3 expression may not always specify Treg cells; in humans, activated T cells transiently upregulate FOXP3 without acquiring a Treg cell phenotype and function (Allan *et al.*, 2007; Gavin *et al.*, 2006; Miyara *et al.*, 2009; Pillai *et al.*, 2007; Tran *et al.*, 2007; Walker *et al.*, 2003; Wang *et al.*, 2007; Ziegler, 2006), although such "promiscuous" Foxp3 expression has not been demonstrated so far in mice (Fontenot *et al.*, 2003, 2005c; Hori *et al.*, 2003; Khattri *et al.*, 2003).

3. DIFFERENTIATION AND FUNCTIONAL ADAPTATION OF $FOXP3^+$ TREG CELLS

3.1. Intrathymic differentiation

Consistent with the notion of intrathymic commitment of Treg cells, Foxp3 induction was shown to take place during thymic differentiation, mostly at the $CD4^+CD8^-$ single-positive (CD4SP) stage, concomitantly with positive and/or negative selection (Fontenot *et al.*, 2005a,c). Furthermore, TCR repertoire analyses have revealed a substantial overlap in CDR3 sequences between Foxp3^{+} CD4SP thymocytes and peripheral Foxp3^{+}CD4^{+} T cells, supporting the idea that the majority of Treg cells originate from the thymus (Hsieh *et al.*, 2006; Pacholczyk *et al.*, 2006; Wong *et al.*, 2007). TCR repertoire analyses have also shown that Foxp3^{+} Treg cells and Foxp3^{-} conventional T cells display largely distinct CDR3 sequences, providing further evidence that commitment to the Treg cell lineage relies on TCR signals distinct from those required for positive selection of conventional T cells (Hsieh *et al.*, 2004; Pacholczyk *et al.*, 2006; Wong *et al.*, 2007). It has been disputed, however, to what extent Treg cell-derived TCRs display heightened avidity toward self-antigens (Hsieh *et al.*, 2004; Pacholczyk *et al.*, 2007). In addition to TCR signals, costimulatory signals through CD28 (Salomon *et al.*, 2000; Tai *et al.*, 2005) and common γ-chain cytokine signals, particularly IL-2 signals (Burchill *et al.*, 2007; Fontenot *et al.*, 2005b; Yao *et al.*, 2007), have been identified as important signaling pathways leading to thymic Foxp3 induction and Treg cell differentiation.

3.2. Extrathymic differentiation

While thymic differentiation of Foxp3^{+} Treg cells has been established, it has also become evident that they can be generated in the periphery from naïve CD4^{+} T cells upon lymphopenia-driven proliferation (Curotto de Lafaille *et al.*, 2004), "tolerogenic" antigen presentation *in vivo* (Apostolou and von Boehmer, 2004; Cobbold *et al.*, 2004; Kretschmer *et al.*, 2005; Mucida *et al.*, 2005) or activation in the presence of TGF-β *in vitro* (Chen *et al.*, 2003). These extrathymically generated Treg (or often called inducible Treg, iTreg) cells have been implicated in maintaining immune homeostasis with microbial flora and innocuous environmental antigens, and in impeding protective immune responses to tumor cells and some microbes (Curotto de Lafaille and Lafaille, 2009). However, their functional importance as well as their numeric contribution to peripheral Treg cell pool under steady state and various perturbed conditions remain a matter of active debate. In addition, full extent of differences and similarities between iTreg cells and intrathymically generated Treg cells is yet to be determined (Bilate and Lafaille, 2011; Curotto de Lafaille and Lafaille, 2009; Feuerer *et al.*, 2009; Josefowicz and Rudensky, 2009). With respect to phenotypic and functional stability, Foxp3 expression and suppressor function in TGF-β-induced iTreg cells has been shown to be unstable and readily lost when deprived of the inductive signals (Floess *et al.*, 2007; Selvaraj and Geiger, 2007). It remains unclear, however, whether this instability is a general feature of iTreg cells, because other studies have shown that iTreg cells generated *in vivo* exhibit stable Foxp3 expression (Nishio *et al.*, 2010; Polansky *et al.*, 2008).

3.3. Functional adaptation

Foxp3$^+$ Treg cells are capable of controlling different classes of immune responses driven by distinct types of effector T cells. Recent studies have demonstrated that Treg cells are able to differentiate further into distinct subsets or substates that are specialized in suppressing different effector Th cell types (Campbell and Koch, 2011; Feuerer *et al.*, 2009). Moreover, Treg cells do so by exploiting the transcription factor machinery utilized by the very Th cells they are regulating. For example, under Th1-polarized settings, a subset of Foxp3$^+$ Treg cells has been shown to acquire T-bet expression in response to IFN-γ signals to control Th1-mediated, but not Th2-mediated, inflammatory responses (Koch *et al.*, 2009). Two recent studies have also shown that, upon immunization with antigens, a subset of Foxp3$^+$ Treg cells acquire Tfh characteristics such as Bcl6 and CXCR5 expression, localize to B cell follicles, and regulate Tfh and germinal center responses in a Bcl6-dependent manner (Chung *et al.*, 2011; Linterman *et al.*, 2011). Likewise, IRF-4 or STAT-3 expression by Foxp3$^+$ T cells is required for selective regulation of Th2 or Th17 responses, respectively (Chaudhry *et al.*, 2009; Zheng *et al.*, 2009). Thus, although Treg cells constitute a unique T-cell lineage, they are capable of changing their phenotype in response to extrinsic cues to adapt to the immune environment for effective immune regulation.

4. REGULATION OF TREG CELL DIFFERENTIATION BY FOXP3

4.1. Role of Foxp3 in Treg cell differentiation

Although studies mentioned above established an indispensable role of Foxp3 in Treg cells, however, precise contribution made by Foxp3 to Treg cell differentiation and function remained unclear. In order to distinguish Foxp3-dependent phenotypic and functional features of Treg cells from Foxp3-independent features conferred by signals that precedes and induces Foxp3 expression, two groups independently generated "GFP-knock-in Foxp3-knock-out" mice in which the function of Foxp3 protein was inactivated but transcriptional activity at the locus could be monitored by in-frame insertions of GFP (Gavin *et al.*, 2007; Lin *et al.*, 2007). Analyses of these mice showed that the resultant GFP$^+$Foxp3^{null} T cells (or so-called "wannabe" or "would-be" Treg cells) shared many features with wild-type Foxp3$^+$ Treg cells including the anergic property and expression of a number of Treg signature genes such as increased expression of CD25 and CTLA4 and decreased expression of IL-2 and IL-7Rα. However, full acquisition of these features as well as suppressive effector function required functional Foxp3 protein. Similar results were also obtained from studies of mice harboring a mutant *Foxp3* allele that results in 10-fold reduction in Foxp3 protein expression (Wan and Flavell, 2007). These results indicate that the function of Foxp3 is to amplify and stabilize the molecular features of Treg cells that are conferred by other developmental cues such as TCR and cytokine signals (Gavin *et al.*, 2007).

This notion is further supported by Hill *et al.* who performed a meta-analysis of transcriptional profiles of natural Foxp3$^+$ Treg cells and other Foxp3$^+$ T cells induced by TGF-β or by retroviral transduction (Hill *et al.*, 2007). Their results showed that Foxp3 is responsible for only a fraction of the transcriptional signatures of Treg cells, while reinforcing the expression of another fraction of the signatures regulated by other signals such as TCR, IL-2, and TGF-β signals. Most importantly, they identified a cluster of genes that are coregulated with, but not activated or repressed by, Foxp3, predicting the existence of a higher level of regulation upstream of Foxp3 that would determine the expression of Foxp3 and its coregulated genes.

To determine the role of Foxp3 in differentiated mature Treg cells, Rudensky and colleagues ablated the *Foxp3* gene in peripheral Foxp3$^+$ Treg cells harboring a loxP-flanked *Foxp3* allele (Williams and Rudensky, 2007). The resultant "ex-Treg" cells not only lost Treg cell function but also acquired Th cell characteristics including the ability to produce effector cytokines, including IFN-γ, IL-4, IL-17, and IL-21. Moreover, upon transfer into lymphopenic mice, they caused severe inflammatory tissue lesions in multiple organs presumably because of their autoreactive TCR repertoire. Thus, these results indicate that continuous Foxp3 expression is required for the maintenance of Treg cell phenotype and for the repression of effector Th cell phenotypes. A recent study has shown that Foxp3 represses Th cell functions at least in part by repressing expression of SATB1, a genome organizer that regulates chromatin structure and gene expression (Beyer *et al.*, 2011). These findings have raised an important issue of whether Treg cells lose Foxp3 expression and be "reprogrammed" into Th cells under physiological or pathological conditions in response to extrinsic cues (see below).

4.2. Mechanisms of Foxp3 induction and maintenance

Considering the critical role played by Foxp3 in controlling Treg cell differentiation and function, it is certainly important to understand how Foxp3 expression is regulated in developing and differentiated Treg cells. Recent studies have identified many transcription factors and *cis*-regulatory elements that positively or negatively regulate Foxp3 expression in thymocytes and in peripheral T cells, and revealed that there are multiple, redundant pathways of Foxp3 induction (Josefowicz and Rudensky, 2009; Merkenschlager and von Boehmer, 2010). Furthermore, they showed that induction and maintenance of Foxp3 expression are two separable processes regulated by distinct *cis*-elements within the *Foxp3* locus (Huehn *et al.*, 2009; Zheng *et al.*, 2010).

Foxp3 transcription is driven by its promoter in response to TCR and IL-2 signaling through binding of NFAT, AP-1, and STAT-5 (Burchill *et al.*, 2007; Mantel *et al.*, 2006; Yao *et al.*, 2007). A recent study has suggested that the promoter activity is negatively regulated by an epigenetic mechanism, namely DNA methylation; the SUMO E3 ligase PIAS1 restricts Foxp3 induction in thymocytes and peripheral T cells by recruiting DNA methyltransferases and heterochromatin protein 1 to the promoter (Liu *et al.*, 2010). Besides the promoter, three evolutionally conserved noncoding sequence (CNS) elements have been identified within the *Foxp3* locus and shown to play important regulatory roles in Foxp3 expression (Zheng *et al.*, 2010). CNS3 acts as a pioneer element essential for induction of Foxp3 expression in thymocytes and peripheral T cells by recruiting c-Rel (Zheng *et al.*, 2010). CNS1 (or enhancer 1), on the other hand, acts as a "TGF-β sensor" through binding of SMAD3 and is indispensable for peripheral, but not thymic, Foxp3 induction (Tone *et al.*, 2008; Zheng *et al.*, 2010). In contrast, CNS2 or also known as Treg-specific demethylated region (TSDR) is dispensable for Foxp3 induction but essential for heritable maintenance of Foxp3 expression in proliferating Treg cells (Huehn *et al.*, 2009; Zheng *et al.*, 2010). An important clue to this notion was provided by the findings that a CpG island contained within TSDR is fully demethylated in natural Treg cells (Floess *et al.*, 2007; Kim and Leonard, 2007). In contrast, TGF-β-induced $Foxp3^+$ T cells as well as activated human $Foxp3^+$ T cells, both of which exhibit unstable Foxp3 expression, displayed no or only limited demethylation at this region (Baron *et al.*, 2007; Floess *et al.*, 2007). Moreover, inhibition of DNA methylation by using a pharmacologic inhibitor or by knock-down or conditional deletion of DNA methyltransferases 1 induced stable Foxp3 expression in a fraction of naïve $CD4^+$ T cells upon activation (Josefowicz *et al.*, 2009; Kim and Leonard, 2007; Polansky *et al.*, 2008). Conversely, germ-line deletion of TSDR resulted in loss of Foxp3 expression in natural Treg cells upon cell divisions. These findings have collectively established that the demethylated state of TSDR critically contributes to the stable Foxp3 expression in natural Treg cells (Zheng *et al.*, 2010).

How then does demethylated TSDR ensure stable *Foxp3* transcription? It has been suggested that Foxp3 itself stabilizes its own transcription through a positive feedback mechanism (Gavin *et al.*, 2007). Consistent with this idea, Zheng *et al.* recently reported that Foxp3 protein, in cooperation with Runx transcription factors, binds to demethylated

TSDR in natural Treg cells but does not bind to its methylated form, suggesting that demethylated TSDR recruits the Runx-Foxp3 transcription factor complex to stabilize its own transcription (Zheng *et al.*, 2010). Furthermore, TSDR contains CpG motifs that may serve as binding sites for CREB/ATF, NFκB, and Ets-1, which act downstream of TCR/CD28 signals (Kim and Leonard, 2007; Mouly *et al.*, 2010; Polansky *et al.*, 2010). Indeed, CREB/ATF and Ets-1 have been shown to bind to TSDR in a demethylation-dependent manner (Kim and Leonard, 2007; Polansky *et al.*, 2010). These results suggest that TCR signal-dependent binding of these transcription factors to demethylated TSDR may be required for the maintenance of Foxp3 expression in Treg cells. Importantly, "wannabe" Treg cells, which transcribe the *Foxp3* locus but do not express Foxp3 protein, showed demethylation of the TSDR, indicating that TSDR demethylation does not rely on Foxp3 protein expression (Zheng *et al.*, 2010). Thus, these findings suggest that stable Foxp3 expression in natural Treg cells is ensured by a two-step process, the Foxp3-independent TSDR demethylation followed by the Foxp3-dependent self-enforcement.

Collectively, these findings indicate that stability of the Treg cell lineage is established at least in part by these "epigenetic" mechanisms. This in turn implies that Foxp3 expression *per se* is not sufficient for commitment of the Treg cell lineage, since induced Foxp3 can be readily lost without engagement of these mechanisms. This is consistent with the notion that the Treg cell lineage is determined by a higher level of regulation upstream of Foxp3 (Hill *et al.*, 2007).

5. CONTROVERSIES OVER STABILITY VERSUS PLASTICITY OF THE TREG CELL LINEAGE

5.1. Evidence for plasticity of Foxp3$^+$ T cells

5.1.1. "ExFoxp3" Th cells

Although Treg cells have been considered to represent a stable and distinct cell lineage committed to suppressive functions, recent studies have challenged this notion and suggested that Foxp3$^+$ Treg cells may retain developmental plasticity and be "reprogrammed" into Th cells in response to certain environmental cues (Bailey-Bucktrout and Bluestone, 2011; O'Shea and Paul, 2010; Zhou *et al.*, 2009a). First suggestion was provided soon after Foxp3 reporter mice became available. When sorted Foxp3$^+$ T cells were stimulated *in vitro* in the presence of Th polarizing cytokines (e.g., IL-4 or IL-6), a fraction of them lose Foxp3 expression and acquire the ability to produce the corresponding Th cytokines (e.g., IL-4 or IL-17) (Komatsu *et al.*, 2009; Xu *et al.*, 2007; Yang *et al.*, 2008). Even in the absence of deliberate polarization, some Foxp3$^+$ T cells were reported to lose Foxp3 expression upon stimulation with prolonged TCR/CD28 signals (Gabrysova *et al.*, 2011) or with certain costimulatory signals (e.g., OX-40 or Tim-1) (Degauque *et al.*, 2008; Vu *et al.*, 2007). Similar *in vitro* observations have also been made in human FOXP3$^+$ T cells: conversion to a Th17 cell phenotype (Koenen *et al.*, 2008) and loss of Foxp3 expression by repetitive TCR stimulation (Hoffmann *et al.*, 2009) have been documented.

Later studies have demonstrated instability of Foxp3 expression *in vivo*. When highly purified peripheral Foxp3$^+$ T cells were transferred alone into T-cell-deficient mice and examined 4–5 weeks later, approximately 50% of the donor cells in the lymph nodes and spleen were found to be negative for expression of Foxp3 and other Treg cell surface markers and failed to exhibit suppressive activity *in vitro* (Duarte *et al.*, 2009; Komatsu *et al.*, 2009). Instead, many of such former Foxp3$^+$ T cells (or so-called "exFoxp3" T cells) produced Th cell-associated cytokines including IFN-γ, IL-2, and IL-17, while some others retained "memory" of previous Foxp3 expression and reexpressed Foxp3 upon *in vitro* stimulation or secondary transfer to lymphopenic mice (Komatsu *et al.*, 2009). The frequency of exFoxp3 T cells was further increased in gut-associated tissues of lymphopenic host mice, particularly in the presence of inflammation (Murai *et al.*, 2009; Tsuji *et al.*, 2009). Interestingly, in Peyer's patches of CD3$\varepsilon^{-/-}$ host mice, many exFoxp3 T cells were found within the B cell follicles, expressed a Tfh cell phenotype and efficiently induced germinal center reactions, thereby promoting IgA production in the intestine (Tsuji *et al.*, 2009).

More recently, genetic lineage-tracing (or fate mapping) approaches have been applied to evaluate stability of Foxp3 expression *in vivo* without relying on cell sorting and adoptive transfers. Zhou *et al.* crossed Foxp3 bacterial artificial chromosome (BAC) transgenic mice expressing a GFP-Cre fusion protein with *ROSA26*^{YFP} Cre-reporter mice (hereafter referred to as Foxp3-GFP-Cre BAC Tg mice) to distinguish Foxp3$^+$ T cells from those that have lost Foxp3 expression (Zhou *et al.*, 2009b). They found that 10–15% of YFP$^+$ cells were negative for Foxp3 as well as GFP expression, and these YFP$^+$GFP$^-$ cells displayed a regulatory, memory cell phenotype with the ability to produce pro-inflammatory cytokines, IFN-γ and IL-17. Moreover, when crossed with transgenic mice expressing diabetogenic BDC2.5 TCR, the frequency of GFP$^-$YFP$^+$ cells was increased in the inflamed pancreas and these cells as well as GFP$^-$YFP$^-$ cells induced pancreatic inflammation and autoimmune diabetes upon transfer into lymphopenic mice, suggesting conversion of autoreactive Treg cells into pathogenic Th cells.

5.1.2. "Hybrid" phenotypes

Under certain circumstances, some Foxp3$^+$ T cells acquire Th cell features while maintaining Foxp3 expression, so exhibiting "hybrid" phenotypes. When peripheral Foxp3$^+$ T cells were stimulated *in vitro* in the presence of dendritic cells activated via a fungal recognition receptor, some expressed IL-17 and RORγt together with Foxp3 (Osorio *et al.*, 2008). Interestingly, these cells were derived from preexisting $Foxp3^+ROR\gamma t^+$ cells but were not induced *de novo* from $Foxp3^+ROR\gamma t^-$ cells. Such $Foxp3^+ROR\gamma t^+IL\text{-}17^+$ cells have been identified *in vivo* in humans (Beriou *et al.*, 2009; Voo *et al.*, 2009) and in mice, particularly in the intestine (Zhou *et al.*, 2008). A recent study has also shown that some $Foxp3^+$ T cells produce IL-17 upon vaccination with CpG in an IL-6-dependent manner and help cross-priming of $CD8^+$ T cells (Baban *et al.*, 2009; Sharma *et al.*, 2009, 2010). Similarly, when stimulated under a Th1 polarization conditions *in vitro*, some $Foxp3^+$ T cells acquired T-bet and IFN-γ expression without losing Foxp3 expression (Wei *et al.*, 2009). *In vivo* evidence for the development of $Foxp3^+T\text{-}bet^+IFN\text{-}\gamma^+$ cells was reported in a lethal model of *Toxoplasma gondii* infection (Oldenhove *et al.*, 2009) and in a nonlethal model of neurotropic hepatitis virus infection (Zhao *et al.*, 2011). Moreover, recent studies have also shown increased frequencies of $Foxp3^+IFN\gamma^+$ cells in autoimmune patients suffering from multiple sclerosis (Dominguez-Villar *et al.*, 2011) or type 1 diabetes (McClymont *et al.*, 2011).

What are such "hybrid" $Foxp3^+$ T cells? Do they represent "reprogrammed" Treg cells? Whereas they retain *in vitro* suppressive activity in most cases, their production of pro-inflammatory cytokines has been proposed to contribute to antimicrobial immunity or pathological inflammation. However, this hypothesis has not been evaluated *in vivo* and it remains unclear whether their cytokine production is a cause or consequence of the inflammatory conditions. Moreover, $Foxp3^+ROR\gamma t^+$ or $Foxp3^+T\text{-}bet^+$ T cells produce smaller amounts of IL-17 or IFN-γ, respectively, at lower frequencies than their $Foxp3^-$ counterparts, raising a question as to their quantitative contribution to overall inflammatory responses. As discussed above, accumulating evidence indicates that Treg cells express and utilize canonical Th cell-associated transcription factors to regulate specific types of effector Th cells (Chaudhry *et al.*, 2009; Chung *et al.*, 2011; Koch *et al.*, 2009; Linterman *et al.*, 2011; Zheng *et al.*, 2009). Hence, it remains to be clarified whether the hybrid phenotypes of $Foxp3^+$ T cells indicate Treg cell reprogramming or a consequence of their functional adaptation to inflammatory environments.

5.1.3. Environmental cues

What are the environmental signals that drive the conversion of $Foxp3^+$ T cells to Th cells? As inflammatory conditions apparently promote it, pro-inflammatory cytokine signals likely play a role (Komatsu *et al.*, 2009; Xu *et al.*, 2007; Yang *et al.*, 2008). In addition, extracellular ATP released perhaps from damaged cells during inflammation may also contribute to the loss of Foxp3 expression and conversion to Th17 cells through activation of purinergic P2X receptors (Schenk *et al.*, 2011).

Conversely, anti-inflammatory signals such as IL-10, TGF-β, and retinoic acid may prevent conversion of Foxp3$^+$ T cells to Th cells. Foxp3$^+$ T cells that are unable to transmit IL-10 signals showed increased conversion to Foxp3$^-$ Th1 cells in the intestine of lymphopenic host mice, although the defective IL-10 signaling alone was not sufficient but the presence of inflammation was also required for this conversion (Murai *et al.*, 2009). Blockade of TGF-β signaling also leads to loss of Foxp3 expression in some Foxp3$^+$ T cells (Komatsu *et al.*, 2009; Marie *et al.*, 2005). Retinoic acid has been reported to cooperate with TGF-β to generate stable iTreg cells *in vitro* (Benson *et al.*, 2007; Coombes *et al.*, 2007; Mucida *et al.*, 2007; Sun *et al.*, 2007) and to prevent IL-6-mediated conversion of natural Foxp3$^+$ T cells into Th17 cells (Zhou *et al.*, 2010).

Extrinsic cues that lead to the development of exFoxp3 T cells in lymphopenic environments are less clear. Because the frequency of exFoxp3 T cells was reduced when Foxp3$^+$ T cells were cotransferred with Foxp3$^-$ T cells (Duarte *et al.*, 2009; Komatsu *et al.*, 2009), it has been suggested that paracrine factors produced from Foxp3$^-$ T cells such as IL-2 may prevent the development of exFoxp3 T cells (Duarte *et al.*, 2009). Indeed, IL-2 treatments of lymphopenic host mice of Foxp3$^+$ T cells resulted in a reduced frequency of exFoxp3 T cells (Duarte *et al.*, 2009) and IL-2-signals facilitated maintenance of Foxp3 expression in TGF-β-induced iTreg cells (Chen *et al.*, 2011). It should be pointed out, however, Foxp3$^-$ T cells as well as IL-2 promote expansion and/or survival of Foxp3$^+$ T cells (Komatsu *et al.*, 2009; Setoguchi *et al.*, 2005). Thus, it is possible that the reduced frequency of exFoxp3 T cells in IL-2-deficient environments may simply reflect impaired maintenance of Foxp3$^+$ T cells rather than increased conversion to exFoxp3 T cells (Hori, 2010, 2011).

5.2. Evidence against the plasticity phenomena

These observations collectively led to the notion that Foxp3$^+$ Treg cells are not irreversibly committed to Treg cell fate but can undergo reprogramming in response to certain extrinsic cues associated with lymphopenia or inflammation. The dogma-challenging nature of this notion has evoked great controversy, particularly because it cannot be easily reconciled with the robustness of Treg function. If Foxp3$^+$ Treg cells change their fate under inflammatory or lymphopenic environments, how then are they able to maintain self-tolerance and prevent or even cure immunopathology under such circumstances (Belkaid and Tarbell, 2009; Demengeot *et al.*, 2006; Mottet *et al.*, 2003)? Supposing that many Foxp3$^+$ Treg cells are selected based on their autoreactive TCR repertoire, reprogramming of Treg cells into effector Th cells would result in catastrophic consequences to the host.

Despite the evidence for environment-induced Treg cell reprogramming, some other recent studies have provided counter-evidence. First, Geiger and colleagues has shown that, in an experimental autoimmune encephalomyelitis model, myelin oligodendrocyte glycoprotein (MOG)-specific Foxp3$^+$ and Foxp3$^-$ T cells infiltrating into the central nervous system display distinct TCR CDR3 sequences and are therefore derived from distinct clones, suggesting that there is no or only limited interconversion between the two populations during autoimmune inflammation (Liu *et al.*, 2009). Second, Rudensky and colleagues have reported a genetic fate mapping study and shown that Foxp3$^+$ T cells are remarkably stable under steady state and various inflammatory conditions (Rubtsov *et al.*, 2010). Their strategy was different from the Foxp3-GFP-Cre BAC Tg system in that they knocked-in the DNA sequence coding for a GFP-Cre-ERT2 triple fusion protein into the endogenous *Foxp3* locus and crossed the knock-in mice with *ROSA26*^{YFP} mice (here termed *Foxp3*$^{GFP\text{-}Cre\text{-}ERT2}$ mice). In this inducible labeling system, Foxp3$^+$ T cells are labeled with YFP only when animals are treated with tamoxifen. When these mice were treated with tamoxifen for 3 days as adults and examined 2 weeks or 5 months later, less than 5% of YFP$^+$ cells were found to be negative for Foxp3 expression, indicating that their Foxp3 expression is stable under the steady state. Furthermore, the frequency of the Foxp3$^-$YFP$^+$ T cells was not increased when the animals were rendered lymphopenic by sublethal irradiation, challenged with *Listeria monocytogenes* infection, or injected with anti-CD40 antibodies that induce Th1-mediated inflammation. In addition, they have also shown that double-sorted, highly purified Foxp3$^+$ T cells expressing diabetogenic BDC2.5 or arthritogenic K/BxN TCRs do not give rise to exFoxp3 Th cells when transferred into prediabetic or prearthritic recipients, respectively. Although they observed an increase in the frequency of exFoxp3 cells when IL-2 was neutralized with specific antibodies, this was partly due to reduced maintenance of Foxp3$^+$YFP$^+$ cells. In addition, the resultant exFoxp3 T cells did not exhibit Th cell characteristics.

5.3. Toward reconciliation

What might be at the root of these apparently conflicting observations? As discussed recently, there are three possibilities (Hori, 2011). First, "exFoxp3" T cells might be an artifact resulting from infidelity of Foxp3 BAC-driven Cre expression or from outgrowth of a few Foxp3$^-$ contaminants in experiments relying on cell sorting of Foxp3$^+$ T cells (Rubtsov *et al.*, 2010). These possibilities are, however, unlikely. In previous studies, congenically marked Foxp3$^-$ T cells spiked into Foxp3$^+$ donor T cells failed to grow out in lymphopenic hosts. In addition, even when Foxp3$^+$ T cells were double sorted into high purity (>99.99%), we still observed the generation of exFoxp3 T cells in lymphopenic and in *in vitro* polarization settings (unpublished observations). Our unpublished results also demonstrated that, in *Foxp3*$^{GFP\text{-}Cre}$ knock-in x *ROSA26*^{RFP} mice, approximately 15% of peripheral RFP^{+}CD4^{+} T cells were Foxp3^{-}, a finding consistent with the Foxp3-GFP-Cre BAC Tg study. Second, the discrepancy might result from different environmental conditions tested in previous studies. Because the *Foxp3*$^{GFP\text{-}Cre\text{-}ERT2}$ study did not evaluate stability of Foxp3 expression under the conditions that were reported to generate exFoxp3 Th cells (i.e., Th polarization conditions, adoptive transfer into T cell-deficient mice), the observed stability might be conditional to the experimental settings tested in this study. Therefore, the results do not formally prove that Foxp3^{+} Treg cells are unable to change their fate in response to certain external perturbations. Finally, the discrepancy between the two fate mapping studies might reflect different time windows during which Foxp3^{+} T cells are labeled with YFP. Because labeling is continuous, Foxp3^{-}YFP^{+} cells found in adult Foxp3-GFP-Cre BAC Tg mice (Zhou *et al.*, 2009b) are an aggregate of cells that have lost Foxp3 expression during ontogeny and accumulated until adulthood. This in turn suggests that the enrichment of exFoxp3 T cells in certain settings might not indicate environment-induced Treg cell reprogramming but reflect environment-induced accumulation (selection) of exFoxp3 T cells that have developed during ontogeny independently of the selecting environment. In contrast, because Foxp3^{+} T cells were pulse-labeled only as adults in the *Foxp3*$^{GFP\text{-}Cre\text{-}ERT2}$ study, their findings do not exclude the possibility that exFoxp3 T cells might develop and accumulate during ontogeny but indicate that the majority (>95%) of adult peripheral Foxp3^{+} T cells exhibits stable Foxp3 expression (Rubtsov *et al.*, 2010).

To resolve the controversies over lineage stability versus plasticity of Treg cells, we have recently proposed that exFoxp3 Th cells do not reflect reprogramming of Treg cells but a minor population of uncommitted (or unprogrammed) Foxp3^{+} T cells (Hori, 2010, 2011). This "heterogeneity model" is based on our observations that the potential to convert to exFoxp3 Th cells in lymphopenic and *in vitro* polarization settings is restricted to a minor fraction of Foxp3^{+} T cells enriched within the CD25^{-} subset (Komatsu *et al.*, 2009). In contrast, most, if not all, of CD25^{+}Foxp3^{+} T cells exhibit resistance to the conversion under those conditions (Komatsu *et al.*, 2009). This model, which postulates that the majority of peripheral Foxp3^{+} T cells are irreversibly committed to the Treg cell fate, is fully consistent with the findings from *Foxp3*$^{GFP\text{-}Cre\text{-}ERT2}$ mice. On the other hand, this model is also compatible with the plasticity phenomena, as the enrichment of exFoxp3 T cells under lymphopenic or inflammatory conditions can be accounted for by conversion and selection (by preferential proliferation, survival, migration, and/or retention) of the minor uncommitted population, rather than by induced reprogramming of committed Treg cells. Indeed, exFoxp3 T cells were shown to proliferate more vigorously than Foxp3^{+} T cells in lymphopenic environments (Komatsu *et al.*, 2009) and under Th polarization conditions *in vitro* (unpublished observations).

This heterogeneity model also "makes sense" in light of the mechanisms of Treg cell lineage commitment and regulation of *Foxp3* gene expression (Hori, 2010, 2011). As discussed above, the Treg cell lineage commitment is not executed by Foxp3 but rather by a higher-order regulatory process that involves stabilization of Foxp3 expression by "epigenetic" mechanisms involving TSDR demethylation and self-enforcement of Foxp3 expression. This view thus allows for the presence of uncommitted state among Foxp3$^+$ T cells. The heterogeneity model remains to be fully substantiated, however, because the nature of such uncommitted Foxp3$^+$ T cells has not been elucidated and the possibility of environment-induced Treg cell reprogramming has not been ruled out.

6. CONCLUDING REMARKS

Stability and adaptability of Foxp3$^+$ Treg cells is certainly an important determinant of the robustness of self-tolerance and immune homeostasis in a changing environment. Whereas it has become apparent that at least some of Foxp3$^+$ T cells can lose Foxp3 expression and adopt Th cell functions in response to environmental perturbations, it has also become evident that Treg cells are capable of adapting to the environment to efficiently regulate diverse classes of immune responses. Understanding of how stability and adaptability of Treg cells is regulated by extrinsic and intrinsic mechanisms is not only one of the most outstanding questions in the field but also an essential prerequisite to establish valid as well as safe protocols of Treg cell therapy for a number of immune-mediated diseases.

ACKNOWLEDGMENTS

This work was supported by Grants-in-Aid for Scientific Research in Priority Areas (19059014) and for Young Scientists (A) (20689012) from the Ministry of Education, Culture, Sports, Science, and Technology of Japan and the Takeda Science Foundation. I thank Ruka Setoguchi for critical reading of the manuscript.

REFERENCES

Allan, S. E., Crome, S. Q., Crellin, N. K., Passerini, L., Steiner, T. S., Bacchetta, R., Roncarolo, M. G., and Levings, M. K. (2007). Activation-induced FOXP3 in human T effector cells does not suppress proliferation or cytokine production. *Int. Immunol.* **19,** 345.

Annacker, O., Burlen-Defranoux, O., Pimenta-Araujo, R., Cumano, A., and Bandeira, A. (2000). Regulatory CD4 T cells control the size of the peripheral activated/memory CD4 T cell compartment. *J. Immunol.* **164,** 3573.

Apostolou, I., and von Boehmer, H. (2004). In vivo instruction of suppressor commitment in naive T cells. *J. Exp. Med.* **199,** 1401.

Apostolou, I., Sarukhan, A., Klein, L., and von Boehmer, H. (2002). Origin of regulatory T cells with known specificity for antigen. *Nat. Immunol.* **3,** 756.

Asano, M., Toda, M., Sakaguchi, N., and Sakaguchi, S. (1996). Autoimmune disease as a consequence of developmental abnormality of a T cell subpopulation. *J. Exp. Med.* **184,** 387.

Baban, B., Chandler, P. R., Sharma, M. D., Pihkala, J., Koni, P. A., Munn, D. H., and Mellor, A. L. (2009). IDO activates regulatory T cells and blocks their conversion into Th17-like T cells. *J. Immunol.* **183,** 2475.

Bailey-Bucktrout, S. L., and Bluestone, J. A. (2011). Regulatory T cells: Stability revisited. *Trends Immunol.* **32,** 301.

Baron, U., Floess, S., Wieczorek, G., Baumann, K., Grutzkau, A., Dong, J., Thiel, A., Boeld, T. J., Hoffmann, P., Edinger, M., Turbachova, I., Hamann, A., *et al.* (2007). DNA demethylation in the human FOXP3 locus discriminates regulatory T cells from activated FOXP3(+) conventional T cells. *Eur. J. Immunol.* **37,** 2378.

Belkaid, Y., and Tarbell, K. (2009). Regulatory T cells in the control of host-microorganism interactions (*). *Annu. Rev. Immunol.* **27,** 551.

Bennett, C. L., Christie, J., Ramsdell, F., Brunkow, M. E., Ferguson, P. J., Whitesell, L., Kelly, T. E., Saulsbury, F. T., Chance, P. F., and Ochs, H. D. (2001). The immune dysregulation, polyendocrinopathy, enteropathy, X-linked syndrome (IPEX) is caused by mutations of FOXP3. *Nat. Genet.* **27,** 20.

Benson, M. J., Pino-Lagos, K., Rosemblatt, M., and Noelle, R. J. (2007). All-trans retinoic acid mediates enhanced T reg cell growth, differentiation, and gut homing in the face of high levels of co-stimulation. *J. Exp. Med.* **204,** 1765.

Beriou, G., Costantino, C. M., Ashley, C. W., Yang, L., Kuchroo, V. K., Baecher-Allan, C., and Hafler, D. A. (2009). IL-17-producing human peripheral regulatory T cells retain suppressive function. *Blood* **113,** 4240.

Beyer, M., Thabet, Y., Muller, R. U., Sadlon, T., Classen, S., Lahl, K., Basu, S., Zhou, X., Bailey-Bucktrout, S. L., Krebs, W., Schonfeld, E. A., Bottcher, J., *et al.* (2011). Repression of the genome organizer SATB1 in regulatory T cells is required for suppressive function and inhibition of effector differentiation. *Nat. Immunol.* **12,** 898.

Bilate, A. B., and Lafaille, J. J. (2011). It takes two to tango. *Immunity* **35,** 6.

Brunkow, M. E., Jeffery, E. W., Hjerrild, K. A., Paeper, B., Clark, L. B., Yasayko, S. A., Wilkinson, J. E., Galas, D., Ziegler, S. F., and Ramsdell, F. (2001). Disruption of a new forkhead/winged-helix protein, scurfin, results in the fatal lymphoproliferative disorder of the scurfy mouse. *Nat. Genet.* **27,** 68.

Burchill, M. A., Yang, J., Vogtenhuber, C., Blazar, B. R., and Farrar, M. A. (2007). IL-2 receptor beta-dependent STAT5 activation is required for the development of Foxp3+ regulatory T cells. *J. Immunol.* **178,** 280.

Burnet, F. M. (1957). A modification of Jerne's theory of antibody production using the concept of clonal selection. *Aust. J. Sci.* **20,** 67.

Campbell, D. J., and Koch, M. A. (2011). Phenotypical and functional specialization of FOXP3+ regulatory T cells. *Nat. Rev. Immunol.* **11,** 119.

Chatila, T. A., Blaeser, F., Ho, N., Lederman, H. M., Voulgaropoulos, C., Helms, C., and Bowcock, A. M. (2000). JM2, encoding a fork head-related protein, is mutated in X-linked autoimmunity-allergic disregulation syndrome. *J. Clin. Invest.* **106,** R75.

Chaudhry, A., Rudra, D., Treuting, P., Samstein, R. M., Liang, Y., Kas, A., and Rudensky, A. Y. (2009). CD4+ regulatory T cells control TH17 responses in a Stat3-dependent manner. *Science* **326,** 986.

Chen, W., Jin, W., Hardegen, N., Lei, K. J., Li, L., Marinos, N., McGrady, G., and Wahl, S. M. (2003). Conversion of peripheral CD4+CD25- naive T cells to CD4+CD25+ regulatory T cells by TGF-beta induction of transcription factor Foxp3. *J. Exp. Med.* **198,** 1875.

Chen, Q., Kim, Y. C., Laurence, A., Punkosdy, G. A., and Shevach, E. M. (2011). IL-2 controls the stability of Foxp3 expression in TGF-beta-induced Foxp3+ T cells in vivo. *J. Immunol.* **186,** 6329.

Chung, Y., Tanaka, S., Chu, F., Nurieva, R. I., Martinez, G. J., Rawal, S., Wang, Y. H., Lim, H., Reynolds, J. M., Zhou, X. H., Fan, H. M., Liu, Z. M., *et al.* (2011). Follicular regulatory T cells expressing Foxp3 and Bcl-6 suppress germinal center reactions. *Nat. Med.* **17,** 983.

Cobbold, S. P., Castejon, R., Adams, E., Zelenika, D., Graca, L., Humm, S., and Waldmann, H. (2004). Induction of foxP3+ regulatory T cells in the periphery of T cell receptor transgenic mice tolerized to transplants. *J. Immunol.* **172,** 6003.

Coombes, J. L., Siddiqui, K. R., Arancibia-Carcamo, C. V., Hall, J., Sun, C. M., Belkaid, Y., and Powrie, F. (2007). A functionally specialized population of mucosal CD103+ DCs induces Foxp3+ regulatory T cells via a TGF-beta and retinoic acid-dependent mechanism. *J. Exp. Med.* **204,** 1757.

Coutinho, A., Caramalho, I., Seixas, E., and Demengeot, J. (2005). Thymic commitment of regulatory T cells is a pathway of TCR-dependent selection that isolates repertoires undergoing positive or negative selection. *Curr. Top. Microbiol. Immunol.* **293,** 43.

Curotto de Lafaille, M. A., and Lafaille, J. J. (2009). Natural and adaptive foxp3+ regulatory T cells: More of the same or a division of labor? *Immunity* **30,** 626.

Curotto de Lafaille, M. A., Lino, A. C., Kutchukhidze, N., and Lafaille, J. J. (2004). CD25-T cells generate CD25+Foxp3+ regulatory T cells by peripheral expansion. *J. Immunol.* **173,** 7259.

Curotto de Lafaille, M. A., Lafaille, J. J., and Graca, L. (2010). Mechanisms of tolerance and allergic sensitization in the airways and the lungs. *Curr. Opin. Immunol.* **22,** 616.

Degauque, N., Mariat, C., Kenny, J., Zhang, D., Gao, W., Vu, M. D., Alexopoulos, S., Oukka, M., Umetsu, D. T., DeKruyff, R. H., Kuchroo, V., Zheng, X. X., *et al.* (2008). Immunostimulatory Tim-1-specific antibody deprograms Tregs and prevents transplant tolerance in mice. *J. Clin. Invest.* **118,** 735.

Demengeot, J., Zelenay, S., Moraes-Fontes, M. F., Caramalho, I., and Coutinho, A. (2006). Regulatory T cells in microbial infection. *Springer Semin. Immunopathol.* **28,** 41.

Dominguez-Villar, M., Baecher-Allan, C. M., and Hafler, D. A. (2011). Identification of T helper type 1-like, Foxp3+ regulatory T cells in human autoimmune disease. *Nat. Med.* **17,** 673.

Duarte, J. H., Zelenay, S., Bergman, M. L., Martins, A. C., and Demengeot, J. (2009). Natural Treg cells spontaneously differentiate into pathogenic helper cells in lymphopenic conditions. *Eur. J. Immunol.* **39,** 948.

Edinger, M., and Hoffmann, P. (2011). Regulatory T cells in stem cell transplantation: strategies and first clinical experiences. *Curr. Opin. Immunol.* **23,** 679.

Feuerer, M., Hill, J. A., Mathis, D., and Benoist, C. (2009). Foxp3+ regulatory T cells: Differentiation, specification, subphenotypes. *Nat. Immunol.* **10,** 689.

Fisson, S., Darrasse-Jeze, G., Litvinova, E., Septier, F., Klatzmann, D., Liblau, R., and Salomon, B. L. (2003). Continuous activation of autoreactive CD4+ CD25+ regulatory T cells in the steady state. *J. Exp. Med.* **198,** 737.

Floess, S., Freyer, J., Siewert, C., Baron, U., Olek, S., Polansky, J., Schlawe, K., Chang, H. D., Bopp, T., Schmitt, E., Klein-Hessling, S., Serfling, E., *et al.* (2007). Epigenetic control of the foxp3 locus in regulatory T cells. *PLoS Biol.* **5,** e38.

Fontenot, J. D., and Rudensky, A. Y. (2005). A well adapted regulatory contrivance: Regulatory T cell development and the forkhead family transcription factor Foxp3. *Nat. Immunol.* **6,** 331.

Fontenot, J. D., Gavin, M. A., and Rudensky, A. Y. (2003). Foxp3 programs the development and function of CD4+CD25+ regulatory T cells. *Nat. Immunol.* **4,** 330.

Fontenot, J. D., Dooley, J. L., Farr, A. G., and Rudensky, A. Y. (2005a). Developmental regulation of Foxp3 expression during ontogeny. *J. Exp. Med.* **202,** 901.

Fontenot, J. D., Rasmussen, J. P., Gavin, M. A., and Rudensky, A. Y. (2005b). A function for interleukin 2 in Foxp3-expressing regulatory T cells. *Nat. Immunol.* **6,** 1142.

Fontenot, J. D., Rasmussen, J. P., Williams, L. M., Dooley, J. L., Farr, A. G., and Rudensky, A. Y. (2005c). Regulatory T cell lineage specification by the forkhead transcription factor foxp3. *Immunity* **22,** 329.

Gabrysova, L., Christensen, J. R., Wu, X., Kissenpfennig, A., Malissen, B., and O'Garra, A. (2011). Integrated T-cell receptor and costimulatory signals determine TGF-beta-dependent differentiation and maintenance of Foxp3+ regulatory T cells. *Eur. J. Immunol.* **41,** 1242.

Gavin, M. A., Clarke, S. R., Negrou, E., Gallegos, A., and Rudensky, A. (2002). Homeostasis and anergy of CD4(+)CD25(+) suppressor T cells in vivo. *Nat. Immunol.* **3,** 33.

Gavin, M. A., Torgerson, T. R., Houston, E., DeRoos, P., Ho, W. Y., Stray-Pedersen, A., Ocheltree, E. L., Greenberg, P. D., Ochs, H. D., and Rudensky, A. Y. (2006). Single-cell analysis of normal and FOXP3-mutant human T cells: FOXP3 expression without regulatory T cell development. *Proc. Natl. Acad. Sci. USA* **103,** 6659.

Gavin, M. A., Rasmussen, J. P., Fontenot, J. D., Vasta, V., Manganiello, V. C., Beavo, J. A., and Rudensky, A. Y. (2007). Foxp3-dependent programme of regulatory T-cell differentiation. *Nature* **445,** 771.

Hill, J. A., Feuerer, M., Tash, K., Haxhinasto, S., Perez, J., Melamed, R., Mathis, D., and Benoist, C. (2007). Foxp3 transcription-factor-dependent and -independent regulation of the regulatory T cell transcriptional signature. *Immunity* **27,** 786.

Hoffmann, P., Boeld, T. J., Eder, R., Huehn, J., Floess, S., Wieczorek, G., Olek, S., Dietmaier, W., Andreesen, R., and Edinger, M. (2009). Loss of FOXP3 expression in natural human CD4+CD25+ regulatory T cells upon repetitive in vitro stimulation. *Eur. J. Immunol.* **39,** 1088.

Hori, S. (2010). Developmental plasticity of Foxp3+ regulatory T cells. *Curr. Opin. Immunol.* **22,** 575.

Hori, S. (2011). Regulatory T cell plasticity: Beyond the controversies. *Trends Immunol.* **32,** 295.

Hori, S., and Sakaguchi, S. (2004). Foxp3: A critical regulator of the development and function of regulatory T cells. *Microbes Infect.* **6,** 745.

Hori, S., Haury, M., Lafaille, J. J., Demengeot, J., and Coutinho, A. (2002). Peripheral expansion of thymus-derived regulatory cells in anti-myelin basic protein T cell receptor transgenic mice. *Eur. J. Immunol.* **32,** 3729.

Hori, S., Nomura, T., and Sakaguchi, S. (2003). Control of regulatory T cell development by the transcription factor Foxp3. *Science* **299,** 1057.

Hsieh, C. S., Liang, Y., Tyznik, A. J., Self, S. G., Liggitt, D., and Rudensky, A. Y. (2004). Recognition of the peripheral self by naturally arising CD25+ CD4+ T cell receptors. *Immunity* **21,** 267.

Hsieh, C. S., Zheng, Y., Liang, Y., Fontenot, J. D., and Rudensky, A. Y. (2006). An intersection between the self-reactive regulatory and nonregulatory T cell receptor repertoires. *Nat. Immunol.* **7,** 401.

Huehn, J., Polansky, J. K., and Hamann, A. (2009). Epigenetic control of FOXP3 expression: The key to a stable regulatory T-cell lineage? *Nat. Rev. Immunol.* **9,** 83.

Itoh, M., Takahashi, T., Sakaguchi, N., Kuniyasu, Y., Shimizu, J., Otsuka, F., and Sakaguchi, S. (1999). Thymus and autoimmunity: Production of CD25+CD4+ naturally anergic and suppressive T cells as a key function of the thymus in maintaining immunologic self-tolerance. *J. Immunol.* **162,** 5317.

Izcue, A., Coombes, J. L., and Powrie, F. (2009). Regulatory lymphocytes and intestinal inflammation. *Annu. Rev. Immunol.* **27,** 313.

Jordan, M. S., Boesteanu, A., Reed, A. J., Petrone, A. L., Holenbeck, A. E., Lerman, M. A., Naji, A., and Caton, A. J. (2001). Thymic selection of CD4+CD25+ regulatory T cells induced by an agonist self-peptide. *Nat. Immunol.* **2,** 301.

Josefowicz, S. Z., and Rudensky, A. (2009). Control of regulatory T cell lineage commitment and maintenance. *Immunity* **30,** 616.

Josefowicz, S. Z., Wilson, C. B., and Rudensky, A. Y. (2009). Cutting edge: TCR stimulation is sufficient for induction of Foxp3 expression in the absence of DNA methyltransferase 1. *J. Immunol.* **182,** 6648.

Kappler, J. W., Wade, T., White, J., Kushnir, E., Blackman, M., Bill, J., Roehm, N., and Marrack, P. (1987). A T cell receptor V beta segment that imparts reactivity to a class II major histocompatibility complex product. *Cell* **49,** 263.

Kawahata, K., Misaki, Y., Yamauchi, M., Tsunekawa, S., Setoguchi, K., Miyazaki, J., and Yamamoto, K. (2002). Generation of CD4(+)CD25(+) regulatory T cells from autoreactive T cells simultaneously with their negative selection in the thymus and from nonautoreactive T cells by endogenous TCR expression. *J. Immunol.* **168,** 4399.

Kendal, A. R., and Waldmann, H. (2010). Infectious tolerance: Therapeutic potential. *Curr. Opin. Immunol.* **22,** 560.

Khattri, R., Cox, T., Yasayko, S. A., and Ramsdell, F. (2003). An essential role for Scurfin in CD4+CD25+ T regulatory cells. *Nat. Immunol.* **4,** 337.

Kim, H. P., and Leonard, W. J. (2007). CREB/ATF-dependent T cell receptor-induced FoxP3 gene expression: A role for DNA methylation. *J. Exp. Med.* **204,** 1543.

Kim, J. M., Rasmussen, J. P., and Rudensky, A. Y. (2007). Regulatory T cells prevent catastrophic autoimmunity throughout the lifespan of mice. *Nat. Immunol.* **8,** 191.

Kim, J., Lahl, K., Hori, S., Loddenkemper, C., Chaudhry, A., deRoos, P., Rudensky, A., and Sparwasser, T. (2009). Cutting edge: Depletion of Foxp3+ cells leads to induction of autoimmunity by specific ablation of regulatory T cells in genetically targeted mice. *J. Immunol.* **183,** 7631.

Kisielow, P., Bluthmann, H., Staerz, U. D., Steinmetz, M., and von Boehmer, H. (1988). Tolerance in T-cell-receptor transgenic mice involves deletion of nonmature CD4+8+ thymocytes. *Nature* **333,** 742.

Klein, L., Khazaie, K., and von Boehmer, H. (2003). In vivo dynamics of antigen-specific regulatory T cells not predicted from behavior in vitro. *Proc. Natl. Acad. Sci. USA* **100,** 8886.

Koch, M. A., Tucker-Heard, G., Perdue, N. R., Killebrew, J. R., Urdahl, K. B., and Campbell, D. J. (2009). The transcription factor T-bet controls regulatory T cell homeostasis and function during type 1 inflammation. *Nat. Immunol.* **10,** 595.

Koenen, H. J., Smeets, R. L., Vink, P. M., van Rijssen, E., Boots, A. M., and Joosten, I. (2008). Human CD25highFoxp3pos regulatory T cells differentiate into IL-17-producing cells. *Blood* **112,** 2340.

Kojima, A., Tanaka-Kojima, Y., Sakakura, T., and Nishizuka, Y. (1976). Prevention of postthymectomy autoimmune thyroiditis in mice. *Lab. Invest.* **34,** 601.

Komatsu, N., and Hori, S. (2007). Full restoration of peripheral Foxp3+ regulatory T cell pool by radioresistant host cells in scurfy bone marrow chimeras. *Proc. Natl. Acad. Sci. USA* **104,** 8959.

Komatsu, N., Mariotti-Ferrandiz, M. E., Wang, Y., Malissen, B., Waldmann, H., and Hori, S. (2009). Heterogeneity of natural Foxp3+ T cells: A committed regulatory T-cell lineage and an uncommitted minor population retaining plasticity. *Proc. Natl. Acad. Sci. USA* **106,** 1903.

Kretschmer, K., Apostolou, I., Hawiger, D., Khazaie, K., Nussenzweig, M. C., and von Boehmer, H. (2005). Inducing and expanding regulatory T cell populations by foreign antigen. *Nat. Immunol.* **6,** 1219.

Lahl, K., Loddenkemper, C., Drouin, C., Freyer, J., Arnason, J., Eberl, G., Hamann, A., Wagner, H., Huehn, J., and Sparwasser, T. (2007). Selective depletion of Foxp3+ regulatory T cells induces a scurfy-like disease. *J. Exp. Med.* **204,** 57.

Le Douarin, N., Corbel, C., Bandeira, A., Thomas-Vaslin, V., Modigliani, Y., Coutinho, A., and Salaun, J. (1996). Evidence for a thymus-dependent form of tolerance that is not based on elimination or anergy of reactive T cells. *Immunol. Rev.* **149,** 35.

Lederberg, J. (1959). Genes and antibodies. *Science* **129,** 1649.

Lin, W., Haribhai, D., Relland, L. M., Truong, N., Carlson, M. R., Williams, C. B., and Chatila, T. A. (2007). Regulatory T cell development in the absence of functional Foxp3. *Nat. Immunol.* **8,** 359.

Linterman, M. A., Pierson, W., Lee, S. K., Kallies, A., Kawamoto, S., Rayner, T. F., Srivastava, M., Divekar, D. P., Beaton, L., Hogan, J. J., Fagarasan, S., Liston, A., *et al.* (2011). Foxp3(+) follicular regulatory T cells control the germinal center response. *Nat. Med.* **17,** 975.

Liston, A., Farr, A. G., Chen, Z., Benoist, C., Mathis, D., Manley, N. R., and Rudensky, A. Y. (2007). Lack of Foxp3 function and expression in the thymic epithelium. *J. Exp. Med.* **204,** 475.

Liu, X., Nguyen, P., Liu, W., Cheng, C., Steeves, M., Obenauer, J. C., Ma, J., and Geiger, T. L. (2009). T cell receptor CDR3 sequence but not recognition characteristics distinguish autoreactive effector and Foxp3(+) regulatory T cells. *Immunity* **31,** 909.

Liu, B., Tahk, S., Yee, K. M., Fan, G., and Shuai, K. (2010). The ligase PIAS1 restricts natural regulatory T cell differentiation by epigenetic repression. *Science* **330,** 521.

Mantel, P. Y., Ouaked, N., Ruckert, B., Karagiannidis, C., Welz, R., Blaser, K., and Schmidt-Weber, C. B. (2006). Molecular mechanisms underlying FOXP3 induction in human T cells. *J. Immunol.* **176,** 3593.

Marie, J. C., Letterio, J. J., Gavin, M., and Rudensky, A. Y. (2005). TGF-beta1 maintains suppressor function and Foxp3 expression in CD4+CD25+ regulatory T cells. *J. Exp. Med.* **201,** 1061.

McClymont, S. A., Putnam, A. L., Lee, M. R., Esensten, J. H., Liu, W., Hulme, M. A., Hoffmuller, U., Baron, U., Olek, S., Bluestone, J. A., and Brusko, T. M. (2011). Plasticity of human regulatory T cells in healthy subjects and patients with type 1 diabetes. *J. Immunol.* **186,** 3918.

Merkenschlager, M., and von Boehmer, H. (2010). PI3 kinase signalling blocks Foxp3 expression by sequestering Foxo factors. *J. Exp. Med.* **207,** 1347.

Miyara, M., Yoshioka, Y., Kitoh, A., Shima, T., Wing, K., Niwa, A., Parizot, C., Taflin, C., Heike, T., Valeyre, D., Mathian, A., Nakahata, T., *et al.* (2009). Functional delineation and differentiation dynamics of human CD4+ T cells expressing the FoxP3 transcription factor. *Immunity* **30,** 899.

Modigliani, Y., Thomas-Vaslin, V., Bandeira, A., Coltey, M., Le Douarin, N. M., Coutinho, A., and Salaun, J. (1995). Lymphocytes selected in allogeneic thymic epithelium mediate dominant tolerance toward tissue grafts of the thymic epithelium haplotype. *Proc. Natl. Acad. Sci. USA* **92,** 7555.

Modigliani, Y., Bandeira, A., and Coutinho, A. (1996). A model for developmentally acquired thymus-dependent tolerance to central and peripheral antigens. *Immunol. Rev.* **149,** 155.

Mottet, C., Uhlig, H. H., and Powrie, F. (2003). Cutting edge: Cure of colitis by CD4+CD25+ regulatory T cells. *J. Immunol.* **170,** 3939.

Mouly, E., Chemin, K., Nguyen, H. V., Chopin, M., Mesnard, L., Leite-de-Moraes, M., Burlen-defranoux, O., Bandeira, A., and Bories, J. C. (2010). The Ets-1 transcription factor controls the development and function of natural regulatory T cells. *J. Exp. Med.* **207,** 2113.

Mucida, D., Kutchukhidze, N., Erazo, A., Russo, M., Lafaille, J. J., and Curotto de Lafaille, M. A. (2005). Oral tolerance in the absence of naturally occurring Tregs. *J. Clin. Invest.* **115,** 1923.

Mucida, D., Park, Y., Kim, G., Turovskaya, O., Scott, I., Kronenberg, M., and Cheroutre, H. (2007). Reciprocal TH17 and regulatory T cell differentiation mediated by retinoic acid. *Science* **317,** 256.

Murai, M., Turovskaya, O., Kim, G., Madan, R., Karp, C. L., Cheroutre, H., and Kronenberg, M. (2009). Interleukin 10 acts on regulatory T cells to maintain expression of the transcription factor Foxp3 and suppressive function in mice with colitis. *Nat. Immunol.* **10,** 1178.

Nishio, J., Feuerer, M., Wong, J., Mathis, D., and Benoist, C. (2010). Anti-CD3 therapy permits regulatory T cells to surmount T cell receptor-specified peripheral niche constraints. *J. Exp. Med.* **207,** 1879.

Nishizuka, Y., and Sakakura, T. (1969). Thymus and reproduction: Sex-linked dysgenesia of the gonad after neonatal thymectomy in mice. *Science* **166,** 753.

Ohki, H., Martin, C., Corbel, C., Coltey, M., and Le Douarin, N. M. (1987). Tolerance induced by thymic epithelial grafts in birds. *Science* **237,** 1032.

Oldenhove, G., Bouladoux, N., Wohlfert, E. A., Hall, J. A., Chou, D., Dos Santos, L., O'Brien, S., Blank, R., Lamb, E., Natarajan, S., Kastenmayer, R., Hunter, C., *et al.* (2009). Decrease of Foxp3+ Treg cell number and acquisition of effector cell phenotype during lethal infection. *Immunity* **31,** 772.

O'Shea, J. J., and Paul, W. E. (2010). Mechanisms underlying lineage commitment and plasticity of helper CD4+ T cells. *Science* **327,** 1098.

Osorio, F., LeibundGut-Landmann, S., Lochner, M., Lahl, K., Sparwasser, T., Eberl, G., and Reis e Sousa, C. (2008). DC activated via dectin-1 convert Treg into IL-17 producers. *Eur. J. Immunol.* **38,** 3274.

Pacholczyk, R., Ignatowicz, H., Kraj, P., and Ignatowicz, L. (2006). Origin and T cell receptor diversity of Foxp3+CD4+CD25+ T cells. *Immunity* **25,** 249.

Pacholczyk, R., Kern, J., Singh, N., Iwashima, M., Kraj, P., and Ignatowicz, L. (2007). Nonself-antigens are the cognate specificities of Foxp3+ regulatory T cells. *Immunity* **27,** 493.

Pillai, V., Ortega, S. B., Wang, C. K., and Karandikar, N. J. (2007). Transient regulatory T-cells: A state attained by all activated human T-cells. *Clin. Immunol.* **123,** 18.

Polansky, J. K., Kretschmer, K., Freyer, J., Floess, S., Garbe, A., Baron, U., Olek, S., Hamann, A., von Boehmer, H., and Huehn, J. (2008). DNA methylation controls Foxp3 gene expression. *Eur. J. Immunol.* **38,** 1654.

Polansky, J. K., Schreiber, L., Thelemann, C., Ludwig, L., Kruger, M., Baumgrass, R., Cording, S., Floess, S., Hamann, A., and Huehn, J. (2010). Methylation matters: Binding of Ets-1 to the demethylated Foxp3 gene contributes to the stabilization of Foxp3 expression in regulatory T cells. *J. Mol. Med. (Berl)* **88,** 1029.

Ramsdell, F., Lantz, T., and Fowlkes, B. J. (1989). A nondeletional mechanism of thymic self tolerance. *Science* **246,** 1038.

Riley, J. L., June, C. H., and Blazar, B. R. (2009). Human T regulatory cell therapy: Take a billion or so and call me in the morning. *Immunity* **30,** 656.

Roberts, J. L., Sharrow, S. O., and Singer, A. (1990). Clonal deletion and clonal anergy in the thymus induced by cellular elements with different radiation sensitivities. *J. Exp. Med.* **171,** 935.

Rubtsov, Y. P., Niec, R. E., Josefowicz, S., Li, L., Darce, J., Mathis, D., Benoist, C., and Rudensky, A. Y. (2010). Stability of the regulatory T cell lineage in vivo. *Science* **329,** 1667.

Sakaguchi, S. (2004). Naturally arising CD4+ regulatory t cells for immunologic self-tolerance and negative control of immune responses. *Annu. Rev. Immunol.* **22,** 531.

Sakaguchi, S., Sakaguchi, N., Asano, M., Itoh, M., and Toda, M. (1995). Immunologic self-tolerance maintained by activated T cells expressing IL-2 receptor alpha-chains (CD25). Breakdown of a single mechanism of self-tolerance causes various autoimmune diseases. *J. Immunol.* **155,** 1151.

Salaun, J., Bandeira, A., Khazaal, I., Calman, F., Coltey, M., Coutinho, A., and Le Douarin, N. M. (1990). Thymic epithelium tolerizes for histocompatibility antigens. *Science* **247,** 1471.

Salomon, B., Lenschow, D. J., Rhee, L., Ashourian, N., Singh, B., Sharpe, A., and Bluestone, J. A. (2000). B7/CD28 costimulation is essential for the homeostasis of the CD4+CD25+ immunoregulatory T cells that control autoimmune diabetes. *Immunity* **12,** 431.

Schenk, U., Frascoli, M., Proietti, M., Geffers, R., Traggiai, E., Buer, J., Ricordi, C., Westendorf, A. M., and Grassi, F. (2011). ATP inhibits the generation and function of regulatory T cells through the activation of purinergic P2X receptors. *Sci. Signal* **4,** ra12.

Selvaraj, R. K., and Geiger, T. L. (2007). A kinetic and dynamic analysis of Foxp3 induced in T cells by TGF-beta. *J. Immunol.* **178,** 7667.

Setoguchi, R., Hori, S., Takahashi, T., and Sakaguchi, S. (2005). Homeostatic maintenance of natural Foxp3(+) CD25(+) CD4(+) regulatory T cells by interleukin (IL)-2 and induction of autoimmune disease by IL-2 neutralization. *J. Exp. Med.* **201,** 723.

Sharma, M. D., Hou, D. Y., Liu, Y., Koni, P. A., Metz, R., Chandler, P., Mellor, A. L., He, Y., and Munn, D. H. (2009). Indoleamine 2,3-dioxygenase controls conversion of Foxp3+ Tregs to TH17-like cells in tumor-draining lymph nodes. *Blood* **113,** 6102.

Sharma, M. D., Hou, D. Y., Baban, B., Koni, P. A., He, Y., Chandler, P. R., Blazar, B. R., Mellor, A. L., and Munn, D. H. (2010). Reprogrammed foxp3(+) regulatory T cells provide essential help to support cross-presentation and CD8(+) T cell priming in naive mice. *Immunity* **33,** 942.

Sun, C. M., Hall, J. A., Blank, R. B., Bouladoux, N., Oukka, M., Mora, J. R., and Belkaid, Y. (2007). Small intestine lamina propria dendritic cells promote de novo generation of Foxp3 T reg cells via retinoic acid. *J. Exp. Med.* **204,** 1775.

Tai, X., Cowan, M., Feigenbaum, L., and Singer, A. (2005). CD28 costimulation of developing thymocytes induces Foxp3 expression and regulatory T cell differentiation independently of interleukin 2. *Nat. Immunol.* **6,** 152.

Takahashi, T., Kuniyasu, Y., Toda, M., Sakaguchi, N., Itoh, M., Iwata, M., Shimizu, J., and Sakaguchi, S. (1998). Immunologic self-tolerance maintained by CD25+CD4+ naturally anergic and suppressive T cells: Induction of autoimmune disease by breaking their anergic/suppressive state. *Int. Immunol.* **10,** 1969.

Thornton, A. M., and Shevach, E. M. (1998). CD4+CD25+ immunoregulatory T cells suppress polyclonal T cell activation in vitro by inhibiting interleukin 2 production. *J. Exp. Med.* **188,** 287.

Tone, Y., Furuuchi, K., Kojima, Y., Tykocinski, M. L., Greene, M. I., and Tone, M. (2008). Smad3 and NFAT cooperate to induce Foxp3 expression through its enhancer. *Nat. Immunol.* **9,** 194.

Tran, D. Q., Ramsey, H., and Shevach, E. M. (2007). Induction of FOXP3 expression in naive human CD4+FOXP3 T cells by T-cell receptor stimulation is transforming growth factor-beta dependent but does not confer a regulatory phenotype. *Blood* **110,** 2983.

Tsuji, M., Komatsu, N., Kawamoto, S., Suzuki, K., Kanagawa, O., Honjo, T., Hori, S., and Fagarasan, S. (2009). Preferential generation of follicular B helper T cells from Foxp3+ T cells in gut Peyer's patches. *Science* **323,** 1488.

von Boehmer, H. (2009). Central tolerance: Essential for preventing autoimmune disease? *Eur. J. Immunol.* **39,** 2313.

Voo, K. S., Wang, Y. H., Santori, F. R., Boggiano, C., Arima, K., Bover, L., Hanabuchi, S., Khalili, J., Marinova, E., Zheng, B., Littman, D. R., and Liu, Y. J. (2009). Identification of IL-17-producing FOXP3+ regulatory T cells in humans. *Proc. Natl. Acad. Sci. USA* **106,** 4793.

Vu, M. D., Xiao, X., Gao, W., Degauque, N., Chen, M., Kroemer, A., Killeen, N., Ishii, N., and Chang Li, X. (2007). OX40 costimulation turns off Foxp3+ Tregs. *Blood* **110,** 2501.

Walker, M. R., Kasprowicz, D. J., Gersuk, V. H., Benard, A., Van Landeghen, M., Buckner, J. H., and Ziegler, S. F. (2003). Induction of FoxP3 and acquisition of T regulatory activity by stimulated human CD4+CD25- T cells. *J. Clin. Invest.* **112,** 1437.

Wan, Y. Y., and Flavell, R. A. (2005). Identifying Foxp3-expressing suppressor T cells with a bicistronic reporter. *Proc. Natl. Acad. Sci. USA* **102,** 5126.

Wan, Y. Y., and Flavell, R. A. (2007). Regulatory T-cell functions are subverted and converted owing to attenuated Foxp3 expression. *Nature* **445,** 766.

Wang, J., Ioan-Facsinay, A., van der Voort, E. I., Huizinga, T. W., and Toes, R. E. (2007). Transient expression of FOXP3 in human activated nonregulatory CD4+ T cells. *Eur. J. Immunol.* **37,** 129.

Wei, G., Wei, L., Zhu, J., Zang, C., Hu-Li, J., Yao, Z., Cui, K., Kanno, Y., Roh, T. Y., Watford, W. T., Schones, D. E., Peng, W., *et al.* (2009). Global mapping of H3K4me3 and H3K27me3 reveals specificity and plasticity in lineage fate determination of differentiating CD4+ T cells. *Immunity* **30,** 155.

Wildin, R. S., Ramsdell, F., Peake, J., Faravelli, F., Casanova, J. L., Buist, N., Levy-Lahad, E., Mazzella, M., Goulet, O., Perroni, L., Bricarelli, F. D., Byrne, G., *et al.* (2001). X-linked neonatal diabetes mellitus, enteropathy and endocrinopathy syndrome is the human equivalent of mouse scurfy. *Nat. Genet.* **27,** 18.

Williams, L. M., and Rudensky, A. Y. (2007). Maintenance of the Foxp3-dependent developmental program in mature regulatory T cells requires continued expression of Foxp3. *Nat. Immunol.* **8,** 277.

Wong, J., Mathis, D., and Benoist, C. (2007). TCR-based lineage tracing: No evidence for conversion of conventional into regulatory T cells in response to a natural self-antigen in pancreatic islets. *J. Exp. Med.* **204,** 2039.

Xu, L., Kitani, A., Fuss, I., and Strober, W. (2007). Cutting edge: Regulatory T cells induce CD4+CD25-Foxp3- T cells or are self-induced to become Th17 cells in the absence of exogenous TGF-beta. *J. Immunol.* **178,** 6725.

Yamaguchi, T., and Sakaguchi, S. (2006). Regulatory T cells in immune surveillance and treatment of cancer. *Semin. Cancer Biol.* **16,** 115.

Yang, X. O., Nurieva, R., Martinez, G. J., Kang, H. S., Chung, Y., Pappu, B. P., Shah, B., Chang, S. H., Schluns, K. S., Watowich, S. S., Feng, X. H., Jetten, A. M., *et al.* (2008). Molecular antagonism and plasticity of regulatory and inflammatory T cell programs. *Immunity* **29,** 44.

Yao, Z., Kanno, Y., Kerenyi, M., Stephens, G., Durant, L., Watford, W. T., Laurence, A., Robinson, G. W., Shevach, E. M., Moriggl, R., Hennighausen, L., Wu, C., *et al.* (2007). Nonredundant roles for Stat5a/b in directly regulating Foxp3. *Blood* **109,** 4368.

Zhao, J., Fett, C., Trandem, K., Fleming, E., and Perlman, S. (2011). IFN-{gamma}- and IL-10-expressing virus epitope-specific Foxp3+ T reg cells in the central nervous system during encephalomyelitis. *J. Exp. Med.* **208,** 1571.

Zheng, Y., Chaudhry, A., Kas, A., deRoos, P., Kim, J. M., Chu, T. T., Corcoran, L., Treuting, P., Klein, U., and Rudensky, A. Y. (2009). Regulatory T-cell suppressor program co-opts transcription factor IRF4 to control T(H)2 responses. *Nature* **458,** 351.

Zheng, Y., Josefowicz, S., Chaudhry, A., Peng, X. P., Forbush, K., and Rudensky, A. Y. (2010). Role of conserved non-coding DNA elements in the Foxp3 gene in regulatory T-cell fate. *Nature* **463,** 808.

Zhou, L., Lopes, J. E., Chong, M. M., Ivanov, I. I., Min, R., Victora, G. D., Shen, Y., Du, J., Rubtsov, Y. P., Rudensky, A. Y., Ziegler, S. F., and Littman, D. R. (2008). TGF-beta-induced Foxp3 inhibits T(H)17 cell differentiation by antagonizing RORgammat function. *Nature* **453,** 236.

Zhou, X., Bailey-Bucktrout, S., Jeker, L. T., and Bluestone, J. A. (2009a). Plasticity of CD4(+) FoxP3(+) T cells. *Curr. Opin. Immunol.* **21,** 281.

Zhou, X., Bailey-Bucktrout, S. L., Jeker, L. T., Penaranda, C., Martinez-Llordella, M., Ashby, M., Nakayama, M., Rosenthal, W., and Bluestone, J. A. (2009b). Instability of the transcription factor Foxp3 leads to the generation of pathogenic memory T cells in vivo. *Nat. Immunol.* **10,** 1000.

Zhou, X., Kong, N., Wang, J., Fan, H., Zou, H., Horwitz, D., Brand, D., Liu, Z., and Zheng, S. G. (2010). Cutting edge: All-trans retinoic acid sustains the stability and function of natural regulatory T cells in an inflammatory milieu. *J. Immunol.* **185,** 2675.

Ziegler, S. F. (2006). FOXP3: Of mice and men. *Annu. Rev. Immunol.* **24,** 209.

CHAPTER 2

Thymic and Peripheral Differentiation of Regulatory T Cells

Hyang-Mi Lee, Jhoanne Lynne Bautista, and **Chyi-Song Hsieh**

Contents

Abstract

The development of regulatory T (Treg) cells is essential for the maintenance of immune tolerance and homeostasis. Here, we review recent studies that have advanced our understanding of Treg cell differentiation. In the thymus, TCR specificity to self-antigen appears to be a primary determinant for Treg cell lineage commitment, with c-Rel being an important factor that links T cell receptor (TCR) engagement and Foxp3 expression, along with cytokines and costimulatory molecules. It is also clear that postthymic

Department of Internal Medicine, Division of Rheumatology, Washington University in St. Louis, St. Louis, Missouri, USA

Advances in Immunology, Volume 112
ISSN 0065-2776, DOI: 10.1016/B978-0-12-387827-4.00002-4

events shape the peripheral Treg cell population. This includes preferential maintenance of Treg cells specific to self-antigens presented in the periphery, as well as the *de novo* generation of Treg cells from conventional $Foxp3^-$ T cells. The process of peripheral Treg cell differentiation shares some features with thymic Treg cell development, but there are notable differences. Together, thymic and peripheral Treg cell differentiation appear to generate an "imprint" of both self- and foreign antigens in the peripheral Treg cell population to provide dominant tolerance.

1. INTRODUCTION

It has become well established that the naturally occurring regulatory subset (Treg) of $CD4^+$ T cells plays an important role in maintaining immune tolerance (Barnes and Powrie, 2009; Josefowicz and Rudensky, 2009; Wing and Sakaguchi, 2010). This is clearly illustrated by the observation that mutations in the forkhead transcription factor Foxp3, important for Treg cell function and development, result in lethal multiorgan autoimmunity. In humans, children are afflicted by the IPEX syndrome (immunodysregulation polyendocrinopathy and enteropathy) (d'Hennezel *et al.*, 2009), whereas the murine equivalent is termed scurfy (Appleby and Ramsdell, 2008; Ziegler, 2006). Moreover, the acute depletion of $Foxp3^+$ Treg cell from normal mice unleashes a rapid diffuse lymphoproliferative syndrome that leads to death within 2 weeks (Kim *et al.*, 2007; Lahl *et al.*, 2007). Treg cells have also been shown to prevent excessive immunopathology to pathogens and limit immune responses to commensal microbiota and tumor cells (Belkaid and Tarbell, 2009; Weaver and Hatton, 2009; Yamaguchi and Sakaguchi, 2006). In order to understand how Treg cells impact these multiple facets of immunity, it is important to determine how Treg cells are generated. Here, we review the current understanding of the cellular and molecular mechanisms that control Treg cell differentiation.

2. THYMIC DEVELOPMENT OF TREG CELLS

2.1. Introduction

2.1.1. Brief history

In order to recognize a wide variety of pathogens, the adaptive immune system generates an enormously diverse repertoire of antigen receptors by somatic gene rearrangement. One consequence of this diversity, however, is that the repertoire includes receptors that recognize self-antigen and have the potential to cause autoimmunity. For T cells, the immune system utilizes an intricate system of development in a specialized organ, the thymus, to educate the immature T cell population to self prior to their release into the periphery. One mechanism of education is the elimination of immature self-reactive cells, a process also known as negative selection (McCaughtry and Hogquist, 2008; Palmer and Naeher, 2009; Siggs *et al.*, 2006). However, it has become clear that this culling is not sufficient to achieve tolerance, and some potentially pathogenic cells escape into the periphery (Bouneaud *et al.*, 2000; Zehn and Bevan, 2006).

The notion that a specialized regulatory T cell population is required to inhibit the cells that escape negative selection was suggested by the classic neonatal thymectomy experiment, in which T cell-dependent autoimmunity occurred if thymectomy occurred at day 3 after birth, but not day 7 (Nishizuka and Sakakura, 1969), resulting from what is now thought to be a delay in thymic Treg cell export (Asano *et al.*, 1996; Fontenot *et al.*, 2005a). Follow-up studies found that depletion of the CD25 subset revealed the autoimmune propensity in CD4 single-positive (SP) thymocytes (Itoh *et al.*, 1999). Thus, it is now believed that another important educational function of the thymus is to select Treg cells to prevent spontaneous autoimmunity (Saoudi *et al.*, 1996).

2.1.2. Timing of thymic Treg cell development

The classic neonatal thymectomy experiment revealed that spontaneous autoimmunity occurred if the surgery occurred on day 3, but not on day 1 or day 7, suggesting a narrow window for this effect (Nishizuka and Sakakura, 1969). As markers for Treg cells were discovered, it was found that Treg cells are delayed in their thymic development compared with conventional T cells (Asano *et al.*, 1996; Fontenot *et al.*, 2005a). Subsequent experiments using neonatal bone marrow chimeras revealed that the developmental delay also occurred in 2- to 3-week-old thymuses, suggesting that this was not an oddity of the neonatal thymus (Lee and Hsieh, 2009). Studies of thymic T cell development using Rag1-GFP as a molecular timer for events past positive selection suggest that a substantial percentage of Foxp3^{+} cells are resident in the thymus for many days or have recirculated from the periphery (McCaughtry *et al.*, 2007). While it is clear that recirculation occurs readily in lymphopenic settings (Bosco *et al.*, 2006), the percentage in nonlymphopenic settings may be considerably lower (Lathrop *et al.*, 2008). Although day 3 thymectomy was serendipitously a very useful experimental protocol for exploring the notion of T cell-mediated immune regulation, the teleologic reason for this developmental delay is unclear. Is it simply a reflection of the additional developmental steps required to upregulate Foxp3? Is this important for self-reactive thymocytes to completely poll the thymic antigen presenting cells (APC) population? Future experiments will be required to address these questions as well as determine the half-life of newly developed thymic Treg cells and the frequency of Treg cells in the thymus that come via recirculation.

2.2. Intrinsic factors involved in thymic Treg cell development

2.2.1. TCR specificity

2.2.1.1. Self-reactivity of thymic Treg cells Since Foxp3$^+$ Treg cells comprise only a small fraction of CD4$^+$ T cells, an important question is whether there are defining features that determine selection into the Treg cell subset. Early studies inferred that Treg cells may be reactive to self-antigens, as the presence of an organ was required for the maintenance or generation of peripheral T cells with the ability to suppress autoimmunity to that tissue (Garza *et al.*, 2000; Seddon and Mason, 1999; Taguchi *et al.*, 1994). Subsequent studies were more direct. For example, it was observed that superantigen increased the frequency of Treg cells with the appropriate TCRβ chain, suggesting that TCR stimulation was involved in thymic Treg cell development (Papiernik *et al.*, 1998). Moreover, TCR transgenic mice on a Rag-deficient background were not found to contain thymic Treg cells, whereas Rag-sufficient mice, able to rearrange endogenous TCRα chains, could generate thymic Treg cells (Itoh *et al.*, 1999; Olivares-Villagomez *et al.*, 2000). These data therefore suggested that only certain TCR specificities could facilitate thymic Treg cell development.

Studies of TCR by cognate antigen double transgenic mice provided direct support for the hypothesis that self-antigen recognition in the thymus is important for directing thymic Treg cell selection. For example, hemagglutinin (HA)-specific TCR transgenic cells were selected to become Treg cells only when HA was transgenically expressed in the thymus (Jordan *et al.*, 2001). Other TCR transgenic models including ovalbumin (OVA)-specific DO11.10 yielded similar results (Kawahata *et al.*, 2002; Lohr *et al.*, 2003). From these data, the notion arose that Treg cell development occurred in the avidity window for self-antigen between positive and negative selection (Fig. 2.1A; Maloy and Powrie, 2001).

However, subsequent data were conflicted regarding whether intrinsic properties of TCR, such as self-reactivity, specify Treg lineage commitment. In hen egg lysozyme (HEL)-specific 3A9 TCR transgenic mice, analysis of the absolute number of Treg cells revealed no significant increase in Treg cell generation when transgenic TCR was coexpressed with HEL (Liston *et al.*, 2004). Similarly, it was found that the proportion of Treg cells increased without corresponding changes in the absolute numbers when pigeon cytochrome *c*-reactive and TCR transgenic cells encountered varying amount of antigen expressed under the control of tetracycline-dependent transgene system (van Santen *et al.*, 2004). This observation suggested that agonist TCR interaction resulted in deletion of non-Treg cells, rather than enhancement of Treg cell selection. However, TCR specificities induced by alternative TCRα

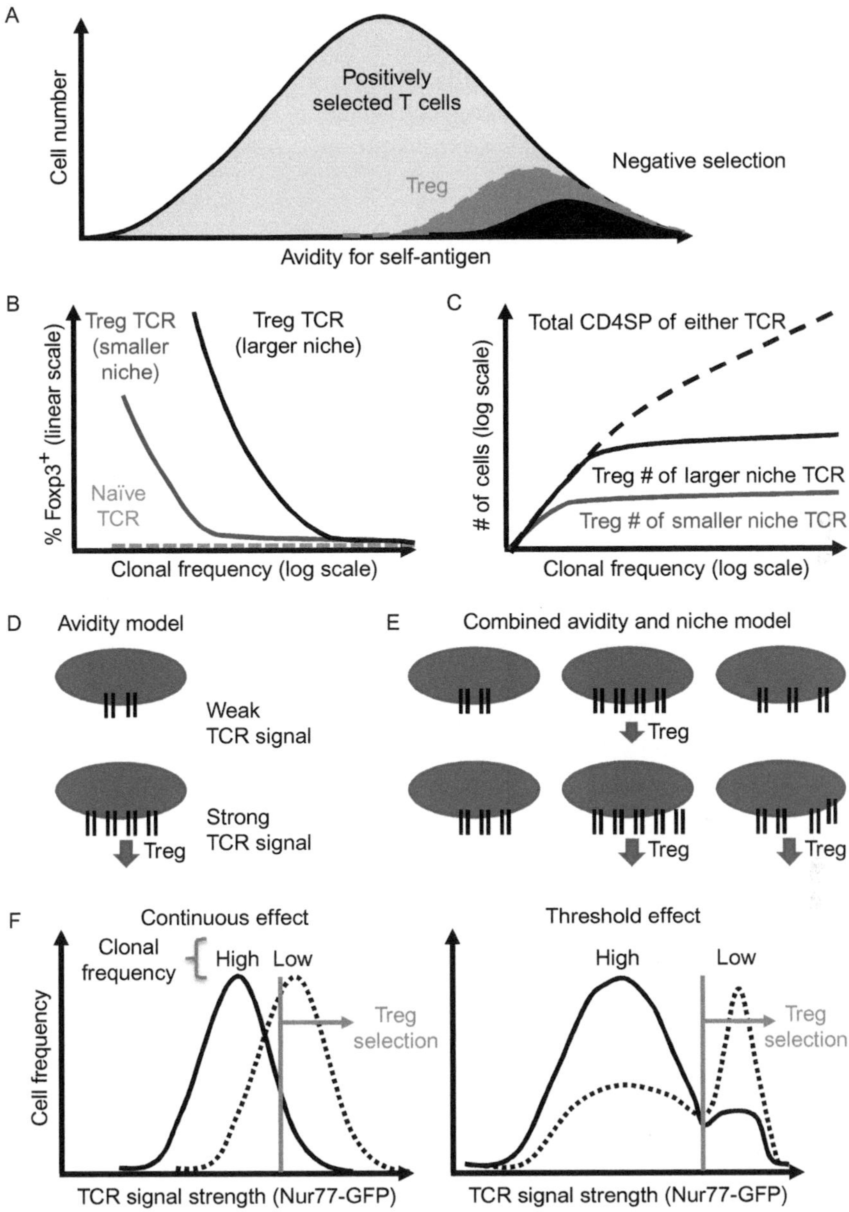

FIGURE 2.1 Role of TCR specificity in thymic Treg cell development. (A) Classic avidity model for thymic Treg cell selection. Thymocytes bearing TCRs with avidity for self-antigen sufficient for positive selection are represented by a Gaussian distribution. Those TCRs which encounter strong self-antigens are either negatively selected (black) or undergo Treg cell differentiation (green). (B, C) Intraclonal competition and thymic Treg cell development. The inverse relationship between clonal frequency and the

chain usage could induce Treg cells and protect them from agonist-mediated deletion, resulting in enhancement of the frequency, but not number, of Treg cells with increasing agonist ligands. Thus, not all TCR transgenic studies supported an agonist model of Treg cell selection.

Analyzes of polyclonal TCR repertoire also reached conflicting conclusions. It was reported that Treg cells utilize a distinct, but overlapping, TCR repertoire from that of non-Treg cells, in favor of a model in which Treg lineage commitment is instructed by TCR specificity (Hsieh *et al.*, 2006; Pacholczyk *et al.*, 2006; Wong *et al.*, 2007b). Moreover, T cells transduced with Treg TCRs often rapidly expanded after adoptive transfer into either normal or lymphopenic hosts, suggesting that Treg-derived TCRs confer reactivity to self-antigens *in vivo* (Hsieh *et al.*, 2004, 2006). In contrast, another study argued that self-reactivity is dispensable for Treg cell selection, based on the observations that TCR repertoire of Treg and non-Treg cell subset was significantly overlapped, and hybridoma cells expressing Treg TCRs failed to respond to self-antigens presented by splenocytes or dendritic cells *in vitro* (Pacholczyk *et al.*, 2007).

Finally, it was suggested that early thymic developmental events condition thymocytes for differentiation into Treg cell lineage before TCR rearrangement, implying that TCR specificity is not a specifying factor for thymic Treg cell induction, although postdifferentiation selection based on antigen specificity is possible (Pennington *et al.*, 2006). Thus, there was considerable controversy regarding the role of TCR specificity in thymic Treg cell selection.

---

frequency of thymic Foxp3^{+} cells is illustrated in (B), with Treg TCRs of two different niche sizes. T cells from expressing naive TCRs often have no ability to induce Treg cell differentiation. The Treg "niche" sizes is illustrated in (C), where the absolute number of Treg cells reaches a plateau much earlier than the absolute number of positively selected CD4SP cells (dashed line). (D, E) Antigen "dosage" and thymic Treg cell generation. The traditional concept of avidity is that TCR signal strength is increased by the number of available ligands per APC. An alternative, and not mutually exclusive possibility, is that titration of the amount of ligand actually increases the "niche" size by increasing the number of APCs that can present threshold levels of ligand to induce TCR-mediated Treg cell differentiation. These two models cannot be easily distinguished via the analysis of Treg cell frequencies at the population level. (F) Intraclonal competition could affect the TCR signaling in an "analog" or "digital" fashion. Using Nur77-GFP as a model for TCR signal strength, an analog change would be represented as a monophasic shift in the plot (left), whereas a digital change would be represented as a biphasic plot (right), with changes in the proportion of the GFP^{hi} and GFP^{lo} cells. The latter model is most consistent with the data presented by Hogquist and colleagues. Treg cell selection is represented as a vertical threshold but could also be dependent on TCR signal strength, which would then be represented by a rightward tilting line. (For interpretation of the references to color in this figure legend, the reader is referred to the Web version of this chapter.)

Data from the past several years, however, have been in favor of TCR-mediated Treg cell selection. First, TCRs from natural Treg, but not naive, cells are able to facilitate thymic Treg cell development in TCR transgenic models (Bautista *et al.*, 2009; Leung *et al.*, 2009). Second, experimental attenuation of major histocompatibility complex (MHC) II expression on medullary thymic epithelial cells (mTECs) resulted in preferential Treg development, rather than negative selection, in a TCR transgenic model involving an Aire promoter-driven antigen (Hinterberger *et al.*, 2010). Third, intravenous peptide could stimulate Treg cell development if given within certain dose ranges (Atibalentja *et al.*, 2009). Finally, direct assessment of TCR activation using a new Nur77-GFP reporter revealed that thymic Treg cells were recently exposed to enhanced levels of TCR stimulation (Moran *et al.*, 2011). Thus, the preponderance of current evidence, discussed in greater detail below, supports an important role for TCR specificity in thymic Treg cell differentiation.

2.2.1.2. TCR transgenic mice with natural Treg TCRs The aforementioned TCR transgenic studies utilized TCRs that were not known to be normally found in the natural Treg cell population. It was therefore possible that the experimental manipulation of expressing the cognate antigen using various promoters may not mimic natural Treg cell development. Recently, several groups independently reported their experiences with TCR transgenic mice expressing natural Treg TCRs. One group isolated the natural Treg TCR from OTII $\alpha\beta$-TCR transgenic cells and only observed negative selection after generating the Treg TCR transgenic line (DiPaolo and Shevach, 2009). It is possible that this particular TCR was expressed as a secondary TCRα chain at low levels in OTII cells and therefore had higher affinity than usual for self-antigens leading to negative selection when expressed in a normal context. Two other groups used TCRs isolated from polyclonal-fixed TCRβ repertoires in which the skewing of the TCR to the Treg cell subset was known (Bautista *et al.*, 2009; Leung *et al.*, 2009). Intriguingly, the frequency of Foxp3$^+$ thymocytes was surprisingly low in the TCR transgenic mice on a Rag-deficient background, which would not have been predicted based on previous studies of TCR by cognate antigen transgenic mice (Jordan *et al.*, 2001; Kawahata *et al.*, 2002; Lohr *et al.*, 2003). In fact, the initial interpretation by both groups was that this observation resulted from an experimental artifact of TCR transgenesis.

Substantial thymic Treg cell development was eventually observed only when the clonal frequency of the TCR transgenic cells was diminished using either mixed bone marrow chimeras or intrathymic injection. In other words, the clonal frequency of TCR transgenic cells was inversely correlated with the frequency of Foxp3$^+$ cells (Fig. 2.1B). Moreover, the number of Treg cells generated reached plateau at high clonal frequency and was much smaller than the total number of $CD4^{+}CD8^{-}$ (CD4SP) cells that could be generated by positive selection (Fig. 2.1C). These observations led to the notion that thymocytes with a given TCR specificity undergo intraclonal competition for a small niche with limited microenvironmental factor(s) important for thymic Treg cell development. Although the factors defining the Treg cell selection niche are unclear, the most compelling hypothesis is that the niche represents a limited amount of antigen for which developing thymocytes must compete in order to undergo Treg cell differentiation.

It is important to note, however, that the limitations imposed by niche size resulting in intra-(or inter) clonal competition may be more applicable to experimental immunology rather than normal T cell development in a fully polyclonal population, where the clonal frequency is likely to be extremely low—on the order of a few to tens of cells per antigen specificity (Moon *et al.*, 2007). High clonal frequencies may explain why transgenic mice with self-reactive TCRs undergo spontaneous autoimmunity on the Rag-deficient background due to a lack of Treg cell development (Nishio *et al.*, 2010; Olivares-Villagomez *et al.*, 1998). In addition to TCR transgenic mice, restricted TCR repertoires may be susceptible to niche size issues depending on the extent of the limitation of diversity (Pacholczyk *et al.*, 2006) and may provide an explanation regarding the increased overlap between Treg and non-Treg cell subsets in some studies (Pacholczyk *et al.*, 2007) compared with others (Hsieh *et al.*, 2006).

The physiologic implication from the observation of a small Treg cell developmental niche, at least based on the natural Treg TCRs reported to date, is that the antigens that drive thymic Treg cell development are probably tissue specific and rare, rather than ubiquitous. This is consistent with the recent description of a transgenic line using a TCR isolated from scurfy mice, in which the TCR not only facilitates thymic Treg cell differentiation but also recognizes a self-antigen in the skin (Killebrew *et al.*, 2011). Moreover, the size of the niche varies per TCR, suggesting that some Treg TCR ligands are more abundant than others. The notion of rare Treg cell-inducing ligands would be consistent with the observation that Aire-dependent tissue-specific antigens presented by mTECs can facilitate thymic Treg cell development (Aschenbrenner *et al.*, 2007). This is also supported by an observation of preferred Treg cell development compared to deletion in low amount of cognate antigen expression (Picca *et al.*, 2009). Uncommon self-ligands may explain why self-reactivity could not be easily detected by direct hybridoma assays (Pacholczyk *et al.*, 2007) but required assessment of *in vivo* proliferation in lymphopenic hosts (Hsieh *et al.*, 2004, 2006). Finally, it may be possible that ubiquitous expression of certain high-affinity ligands preferentially results in negative selection and explains the lack of Treg cell differentiation in some TCR transgenic models but not others. Direct proof of these hypotheses will benefit from the isolation of natural Treg TCR ligands, which has not yet been reported to our knowledge. Thus, the preponderance of current data suggests that thymic Treg cell selection is driven by TCR encounter with rare self-antigens.

2.2.1.3. Affinity and avidity As TCR specificity is a primary determinant for thymic Treg cell selection, an important question is to characterize the parameters of TCR:ligand engagement required for this process. Initial experiments with TCR transgenic models suggested that a high-affinity TCR interaction is required for thymic Treg cell selection. Comparison of two HA-specific TCR transgenic lines with 1000-fold differences in sensitivity to HA revealed that Treg cell development only occurred with the high-affinity TCR (Jordan *et al.*, 2001). Increasing the amount of antigen, that is, avidity, could not compensate for the difference in affinity, as levels of antigen sufficient to induce negative selection were still unable to generate Foxp3^{+} cells. A different study of Treg cell differentiation in thymic organ cultures used altered peptide ligands with varying affinities to the hemoglobin-specific N3.L2 TCR (Relland *et al.*, 2009). They observed that strong and weak agonist, but not antagonist, peptides could induce Treg cell differentiation. Thus, these data suggest that there may be an affinity threshold, akin to negative selection (Naeher *et al.*, 2007), for which TCR recognition of agonist ligands is required for thymic Treg cell differentiation.

The role of the avidity in Treg cell differentiation has been also addressed using transgenic mice and *in vitro* thymic organ cultures. Using a panel of transgenic lines with different expression levels of HA in the thymus, it was observed that Treg cell number was enhanced at low expression levels, whereas high levels resulted predominantly in negative selection (Picca *et al.*, 2009). A similar monophasic response was observed *in vitro* using a titration of agonist peptides into fetal thymic organ cultures (FTOC) of NOD TCR transgenic cells, in which Treg cell numbers were positively related to peptide dose at low concentration but diminished at high antigen concentration due to extensive negative selection (Feuerer *et al.*, 2007). This phenomenon was also replicated *in vivo* using intravenous peptide administration (Atibalentja *et al.*, 2009). Finally, a recent study using transgenic mice in which the level of antigen presentation was decreased by the downmodulation of MHC class II expression in Aire-expressing mTEC cells via an Aire promoter-driven shRNA against C2TA, the master regulator of MHC class II expression (C2TA^{kd}), also showed a shift from negative selection to Treg cell development (Hinterberger *et al.*, 2010). Thus, the experimental evidence favors a model by which Treg cell development is most efficient within an avidity window below that of negative selection (Fig. 2.1A).

While the current data provide a framework for understanding the role of avidity and affinity in thymic Treg cell selection, a number of questions remain. One issue is the interpretation of experiments that increase the amount of antigen as a means to understand avidity. The traditional notion of avidity is that an increased number of ligands per APC leads to enhanced TCR signaling and Treg cell selection at the individual cell level (Fig. 2.1D). However, it is possible that the increased amount of antigen also results in more APCs loaded with sufficient ligand, increasing the size of the antigenic niche (Fig. 2.1E). As these studies were performed at high clonal frequencies using TCR transgenic mice, increasing the amount of antigen may lead to increased percentages of Foxp3$^+$ more via enlarging the niche, rather than enhancing the avidity of TCR interactions.

In addition to Treg cell selection, intraclonal competition also affects positive selection, albeit to a lesser degree (Canelles *et al.*, 2003; Huesmann *et al.*, 1991). It is therefore reasonable to expect that negative selection is also affected by intraclonal competition for an antigenic niche. However, the relationship between clonal frequency and negative selection versus Treg cell differentiation is unknown. Although the Treg TCRs studied to date do not appear to undergo obvious negative selection (Bautista *et al.*, 2009), it is expected that some Treg TCRs will result in negative selection. Future studies will be required to address how clonal frequency affects the balance between negative selection and Treg cell differentiation.

Another issue is the use of ubiquitous promoters to express the cognate antigens, peptide injection into mice, or addition of peptides to FTOC, which presumes that Treg cell selecting ligands and the APC environment of the thymus are monolithic. Experimental evidence suggests that Treg cell selection utilizes a limited antigen niche (Bautista *et al.*, 2009; Leung *et al.*, 2009), making ubiquitous antigens less likely to model thymic Treg cell development, although the range of antigens that select thymic Treg cells is by no means fully explored. Moreover, the distribution of selecting ligand on the various thymic APC subsets may or may not represent that of natural Treg cell ligands. For example, expression of antigen in the cortex might favor negative selection rather than Treg cell development. Thus, although the current data show that affinity and avidity play important roles in thymic Treg cell selection as proof of principle, future experiments will be required to understand in a quantitative fashion how reactivity to self-antigens dictates cell-fate choice between Treg cell differentiation, negative selection, and release from the thymus as an effector cell with autoimmune potential.

2.2.1.4. A signaling threshold for thymic Treg cell development? One reason that it has been difficult to address the role of self-antigen reactivity in Treg cell selection is the lack of effective *in vivo* assessment for TCR stimulation. While T cell activation markers such as CD69 and CD25 are associated with the process of Treg cell development (Lio and Hsieh, 2008), they did not provide compelling evidence due to their transient expression or their upregulation as a Treg cell marker, respectively. Recently, a new marker for TCR activation was reported using a Nur77 promoter-driven GFP which revealed several interesting aspects of thymic Treg cell differentiation (Moran *et al.*, 2011). In a polyclonal population, they found that GFP expression was high on Foxp3^{+} compared with Foxp3^{-} cells, confirming the notion that Treg cells underwent TCR-mediated selection to agonist ligand.

The authors also analyzed the role of intraclonal competition using G113 Treg TCR transgenic cells that were known to be restricted by a small developmental niche (Bautista *et al.*, 2009). One prediction was that intraclonal competition would result in a downward shift of a monophasic TCR signal measured by Nur77-GFP, thereby decreasing the proportion in the "avidity" window of Treg cell development (Fig. 2.1F). However, what was observed was that G113 cells existed in a biphasic plot with distinguishable Nur77-GFP^{hi} and ^{lo} populations, which shifted in proportion depending on the clonal frequency of G113 thymocytes. Assuming that the Treg cell differentiation program itself does not affect Nur77-GFP levels, these data suggest that intraclonal competition operates by limiting the number of T cells obtaining a high-level TCR signal, rather than decreasing the overall signal strength that the T cell population receives. Niche size would then reflect the number of APCs that have sufficient antigen in terms of both affinity and avidity to induce "high-level" TCR signaling, which remains to be biophysically defined in the context of Nur77-GFP. Thus, this suggests that Treg cell differentiation is dependent on a certain threshold of TCR signaling, arguing for a "digital," rather than "analog" interpretation of TCR signals for thymic Treg cell differentiation (Fig. 2.1F).

2.2.2. Cytokines and costimulation

Although many of the initial studies on the role of cytokines revolved primarily on peripheral Treg cell differentiation, it has become increasingly clear that cytokines can play a role in thymic Treg cell differentiation (Fig. 2.2).

*2.2.2.1. **Interleukin-2*** Interleukin-2 (IL-2) was originally discovered as a T cell growth factor crucial for the *in vitro* propagation of T cell clones (reviewed in Oppenheim, 2007). It was therefore quite surprising that mice deficient in IL-2 or its receptor subunits (CD25 or CD122) developed spontaneous autoimmunity instead of the predicted immunodeficiency (Kramer *et al.*, 1995; Schorle *et al.*, 1991; Suzuki *et al.*, 1995, 1999). A seminal observation that explained the autoimmune phenotype was that the dominant function of IL-2 is to promote the survival and function of Treg cells

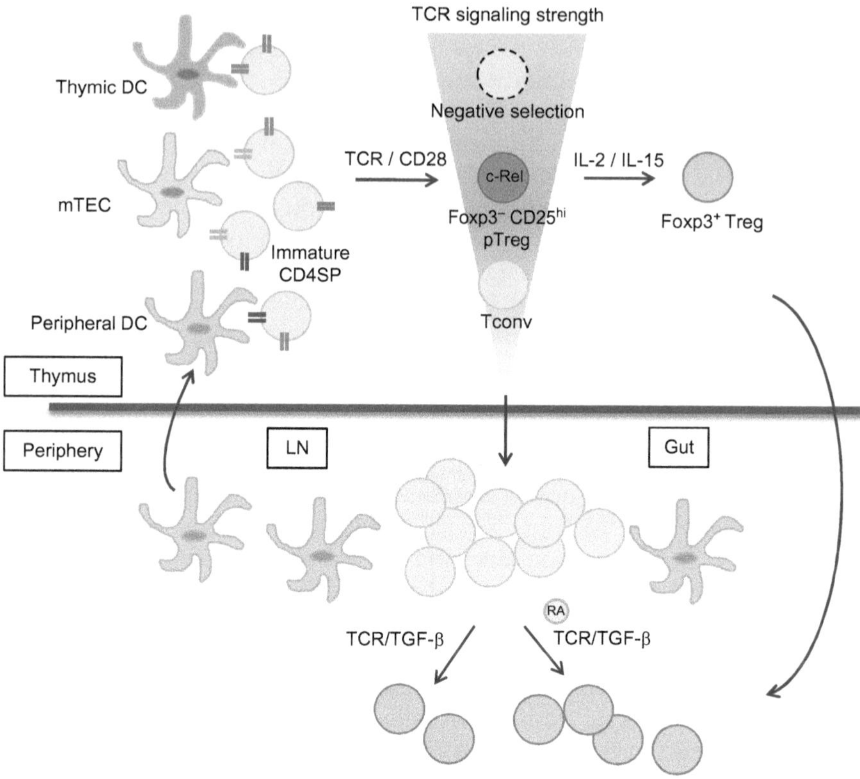

FIGURE 2.2 Overview of Treg cell differentiation. After positive selection, immature CD4SP thymocytes encounter a wide array of self-antigen presented by various APCs such as mTEC, thymic DC, and peripheral-derived DC. Cells receiving very strong TCR stimulation via large amounts of high-affinity ligand or ubiquitous nonagonist ligands are deleted by apoptosis. Thymocytes expressing TCRs that interact with agonist affinity self-antigen are hypothesized to become $Foxp3^{-}CD25^{hi}$ cytokine-responsive Treg precursor cells, which utilize IL-2/IL-15 for induction of Foxp3 expression. TCRs with low self-reactivity in the range of positive selection are exported as conventional naive T cells to the periphery. In contrast to the thymic process, TGF-β appears to play a more important role in peripheral induction of Foxp3. The gut environment is also enhanced in specific factors such as RA that facilitates conversion of naïve T cells into Treg cells. (For color version of this figure, the reader is referred to the Web version of this chapter.)

(Furtado *et al.*, 2002). In retrospect, this might have been predicted based on the observation that CD25, the α-subunit of the high-affinity IL-2R, is constitutively expressed on most $Foxp3^{+}$ Treg cells (Sakaguchi *et al.*, 1995). Recent reports have also shown the remarkable utility of exogenous administration of IL-2 via antibody complexes in enhancing Treg cell numbers and the treatment of murine models of autoimmune disease (Webster *et al.*, 2009). Thus, IL-2 has an important role in Treg cell biology.

Initial studies examining the role of IL-2 in thymic Treg cell differentiation suggested that the IL-2 was not essential, as the frequency of thymic $Foxp3^+$ cells was only modestly reduced in IL-2- or CD25-deficient mice by approximately 50% of that of wild-type mice (Fontenot *et al.*, 2005b). Moreover, a TCR transgenic model suggested that IL-2 is dispensable for agonist-induced Treg (iTreg) differentiation but is required for the survival of mature Treg cells in the periphery (D'Cruz and Klein, 2005). It was therefore felt that IL-2 played a minor role, if at all, in thymic Treg cell development.

More recent studies have favored a role for IL-2 in thymic Treg cell development. In one proposed "two-step" model (Fig. 2.2), the process of thymic Treg cell differentiation can be divided into at least two stages based on the requirement for TCR signaling (Lio and Hsieh, 2008). The first TCR-dependent step also involves CD28 signals resulting in the generation of a putative $CD25^+Foxp3^-$ Treg cell precursor, followed by a TCR-independent step in which cytokines such as IL-2, and to a lesser extent, IL-15 and IL-7, can induce Foxp3 expression. This model is supported by observations from the expression of constitutively active form of STAT5b, which increases the frequency of Treg cells in part by enlarging the Treg TCR repertoire to include TCRs which are more commonly found in the non-Treg cell subset (Burchill *et al.*, 2008). This model is also consistent with data from a study of agonist-driven Treg cell development using TCR transgenic cells (Wirnsberger *et al.*, 2009).

It is important to point out that this model is only a framework for understanding thymic Treg cell differentiation. Although there is a fair amount of supporting data, it has not yet been rigorously established that all Treg cell differentiation must proceed through a $CD25^+Foxp3^-$ cellular intermediate. It may be possible that for some cells, upregulation of Foxp3 is a rapid event that cannot be easily visualized in terms of discrete steps. Future studies will be required to refine or disprove this "two-step" model.

A couple of observations may reconcile the conflicting interpretations regarding the role of IL-2 in thymic Treg cell differentiation. For example, it was reported that the frequency of $Foxp3^+$ cells was fivefold decreased in CD25-deficient, compared with CD25-sufficient, cells in mixed bone marrow chimeras, in comparison with the twofold decrease in knockout animals (Fontenot *et al.*, 2005b). Thus, the importance of IL-2 is more clearly demonstrated in a competitive environment with normal thymocytes. Moreover, it is likely that the role of IL-2 is redundant, and its absence can be compensated for by other common γ-chain cytokines in the thymus. Analysis of common γ-chain receptor-deficient, IL-2, -7, -15-deficient, or STAT5-deficient mice revealed further decreases in the frequency of thymic Treg cells (Bayer *et al.*, 2008; Fontenot *et al.*, 2005b; Vang *et al.*, 2008; Yao *et al.*, 2007). Although IL-2 likely plays a role in thymic Treg cell survival, the available data suggest that IL-2 and related cytokines also play a role in thymic Treg cell differentiation.

One interesting unresolved question is the source of the IL-2. In mixed bone marrow chimera studies, it was observed that provision of normal bone marrow was sufficient to complement the decreased frequency of Treg cells from IL-2-deficient bone marrow, suggesting that the IL-2 could be obtained from other cells, including T cells (Tai *et al.*, 2005). However, it is also possible that IL-2 can originate from APCs (Granucci *et al.*, 2001). In this case, IL-2 could also act as a local factor influencing the niche size for a particular TCR specificity, especially if IL-2 production is restricted to specific APC subsets and microenvironments. Further studies will be required to determine the source of IL-2 and other cytokines involved in thymic Treg cell differentiation and maintenance.

*2.2.2.2. **Transforming growth factor-β*** There is substantial data supporting the notion that transforming growth factor (TGF-β) is important for the peripheral conversion of naïve T cells into Treg cells (Chen *et al.*, 2003; Fantini *et al.*, 2004). However, its function in thymic Treg cell lineage commitment is controversial. It was initially reported that the absence of TGF-βI or the T cell-specific deletion of TGF-β receptor II (TGF-βRII) resulted in a fairly normal frequency of thymic Treg cells (Li *et al.*, 2006; Marie *et al.*, 2005, 2006). It was then reported that the absence of either TGF-βRI or TGF-βRII in T cells causes reduced generation of thymic Treg cells, but only at early time points (Liu *et al.*, 2008; Ouyang *et al.*, 2010a). At later time points, IL-2-dependent expansion of Treg cells appears to compensate for lack of TGF-β signaling. Although it was originally interpreted as a lack of Treg cell development, it has been suggested that the absence of TGF-β signals results in poor Treg cell survival, as Bim deficiency could rescue Treg cell numbers in the absence of TGF-β signaling (Ouyang *et al.*, 2010a). These data suggest that TGF-β contributes to thymic Treg cell development by protecting thymocytes from cell death induced by presumably strong TCR engagement with self-antigen (Ouyang *et al.*, 2010a). Finally, an *in vivo* deletional analysis of transcription factor binding sites in the Foxp3 locus suggested that the TGF-β-dependent SMAD3 binding site located at the interspecies conserved noncoding sequence-1 (CNS1, see Fig. 2.3) Tone *et al.* (2008) was dispensable for thymic, but not peripheral, Treg cell development (Zheng *et al.*, 2010). Taken together, these data suggest that TGF-β may not play an important role in the process of thymic Treg cell differentiation itself but may act instead on the survival of already generated Treg cells.

*2.2.2.3. **Costimulation*** Given that TCR specificity is a pivotal factor driving differentiation of Treg cells, it is reasonable to suspect that costimulatory signals are also involved in Treg cell selection. In fact, the requirement for costimulation by CD28 was one of the first signals identified for Treg cell homeostasis (Salomon *et al.*, 2000). Subsequently, it was

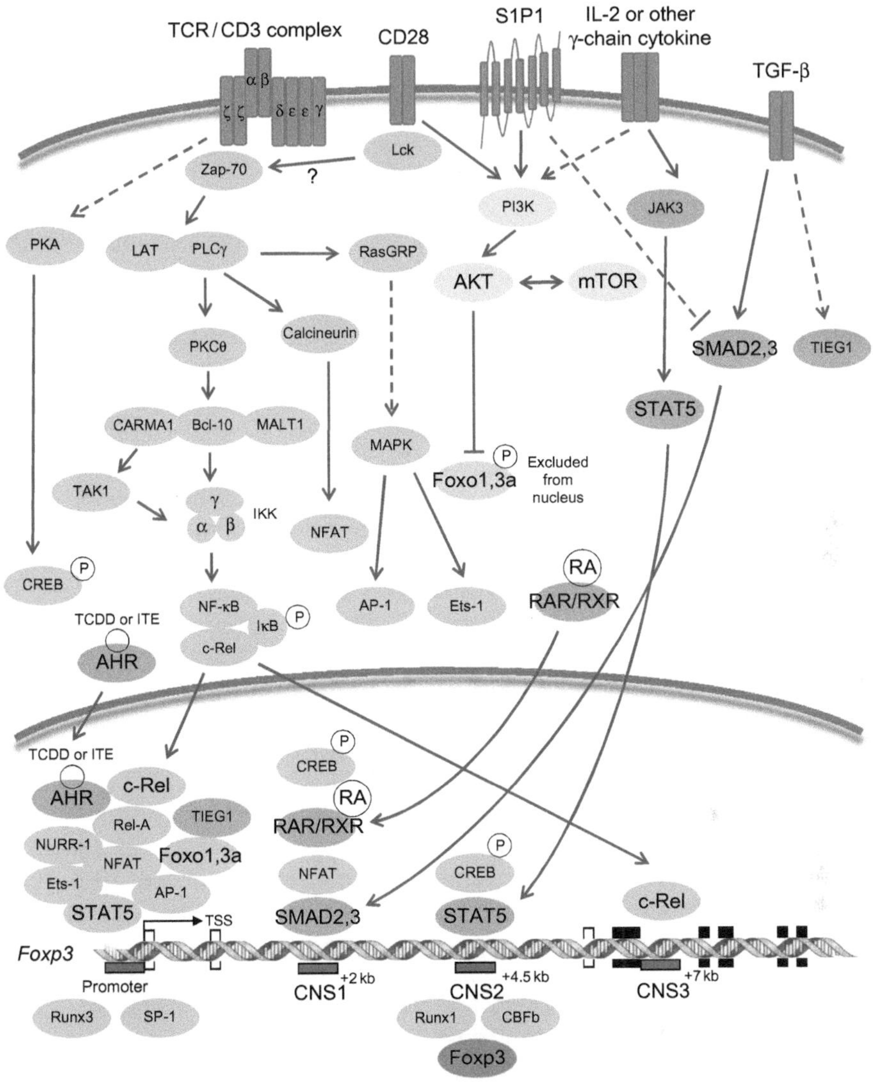

FIGURE 2.3 A model for the molecular mechanisms that induce Foxp3 expression. TCR stimulation (red) leads to NF-κB activation which induces Foxp3 expression by unlocking the locus via the binding of c-Rel to CNS3, and driving protein expression from the promoter via a number of transcription factors, including NFAT, NF-κB family members, AP-1, and so forth. Cytokine signaling (green) from IL-2 results in the phosphorylation of STAT5, which is translocated into the nucleus. On the other hand, stimulation of the AKT pathway (blue) by strong TCR and costimulation signals as well as IL-2 and S1P1 prevents Treg cell differentiation due to the phosphorylation of Foxo proteins and their subsequent exclusion from the nucleus. Interaction of various TCR- and cytokine-derived stimuli induces Treg cell differentiation by interacting together in the proposed

shown that CD28 provided a cell-intrinsic signal for thymic Treg cell generation, rather than enhancing the production of cytokines (Tai *et al.*, 2005). Deficiency in CD28 or its corresponding ligands, CD80/CD86 (B7-1/2), results in an approximately 80% decrease in the frequency of Foxp3$^+$ T cells (Lio *et al.*, 2010; Tai *et al.*, 2005). Thus, CD28 plays an important role in thymic Treg cell selection.

Recent studies have examined the mechanism by which CD28 promotes Treg cell generation. It was shown that the Lck-binding motif PYAP of CD28 is partially involved in TCR-dependent Treg cell precursor generation (Lio *et al.*, 2010; Vang *et al.*, 2010), consistent with previous observation that Lck, but not PI3K or Itk, binding motif is required for efficient Treg cell generation (Tai *et al.*, 2005). The involvement of CD28 in this TCR-dependent process, as expected, suggested that CD28 could act as an amplifier of TCR signaling, thereby increasing the overall pool of TCRs that recognize self- and facilitate Treg cell differentiation. However, TCR repertoire studies were inconsistent with this hypothesis, as they did not reveal a marked difference between CD28-deficient and -sufficient Treg cells (Lio *et al.*, 2010). A study by another group utilized a TCR transgenic model, confirming that CD28 signaling was involved in the TCR-dependent generation of Treg cell precursors (Hinterberger *et al.*, 2011). In intrathymic transfer experiments of TCR transgenic Treg cell precursors, they were able to show that the lack of CD28 costimulation led to their preferential loss. Taken together, the current hypothesis is that CD28 provides a unique signal that facilitates the survival of Treg cell precursors and increases the efficiency of their progression to become Foxp3$^+$ (Hinterberger *et al.*, 2011), but the mechanism by which this occurs remains to be determined.

Other costimulatory molecules have also been implicated. In a TCR transgenic model expressing a receptor specific for myelin basic protein, it was reported that CTLA-4-deficient mice had a higher frequency of thymic Treg cells (Verhagen *et al.*, 2009). Interestingly, they observed Treg cells in the cortex in the CTLA4-deficient, but not sufficient, TCR transgenic mice, corresponding with an enhanced frequency of Foxp3$^+$ cells in the CD4$^+$CD8$^+$ (DP) subset. Future studies will be required to

enhanceosome complex centered around c-Rel in the Foxp3 promoter. In the periphery, TGF-β appears to play a pivotal role in conversion of naive T cells to Treg cells in cooperation with IL-2, via the binding of NFAT and SMAD2,3 to CNS1, as well as the induction of TIEG1 which then binds the promoter. Other soluble factors such as RA bind intracellular receptors RAR/RXR which bind to CNS1 and facilitate Foxp3 induction. The maintenance of Foxp3 expression involves binding of the Runx1–CBFb–Foxp3 complex to demethylated CNS2 in the Foxp3 locus. Signaling components with dashed line indicate that their role is controversial or not yet studied. (For interpretation of the references to color in this figure legend, the reader is referred to the Web version of this chapter.)

determine whether CTLA4 expressed on thymic Treg cells inhibits the TCR stimulation of more immature thymocytes, or whether CTLA4 participates directly in the Treg cell differentiation process.

CD154 (CD40L) interactions with CD40 have also been implicated in thymic Treg cell differentiation, with a threefold decrease in thymic Treg cell frequency observed in mice deficient in either of those molecules (Spence and Green, 2008). The mechanism was not attributed to signals related to that of CD28:B7-1/2 discussed above, but rather due to a decrease in proliferation of $Foxp3^+$ Treg cells. While TCR repertoire shifts were not studied, these data suggest that CD154/CD40 are not primarily involved in Treg cell selection but act to maintain cellular homeostasis post-Treg cell development.

In summary, these data suggest an important role for costimulation in thymic Treg cell selection. However, the teleologic role of signals such as CD28 remains unclear. One might predict that APCs which poorly express B7-1/2 may primarily facilitate negative selection, rather than Treg cell differentiation. Why it would be useful for immune tolerance to delete against certain self-antigens presented on some APCs versus Treg cell induction to different antigens on other APCs is unclear. Future studies will be required to address this question.

2.2.3. Molecular mechanism of thymic Treg cell development

Currently, available data strongly suggest that TCR specificity plays an instructive role in thymic Treg cell development. The mechanisms by which TCR signaling drive other thymic differentiation processes such as positive and negative selection have been studied. For example, it has been proposed that negative selection results from strong but transient ERK activation, whereas positive selection requires sustained low-level ERK activation (Starr *et al.*, 2003). In addition, it was observed that differential kinetics and subcellular localization of Ras/MAPK signaling distinguish positive selection from negative selection (Daniels *et al.*, 2006). Whether these observations are also applicable to Treg cell selection is unknown. Although TCR signaling for thymic Treg cell differentiation has not been studied in as great detail, we will review the current data regarding the molecular mechanisms that result in Foxp3 gene induction (Fig. 2.3).

2.2.3.1. Proximal TCR signaling Zap70 (zeta-associated protein of 70 kDa), a tyrosine kinase associated with CD3 and zeta-subunits of TCR complex, plays a critical role in TCR signaling in both thymocytes and mature T cells (Chan *et al.*, 1992). A study of a series of Zap70 variants observed that a mild loss of function resulted in a normal proportion of Treg cell (Siggs *et al.*, 2007). A subsequent study found that both the frequency and the number of Treg cells were significantly reduced in mice carrying Y315A and Y319A mutations, which have defects in recruiting Vav1,

Lck, or PLCγ1 (phospholipase Cγ1). Another mutant mouse line called SKG, which has a point mutation in the C-terminal SH2 domain of Zap70, also showed similar reductions in the production of thymic Treg cells (Hsu *et al.*, 2009). Although these data demonstrate that ZAP70 is involved in the TCR signals that result in thymic Treg cell development, it is not clear whether these signals are specific for Treg cell differentiation, as the ZAP70 mutants contain defects in other thymic TCR-dependent events such as positive selection.

By contrast, analysis of LAT (linker of activated T cells) suggests that certain TCR downstream signals may be relatively more important for Treg cell differentiation than other thymic TCR-dependent processes. In particular, the Y136F mutation in LAT resulted in complete absence of Treg cells while preserving positive selection (Koonpaew *et al.*, 2006). This mutation specifically blocks the phosphorylation of tyrosine 136, abolishing the recruitment of PLCγ1. Thus, these data suggest that only a subset of TCR-derived downstream signals is essential for thymic Treg cell differentiation.

2.2.3.2. NF-κB and AKT Another important clue to the molecular mechanism which translates TCR reactivity to self into Treg cell development came from knockouts of NF-κB signaling components. It was observed that deficiency in PKCθ, CARMA1, Bcl-10, IKKβ, or TAK1 showed marked reduction in Treg cell numbers (Barnes *et al.*, 2009; Gupta *et al.*, 2008; Medoff *et al.*, 2009; Molinero *et al.*, 2009; Schmidt-Supprian *et al.*, 2003, 2004; Wan *et al.*, 2006). Recent studies have confirmed the important role of NF-κB signaling in thymic Treg cell differentiation (Long *et al.*, 2009; Ruan *et al.*, 2009). *In vivo* enhancement of NF-κB signaling via the transgenic expression of constitutively active IKKβ could rescue the defect in Treg cell development resulting from proximal defects in NF-κB in CARMA1- and TAK1-deficient mice (Long *et al.*, 2009). Importantly, enhanced NF-κB signaling is sufficient to bypass the requirement for TCR stimulation by self-antigens for Treg cell differentiation in OTII and P14 TCR transgenic lines, which otherwise do not generate Foxp3$^+$ cells. In total, these data offer strong support for the notion that Treg cell differentiation is mediated by TCR-induced NF-κB signaling.

There are many NF-κB family members which may be involved in inducing transcription of Foxp3. Although complexes involving RelA (p65), NF-kB1 (p50), and c-Rel have been implicated, the aggregate data suggest that c-Rel is the critical NF-κB family member (Isomura *et al.*, 2009; Long *et al.*, 2009; Ruan *et al.*, 2009; Zheng *et al.*, 2010). c-Rel appears to bind to noncoding sequence-3 (CNS3) of the Foxp3 locus, as deletion of that sequence has a marked defect in thymic Treg cell development (Zheng *et al.*, 2010). Thus, it has been proposed that c-Rel may be the pioneering transcription factor that opens the Foxp3 locus in response to TCR stimulation.

In contrast to NF-κB, AKT signaling appears to have a negative effect on thymic Treg cell differentiation. Expression of constitutively active form of AKT inhibits thymic Treg cell development (Harada *et al.*, 2010). As PI3K/AKT signaling inactivates Foxo1, 3, and 4 by phosphorylation, it was suggested that the inhibitory effect of AKT signaling on Treg cell development involves Foxo transcription factor family based on the observation that Foxo1/Foxo3 can activate Foxp3 expression by binding to Foxp3 promoter, and that Foxo1/Foxo3 double-deficient mice have greatly reduced number of Treg cells (Harada *et al.*, 2010; Ouyang *et al.*, 2010b). Thus, signaling via NF-κB and AKT reciprocally regulate Treg cell differentiation after TCR engagement (Fig. 2.3).

Although it is clear that AKT can play a role in thymic Treg cell differentiation, the conditions under which AKT is preferentially activated compared with NF-κB in the thymus remain to be determined. Alternatively, there may not be a dynamic role for AKT in thymic Treg cell generation, as it is not clear why self-reactivity should sometimes activate AKT to generate effectors that escape negative selection, but other times favor NF-κB without much AKT activation, leading to thymic Treg cell generation. By contrast, dynamic modulation of AKT in the periphery may be important for determining the balance between effector and Treg cell generation. In this model, unopposed NF-kB signaling would be sufficient for induction of Foxp3 in thymocytes (Long *et al.*, 2009), thereby selecting thymocytes into the Treg cell subset entirely on the basis of antigen recognition in the thymus to presumed self-antigens. This would be consistent with a study showing that cell-intrinsic T cell developmental maturity plays a major role in determining the efficiency by which TCR signaling induces Foxp3 (Wirnsberger *et al.*, 2009). Future experiments will be required to determine whether dynamic regulation of AKT in thymic versus peripheral T cells represents the molecular basis for the relationship between T cell maturation and propensity to undergo TCR-driven induction of Foxp3.

2.3. Extrinsic factors for the thymic Treg cell differentiation

2.3.1. APC subsets

The notion that self-antigen specificity selects the thymic Treg cell population engenders the question of which APCs are presenting what self-antigens. For example, it is known that cTEC and mTEC present different antigenic repertoires due to the use of distinct antigen processing enzymes (Nakagawa *et al.*, 1998). Moreover, Aire expression in mTECs has been shown to induce the expression of tissue-specific antigens in the thymus (Anderson *et al.*, 2002). Other differences between APCs such as costimulation may also affect thymic Treg cell differentiation. Thus, it is important to understand the contribution of the various thymic APC subsets in the generation of the thymic Treg cell population.

Initial studies suggested that a normal frequency of $CD25^+$ Treg cells can be generated when MHC class II is exclusively expressed in the cortex of the thymus (Bensinger *et al.*, 2001). Consistent with cortical Treg cell development, some studies showed a considerable frequency of Treg cells at the $CD4^+CD8^+$ (DP) stage (Liston *et al.*, 2008; Ribot *et al.*, 2007). However, other studies favor the notion that Treg cell differentiation mainly occurs at the CD4SP stage (Lee and Hsieh, 2009; Schallenberg *et al.*, 2010). The temporal analysis of Treg cell generation during the neonatal period as well as in young adults using neonatal bone marrow injection demonstrated that DP cells expressing Foxp3 are rare and the development of $Foxp3^+$ cells mostly starts at immature CD4SP stage. Although a quantitative assessment of the relative contribution of cortical versus medullary APCs to thymic Treg cell differentiation is not established, it appears that most Treg cells are generated in the medulla.

The medullary APC population is comprised of mTEC and hematopoietic cells, mainly DCs. Initial studies suggested that mTEC cells were the primary APCs that mediated thymic Treg cell differentiation. It was observed that Treg cell development was normal even when the bone marrow (BM)-derived thymic APCs were rendered MHC class II-deficient in bone marrow chimeras (Aschenbrenner *et al.*, 2007; Liston *et al.*, 2008), demonstrating that mTECs are sufficient to generate normal Treg cell frequencies. Treg cell development also occurred normally in bone marrow chimeras of TCR transgenic cells into antigen-expressing hosts (Jordan *et al.*, 2001). Similarly, a recent transgenic model examining interactions between OTII and OVA expressed by the Aire promoter was highly susceptible to knockdown of MHC class II presentation in Aire-expressing cells using a transgenic shRNA to CIITA ($C2TA^{kd}$), suggesting that mTEC cells were the primary APC responsible for Treg cell differentiation in this model (Hinterberger *et al.*, 2010).

Although antigen presentation by BM-derived APCs is not required for the generation of normal thymic Treg cell numbers, there is evidence that they can also play a role in Treg cell differentiation. For example, transgenic expression of cognate antigen on thymic DCs resulted in Treg cell differentiation by antigen-specific thymocytes (Apostolou *et al.*, 2002; Proietto *et al.*, 2008). Moreover, a 90% decrease in the level of MHC class II on Aire-expressing mTEC cells using the $C2TA^{kd}$ approach described above did not affect the frequency of thymic Treg cells. Finally, even though total B7 deficiency results in a marked decrease in thymic Treg cells, expression of B7 on either hematopoietic or thymic stromal cells is sufficient to generate normal numbers of Treg cells (Roman *et al.*, 2010; Spence and Green, 2008). Taken together, these data suggest that the BM-derived APCs are capable of facilitating thymic Treg cell development and can substitute for mTEC APC function for generating normal Treg cell numbers.

There are a number of BM-derived APC subsets that could be involved in thymic Treg cell differentiation, including SIRPα^+ conventional DCs, SIRPα^- thymus-derived DCs, and plasmacytoid DCs. *In vitro* studies suggested that SIRPα^+ DCs have the highest capacity for Treg induction, whereas the efficiency of Treg cell generation appeared to be independent of the DC subtypes in another study (Proietto *et al.*, 2008; Wirnsberger *et al.*, 2009). However, clear *in vivo* data regarding the relative importance of each of these subsets have not been generated.

In summary, the current experimental evidence suggests that cTECs, mTECs, and BM-derived APCs are individually sufficient to generate normal frequencies of thymic Treg cells. However, it is unclear whether these different APC subsets are truly redundant in terms of the antigenic peptides presented, or more likely, select for different Treg TCR repertoires. The fact that these three disparate APC subsets are redundant in terms of Treg cell generation hints at the possibility that there may be an overall limitation in the size of the Treg cell compartment, in addition to the antigen-specific niches described above.

Based on the notion that Treg cells develop to uncommon, presumably tissue-specific antigens, we would predict that all of the APCs subsets, and, in particular, BM-derived APCs versus mTECs, display unique sets of antigenic self-peptides that serve to broaden the spectrum of antigens that select for thymic Treg cells. While TCR by antigen transgenic models can provide proof of principle of the role a given APC, it is difficult to determine the relative contribution of the APC to the generation of the thymic Treg cell population. Thus, future experiments will need to analyze not only the overall Treg cell frequency but also the TCR repertoire of cells that undergo thymic Treg cell differentiation after antigenic encounter with defined APC subsets.

2.3.2. Role of Aire

It has been established that mature mTECs present tissue-specific antigens in Aire-dependent manner (Anderson *et al.*, 2005; Liston *et al.*, 2003), suggesting that Treg cells could be selected on Aire-dependent antigens. Aire may also affect thymic architecture related to a defect in mTEC maturation (Dooley *et al.*, 2008; Gillard *et al.*, 2007; Yano *et al.*, 2008), which can result in diminished superantigen-mediated deletion in Aire$^{-/-}$ mice (Yano *et al.*, 2008). However, the frequency of Treg cells in Aire-deficient mice is generally reported to be within the normal range (Kuroda *et al.*, 2005), although one study suggested that the Treg cell frequency was reduced by 50% (Lei *et al.*, 2011). Thus, an obvious role of Aire in thymic Treg cell generation has not been reproducibly reported.

However, it seems unlikely that Treg cell development is completely Aire independent given its role in mTEC cell development and tissue-specific antigen expression. The analysis of Treg cell frequencies may not be particularly sensitive to even fairly large changes in thymic antigen presentation, as evidenced by the normal Treg cell frequencies in the various APC manipulations described above (Aschenbrenner *et al.*, 2007; Bensinger *et al.*, 2001; Hinterberger *et al.*, 2010; Liston *et al.*, 2008). Interestingly, analysis of a fixed TCRβ:limited TCRα transgenic model revealed no obvious differences in the Treg TCR repertoires of Aire-deficient and -sufficient mice (Daniely *et al.*, 2010). While this needs to be confirmed by other studies using different limited TCR repertoire mice, these data suggest that Aire-dependent antigens affect only a very small fraction of the Treg cell population, consistent with the lack of autoimmunity in the B6 strain (Jiang *et al.*, 2005).

3. PERIPHERAL TREG CELL DEVELOPMENT

3.1. Introduction

3.1.1. Overview

Thymic negative selection and Treg cell differentiation educate the immune system to be tolerant of self-antigens. However, it is also important to develop tolerance to nonpathogenic foreign antigens normally in contact with the body, as inappropriate immune responses to food, commensal bacteria, and inhaled antigens can cause food allergies, inflammatory bowel disease, and asthma, respectively. Recent papers have highlighted the requirement for peripherally generated iTreg cells for maintaining immune tolerance to foreign antigens. In a TCR transgenic model, oral antigen feeding could induce the generation of iTreg cells which protected against experimental asthma (Mucida *et al.*, 2005). Another TCR transgenic model used an osmotic pump to induce iTreg cells which generated tolerance to HY mismatched grafts (Verginis *et al.*, 2008). Finally, in a neonatal transfer model for treating autoimmunity due to Foxp3 deficiency, both thymic and peripherally derived Treg cells were required to establish immune tolerance (Haribhai *et al.*, 2011). Thus, these data suggest that iTreg cells and thymically derived Treg cells (also termed nTreg) are not synonymous for immune tolerance, and either performed different functions or, more likely, utilized nonoverlapping TCR repertoire (Haribhai *et al.*, 2011). Here, we review the recent studies on the factors and underlying mechanisms for Treg cell differentiation in the periphery.

3.1.2. Contribution of thymic versus peripheral development to the peripheral Treg cell population

As thymic and peripherally generated Treg cells arise at different locations and T cell maturation stages, an important question is to understand their relative contributions to the peripheral Treg cell population.

For example, the site of development could determine the antigens, either foreign or self, that the Treg cell likely recognize. Additionally, there may be differences in Treg cell function (Feuerer *et al.*, 2010) and stability (Chen *et al.*, 2011) that is related to the origin of Treg cell differentiation. The relative contribution would also provide an indication of their relative importance for maintaining tolerance.

It was suggested based on adoptive transfer of Foxp3$^-$ cells into nonlymphopenic hosts that peripheral conversion could account for approximately 4–7% of Treg cells (Lathrop *et al.*, 2008). However, one caveat is that peripheral T cells which had already undergone peripheral conversion would not be accounted for in this transfer model. Also, transfer into a nonlymphopenic host results in competition against the preexisting Treg cells. Thus, these numbers likely represent the lower limit of the frequency of peripherally converted cells, rather than a *bone fide* estimate.

TCR repertoire studies comparing thymic versus peripheral Treg cells have demonstrated a substantial degree of similarity between them, supporting the notion that the majority of peripheral Treg cells originate via thymic Treg cell differentiation (Lathrop *et al.*, 2008; Pacholczyk *et al.*, 2006; Wong *et al.*, 2007a). However, these studies are correlative and do not directly trace the developmental origin of the Treg cell expressing a given TCR. It is therefore possible that T cells undergo peripheral conversion using TCRs which are also found on thymic Treg cells. In fact, peripheral conversion has been observed using adoptive transfer of naive TCR transgenic cells expressing TCRs that facilitate thymic Treg cell development (Bautista *et al.*, 2009). This may represent a second chance for self-reactive T cells to upregulate Foxp3. Thus, these TCR repertoire data strongly suggest, but do not prove, that the majority of Treg cells in the periphery arise due to thymic Treg cell differentiation.

An intriguing study by Shevach's group suggested that Helios, a member of the Ikaros transcription factor family, is a marker for thymic Treg cells. Based on their studies, the extent of peripheral conversion can be up to 30% (Thornton *et al.*, 2010). The role that Helios itself plays in Treg development is still unclear. ChIP assays show that it can bind the Foxp3 promoter, but overexpression of this protein does not result in induction of Foxp3, but rather apoptosis (Getnet *et al.*, 2010). Also, analysis of knockout mice has not revealed an essential function for Helios in T cells (Cai *et al.*, 2009). Follow-up studies suggested that peripheral conversion may generate the dominant proportion of Treg cells in the colon, with only ~20% Helios^{hi} thymic Treg cells (Atarashi *et al.*, 2011). This is consistent with adoptive transfer studies of Foxp3$^-$ cells showing increased conversion in the gut (Coombes *et al.*, 2007), as well as analyzes of deletions of the Foxp3 locus thought to be important specific only for peripheral Treg cell development, which show a 50% decrease in gut Treg cells (Zheng *et al.*, 2010). Finally, an analysis of colonic TCR repertoires suggested that a substantial proportion of colonic Treg cells arose due to peripheral conversion in response to commensal bacteria (Lathrop *et al.*, 2011). Interestingly, the frequency of Helios^{hi} Treg cells in the colon of germ-free mice was quite high (~80%), implying that the presence of commensal bacteria facilitated a redistribution from thymically to peripherally derived Treg cells (Atarashi *et al.*, 2011). Thus, there is a substantial body of data that are consistent with the notion that Helios is a marker for thymic Treg cells.

However, a recent report examining transfer of Foxp3^{-} cells into Foxp3-deficient mice revealed that the converted Treg cells were mainly Helios^{hi}, implying that Helios may not be a perfect marker for iTreg cells under all conditions (Haribhai *et al.*, 2011). It is also worth noting that Treg cells generally thought of as "peripherally converted," such as those isolated from the gut lamina propria and adipose tissue, as well as Treg cells obtained by lymphopenia-driven homeostatic proliferation, express mRNA for Helios as well (Apostolou and von Boehmer, 2004; Feuerer *et al.*, 2010). How mRNA levels are correlated to flow cytometric analysis of Helios at a per-cell level is unclear. Future studies will be required to determine whether Helios or other markers are sensitive and specific markers for thymically derived Treg cells.

In summary, the preponderance of available data suggests that the majority of the peripheral Treg cell population arises via thymic Treg cell differentiation, although certain sites such as the gut may be dominated by Treg cells that arise from peripheral conversion. In addition to Treg cell differentiation, the peripheral antigenic landscape likely affects selective expansion or contraction of both thymic and peripheral Treg TCR clonotypes (Lathrop *et al.*, 2008), presumably dependent on a peripheral antigenic niche (Nishio *et al.*, 2010). While Helios appears to be a promising marker for thymic Treg cells, more work is needed to fully validate the conditions under which it is both sensitive and specific.

3.2. Intrinsic factors

3.2.1. Role of TCR specificity

Seminal studies using TCR transgenic mice provided proof of principle that naive T cells can differentiate into Treg cells in the periphery in response to foreign antigens (Apostolou and von Boehmer, 2004). Subsequent TCR transgenic studies showed that oral feeding, low-dose antigen delivery using an osmotic pump, and APC-directed cell delivery using epitope labeling of anti-dendritic cell antibodies could facilitate peripheral Treg cell generation (Apostolou and von Boehmer, 2004; Kretschmer *et al.*, 2005; Mucida *et al.*, 2005; Verginis *et al.*, 2008). Thus, similar to thymic Treg cell development, peripheral conversion *in vivo* is a TCR-dependent event.

The notion that certain TCRs are predisposed to undergo peripheral Treg cell development was suggested based on polyclonal studies using serial transfer of non-Treg cells in lymphopenic mice (Curotto de Lafaille *et al.*, 2004). The authors observed that the frequency of Treg cells that were generated by peripheral conversion diminished with each subsequent transfer of $CD25^-$ cells. One interpretation of this data is that the $CD25^-$ population was being depleted of TCR specificities that facilitate peripheral conversion in lymphopenic hosts. This notion was supported by studies using limited TCR repertoire mice with a fixed TCRβ chain, in which the $CD4^+Foxp3^-$ cells that became $Foxp3^+$ after transfer into a lymphopenic hosts utilized a different, but overlapping, TCR repertoire than the cells that remained $Foxp3^-$ (Lathrop *et al.*, 2008). Thus, peripheral conversion appears to depend on TCR specificity.

However, the reason why some TCRs facilitate peripheral conversion versus effector cell development is unclear. The Allison group examined the role of TCR affinity and avidity in peripheral Treg cell development using TCR transgenic cells stimulated via a single intravenous injection of a series of different peptides with varying affinities for TCR. They found that Foxp3 induction occurred optimally at concentrations several orders of magnitude below what was ideal for T cell proliferation for each peptide, consistent with previous studies (Gottschalk *et al.*, 2010). Induction of iTreg cells was dose dependent based on TCR affinity to peptide as expected, with high-affinity ligands requiring lower doses. However, high doses of weak agonist peptides were unable to generate iTregs of equivalent stability in comparison with agonist ligands, reminiscent of TCR transgenic studies of peptide affinity in the thymus (Jordan *et al.*, 2001). Whether this single intravenous injection model represents a common physiologic situation for iTreg cell induction is unknown. Nonetheless, in this short-term antigenic stimulation, a low density of a high-affinity ligand represents the optimal signal for iTreg cell generation, suggesting that some TCRs, based on affinity and availability of ligand, will be more efficient at facilitating peripheral conversion to the self- and foreign antigens normally found on peripheral APCs.

3.2.2. Role of cytokines

3.2.2.1. TGF-β An important observation regarding peripheral Treg cell development was that TGF-β is a crucial cytokine for Foxp3 induction *in vitro* (Chen *et al.*, 2003) (Fig. 2.2). *In vivo*, a substantial set of data suggests that TGF-β is important for Treg cell development, survival (Li *et al.*, 2006), and function (Marie *et al.*, 2005). Because TGF-β appears to affect multiple aspects of Treg cell biology, analysis of Treg cell numbers and frequencies *in vivo* may not strictly correlate with iTreg cell development. However, the deletion of CNS1 in the Foxp3 locus (Fig. 2.3), which includes the binding site for SMAD3 downstream of TGF-β signaling, resulted in a decreased number of gut Treg cells (Zheng *et al.*, 2010). As the effect of CNS1 does not appear to affect Treg cells in the thymus, these data offer strong support for the notion that TGF-β is important for iTreg cell generation *in vivo*.

*3.2.2.2. **Retinoic acid*** An important series of papers suggested that retinoic acid (RA), a vitamin A metabolite, plays an important role in facilitating iTreg cell generation (Benson *et al.*, 2007; Coombes *et al.*, 2007; Hill *et al.*, 2008; Sun *et al.*, 2007). Production of RA was associated with $CD103^+$ DCs isolated from the gut or its draining lymph nodes. In conjunction with the observation that levels of TGF-β are higher in gut tissue, these data may explain why iTreg cell generation is enhanced in the gut (Atarashi *et al.*, 2011; Lathrop *et al.*, 2011; Sun *et al.*, 2007). However, a recent report suggested that RA may play a more global role in Th subset differentiation, arguing against the notion that RA primarily acts to enhance iTreg cell differentiation (Hall *et al.*, 2011). While future experiments will be required to reconcile the observations made using Vitamin A depletion *in vivo*, it is very clear that RA is a potent *in vitro* enhancer of iTreg cell differentiation.

*3.2.2.3. **IL-2*** IL-2 likely plays a role in iTreg cell generation, similar to its role in the thymus. *In vitro*, addition of IL-2 is important for Foxp3 induction (Davidson *et al.*, 2007). *In vivo*, mice deficient in IL-2, its receptors (Fontenot *et al.*, 2005b), and signaling components STAT5a/b (Yao *et al.*, 2007) have very low numbers of Treg cells in both the thymus and the periphery. IL-2 was also required to observe peripheral Treg cell generation after adoptive transfer of antigen-specific naïve T cells into lymphopenic hosts that expressed the cognate antigen as a ubiquitous secreted protein (Knoechel *et al.*, 2005). Finally, a recent report has suggested that IL-2 responsive $Foxp3^-$ peripheral Treg precursor cells are also present in the periphery, which represents a similarity between thymic and peripheral Treg cell differentiation (Schallenberg *et al.*, 2010). It is important to note that IL-2 also plays an important role in Treg cell survival and proliferation (D'Cruz and Klein, 2005; Fontenot *et al.*, 2005b), as well as Treg cell stability (Zhou *et al.*, 2009). Thus, while current data suggest that IL-2 plays a role in peripheral Treg cell generation, the relative importance of IL-2 on Treg cell generation versus post-Foxp3 homeostasis is not clear.

3.2.3. Molecular signals in peripheral Treg cell development

*3.2.3.1. **TCR-derived signals*** An important event associated with TCR signaling is the opening of the Foxp3 locus (Hori, 2010). It has been proposed that this involves conserved noncoding sequence-3 (CNS3), located immediately downstream of exon 1 in the Foxp3 locus (Fig. 2.3).

Using immunoprecipitation studies, it was suggested that a c-Rel homodimer or c-Rel-p50 dimer binds to CNS3 (Long *et al.*, 2009; Zheng *et al.*, 2010). The importance of this interaction is supported by the observation that deficiency in c-Rel (Ruan *et al.*, 2009) or CNS3 (Zheng *et al.*, 2010) results in a profound reduction in the frequency of thymic and peripheral Treg cells. Thus, it was hypothesized that c-Rel acts as a pioneer transcription factor that induces global changes in chromatin structure after binding to CNS3 to open up the Foxp3 locus (Long *et al.*, 2009; Zheng *et al.*, 2010), akin to its role on the IL-2 locus (Rao *et al.*, 2003).

It has also been proposed that c-Rel along with p65, is important for the formation of a Treg-specific enhanceosome in the Foxp3 promoter (Ruan *et al.*, 2009). The enhanceosome would act as a "coincidence sensor" to detect the presence of non-Treg-specific transcription factors such as NFATc2, AP1, CREB, and Smad that come together at the promoter and distal enhancers during Treg-inducing conditions. In this way, the Treg cell-specific enhanceosome can act as the molecular switch for turning on Foxp3 expression while incorporating non-Treg cell factors in Foxp3 gene expression.

TCR stimulation can also activate the PI3K–AKT–mammalian target of rapamycin (mTOR) signaling axis (Fig. 2.3). One of the earliest clues that this pathway is involved in Treg cell differentiation was that addition of rapamycin, an inhibitor of mTOR, facilitated peripheral Treg cell development (Battaglia *et al.*, 2005). The mechanism by which AKT affects Treg cell differentiation appears to be mediated via phosphorylation of Foxo1 or Foxo3a, as described above for thymic Treg cell development. Because there is some redundancy between Foxo1 and Foxo3a, only double-deficient mice shown reduced Treg cell numbers (Kerdiles *et al.*, 2010). However, once Foxp3 is induced in the Foxo1$^{-/-}$ Foxo3a$^{-/-}$ Treg cells, the expression level of Foxp3 is normal. Thus, AKT blocks Treg cell differentiation by inhibiting Foxo proteins, which facilitate Foxp3 induction but are not required for the maintenance of Foxp3 expression (Kerdiles *et al.*, 2010).

Consistent with the observation that AKT overexpression negatively impacts iTreg cell development (Haxhinasto *et al.*, 2008), regulators of this pathway also affect Foxp3 induction. For example, deletion of Cbl-b (Casitas B-lineage lymphoma-b), a RING-type E3 ubiquitin ligase involved in degradation of p85, a regulatory subunit of PI3K, results in enhanced AKT activity (Fang and Liu, 2001). Cbl-b-deficient T cells are defective in iTreg cell generation *in vitro* (Wohlfert *et al.*, 2006) and *in vivo* (Harada *et al.*, 2010), via Foxo-dependent mechanisms (Harada *et al.*, 2010; Kerdiles *et al.*, 2010). Another negative regulator of signals upstream of AKT is phosphatase and tensin homolog (PTEN), a phosphatase that breaks down PIP3 thereby inactivating AKT. Accordingly, PTEN-deficient cells were found to be defective in their ability to upregulate Foxp3 in the presence of TGF-β *in vitro* (Sauer *et al.*, 2008). Conversely, T cells deficient in mTOR are skewed toward upregulating Foxp3 instead of differentiating into Th1, Th2, or Th17 after TCR stimulation (Delgoffe *et al.*, 2009). Along the same vein, engagement of PD-1 (programmed death 1) on T cells with its ligands, PD-L1 and PD-L2 on APCs, has been shown to be important for peripheral Treg conversion *in vitro* and *in vivo* via upregulation of PTEN (Francisco *et al.*, 2009). Thus, the AKT–mTOR pathway is important for determining peripheral T cell fate (Fig. 2.3).

Although AKT has been proposed to play a role in both thymic and peripheral Treg cell differentiation (Haxhinasto *et al.*, 2008), it may have a greater role in the periphery. While the TCR signaling window for peripheral Foxp3 induction can be widened somewhat by opposing the AKT–mTOR pathway using inhibitors of PI3K, AKT, or mTOR (Sauer *et al.*, 2008), disruption of this same pathway via the deletion of p110 results in increased negative selection in the thymus (Sasaki *et al.*, 2000). Moreover, T cell-specific deficiencies in the AKT signaling pathway such as PTEN (Walsh *et al.*, 2006) or mTOR (Delgoffe *et al.*, 2009) have normal numbers of Foxp3$^+$ thymocytes. Thus, these data suggest that the AKT–mTOR pathway may not play an important role in thymic Treg cell differentiation.

A greater role for AKT–mTOR in peripheral Treg cell differentiation would be consistent with the observation that the propensity for TCR stimulation to lead to Foxp3 induction decreases with the maturation state of the T cells (Wirnsberger *et al.*, 2009). Moreover, recent thymic emigrants appear to more readily undergo peripheral conversion in comparison with older naïve T cells in the periphery (Schallenberg *et al.*, 2010). In contrast to the ability of T cells to undergo Treg cell development, the propensity to proliferate and undergo T helper development increases with T cell maturation, which would be consistent with a greater role for AKT. One hypothesis is that strong TCR signals in the thymus preferentially induce NF-κB and Treg cell development, whereas those signals in the periphery would induce both NF-κB and AKT, leading to proliferation and effector cell development. Weak TCR signals in the periphery would induce only NF-κB, resulting in Treg cell development. This is consistent with the observation that, *in vivo*, conversion to iTreg is least efficient in the cells that underwent the most robust proliferation (Kretschmer *et al.*, 2005) and that suboptimal agonist peptide stimulation is most efficient for peripheral Treg cell differentiation (Gottschalk *et al.*, 2010). Further experiments will be required to confirm the hypothesis that modulation of AKT–mTOR is responsible for the variation in TCR-dependent cell-fate decisions with T cell maturation.

3.2.3.2. Costimulation-derived signals While the use of agonist CD28 antibodies has long been a mainstay of *in vitro* Treg generation protocols (Chen *et al.*, 2003; Fantini *et al.*, 2004; Kretschmer *et al.*, 2005), the role of

CD28 in iTreg development is still contested. On one hand, a number of studies suggested that similar to thymic development, CD28 is essential for peripheral conversion. For example, *in vivo* transfer of $CD4^{+}CD25^{-}$ T cells into B7-deficient mice failed to generate suppressive $CD4^{+}CD25^{+}$ Treg cells (Liang *et al.*, 2005). A similar study involving the adoptive transfer of naïve T cells from CD28-, ICOS-, and CD28/ICOS-deficient mice suggested that peripheral conversion requires CD28 but not ICOS (Guo *et al.*, 2008). *In vitro*, peripheral Treg cell generation from CD28-deficient T cells in the presence of anti-CD3 TCR stimulation and TGF-β can be corrected upon addition of exogenous IL-2 (Guo *et al.*, 2008), suggesting that the role of CD28 costimulation was primarily to enhance IL-2 production by T cells. On the other hand, other studies have shown that CD28 engagement antagonizes peripheral conversion (Kim and Rudensky, 2006). *In vitro*, increasing the concentration of anti-CD28 antibodies resulted in fewer $Foxp3^{+}$ cells from naïve T cells stimulated with anti-CD3, IL-2, and TGF-β (Semple *et al.*, 2011). Moreover, DCs expressing high levels of CD80/CD86 (B7-1/B7-2) also showed diminished iTreg induction (Benson *et al.*, 2007). This is consistent with the observation that CTLA-4 is needed early after *in vitro* TCR activation to allow TGF-β to efficiently induce Foxp3 expression, presumably via downregulation of the CD28 ligand, B7, on the APC (Zheng *et al.*, 2006). In summary, CD28 does not appear to play as crucial of a role in peripheral Treg cell differentiation in comparison with its role in the thymus. Some of the contradictory results may be related to the relative importance in the particular model system of CD28 on IL-2 production versus AKT activation, which would be predicted to have potentially opposing effects on iTreg cell development.

3.2.3.3. ***Cytokine-derived signals*** As discussed above, a crucial cytokine for iTreg cell differentiation is TGF-β. The primary mechanism appears to be via activation of SMAD3, which binds to a region in the Foxp3 locus, CNS1, downstream of the promoter and UTR-2a and -2b, but upstream of exon 1 (Kim and Leonard, 2007). Interestingly, SMAD3 by itself does not stimulate transcription but likely works synergistically with NFAT (Tone *et al.*, 2008) and pCREB (Ruan *et al.*, 2009) at the CNS1 site as a TGF-β sensitive enhancer. This is consistent with reports of optimal TCR signaling being sufficient for nominal induction of Foxp3, with TGF-β acting as a potent enhancer of its expression (Sauer *et al.*, 2008). Importantly, deletion of this region does not appear to affect thymic Treg cell generation but does impair TGFβ-driven Foxp3 induction *in vitro* as well as reduce the Treg cell numbers in both mesenteric lymph nodes and gut-associated lymphoid tissue (GALT) (Zheng *et al.*, 2010). Thus, it appears that SMAD3 binding to CNS1 is important for TGF-β-mediated induction of iTreg cells.

TGF-β has also been suggested to enhance peripheral conversion via the upregulation of TGF-β inducible early gene-1 (TIEG1), a member of the SP-1-like family of transcription factors, that binds to the SP-1 site in the Foxp3 promoter to positively regulate Foxp3 expression (Venuprasad *et al.*, 2008). An added layer of regulation is provided by the E3 ubiquitin ligase, Itch, which monoubiquitinates TIEG1 to facilitate its export into the nucleus. TGF-β may also facilitate iTreg cell development via its antiproliferative capacity via an unknown mechanism, based on observations from a number of *in vitro* and *in vivo* studies that extensive proliferation results in a low frequency of $Foxp3^+$ cells (Gottschalk *et al.*, 2010; Kretschmer *et al.*, 2005; Rubtsov and Rudensky, 2007). Finally, TGF-β may facilitate Treg cell differentiation by negatively regulating alterative T cell fates by blocking the expression of T-bet for Th1 (Gorelik *et al.*, 2002) and GATA3 for Th2 (Maruyama *et al.*, 2011) development, respectively. Thus, TGF-β likely has multiple roles in iTreg cell differentiation.

RA likely facilitates iTreg cell generation via several nonmutually exclusive pathways—both direct and indirect. RA binding to RARα can block cytokine production by effector T cells. By curtailing the synthesis of effector cytokines like IL-4, IL-21, and IFN-γ by memory/effector CD4+ T cells, RA permits TGF-β to act unopposed for inducing Foxp3 expression (Hill *et al.*, 2008). Also RA binding to RARα and RXR induces their import into the nucleus and binding to CNS1 and to a lesser degree, the Foxp3 promoter. This binding leads to increased histone acetylation at CNS1, which improves accessibility for SMAD3, thereby increasing the efficiency of Foxp3 induction. RA also participates in the reciprocal regulation of iTreg versus Th17 by blocking the expression of IL-6Rα and IL-23R. Finally, RA also enhances TGF-β-driven expression and phosphorylation of SMAD3 (Xiao *et al.*, 2008). Thus, RA appears to have numerous direct and indirect effects on iTreg cell differentiation.

3.2.3.4. Stabilization of Foxp3 expression An important mechanism controlling Foxp3 gene expression appears to be epigenetic via methylation. DNA methyltransferase 1 (Dnmt1) appears to be the primary methylase responsible for transcriptional repression of the Foxp3 locus, as TCR activation alone in the absence of DnmtI is sufficient to induce Foxp3 (Josefowicz *et al.*, 2009), similar to the its role in epigenetic regulation of cytokine genes and other aspects of T cell subset differentiation (Wilson *et al.*, 2005). Thus, methylation of the Foxp3 locus appears to be important for preventing inappropriate Treg cell differentiation.

It has been observed that one region of Foxp3, termed the Treg cell-specific demethylated region (TSDR) and also called CNS2, is completely demethylated during the process of thymic Treg cell differentiation (Floess *et al.*, 2007). Based on reporter assays, it appears that four CpG islands in this region are demethylated to provide access to a number of transcription factors, including NF-κB, CREB/ATF (Kim and Leonard, 2007), Ets-1 (Mouly *et al.*, 2010; Polansky *et al.*, 2010), and pSTAT5 (Zorn *et al.*, 2006). While both *in vivo*-generated iTregs and nTregs can stably demethylate the TSDR, iTregs generated *in vitro* with strong TCR stimulation in the presence of TGF-β are incompletely demethylated, and are subsequently unstable when TGF-β is withdrawn (Floess *et al.*, 2007). Although correlative, these data suggest that TSDR/CNS2 methylation status affects the stability of, but not necessarily access to, the Foxp3 locus.

Despite the number of transcription factors that appear to be regulated by methylation of CNS2, deletion of this region did not affect thymic or TGF-β-driven Foxp3 induction (Zheng *et al.*, 2010). Instead, CNS2-deficient mice have a gradual reduction in Treg numbers that becomes more apparent as the mice age. *In vitro* activation of CNS2-KO Treg cells with anti-CD3, anti-CD28, and IL-2 resulted in a decrease in the percentage of Foxp3$^+$ cells as well as the amount of Foxp3 per cell with each round of cell division. Given that complete demethylation of CNS2 is a characteristic of stable Treg cells, these data suggest that CNS2 appears important for the stable maintenance of Foxp3 expression.

Interestingly, many of the transcription factors that bind to CNS2 may not be involved in CNS2-dependent stabilization of Foxp3 expression. While a dominant-negative form of CREB affected Foxp3-expressing immortalized cells (HLTV-1 transformed MT-2 cells) (Kim and Leonard, 2007), it did not affect the expression of Foxp3 in normal peripheral Treg cells (Zheng *et al.*, 2010). Also, Stat5-deficient Treg cells were as stable as WT when transferred into Treg cell-depleted recipients (Yao *et al.*, 2007). Instead, Foxp3 itself binding in a complex with Runx1-CBFb to demethylated CNS2 was found to be the key to the stability of its expression (Zheng *et al.*, 2010). However, Foxp3 bound to CNS2 is not observed in iTreg cells generated by TGF-β *in vitro* (Zheng *et al.*, 2010) (Fig. 2.3). Presumably, the transcription factor complex on CNS2 imparts permissive epigenetic modifications like the trimethylation of histone H3 at lysine 4 (H3-K4me3) at the promoter while opposing inhibitory modifications like the trimethylation of H3 at lysine 9 (H3-K9me3) (Rudra *et al.*, 2009). These epigenetic modifications would then operate as a feed forward loop for Foxp3 to stabilize its own expression.

3.3. Extrinsic factors

3.3.1. APCs

The notion that APC populations may be tolerogenic and not proinflammatory arose from studies directly delivering antigens to DCs via the fusion of the antigenic peptide and a monoclonal antibody specific for DEC-205 (Hawiger *et al.*, 2001). DEC-205 is an adsorptive endocytosis receptor expressed ubiquitously on DCs from various peripheral lymphoid organs but is not directly responsible for transmitting the tolerogenic signals to responding T cells (Kronin *et al.*, 2000). The role of DEC-205 in iTreg cell development was later confirmed when antigen-specific Foxp3$^+$ T cells were obtained using the same strategy of fusing the antigen to the monoclonal antibody specific to DEC-205 (Kretschmer *et al.*, 2005). *In vitro*, DCs also appear to play an important role on TGF-β-mediated iTreg cell generation (Yamazaki *et al.*, 2007). Thus, DCs likely play an important role in iTreg cell generation.

There are a number of studies suggesting that certain DC subsets are crucial for iTreg cell generation or tolerance. For example, it was reported that plasmacytoid dendritic cells are important for Treg cell development to alloantigens in a transplantation model (Ochando *et al.*, 2006). Steady-state migratory RelB$^+$ langerin$^+$ dermal DCs have been suggested to facilitate Treg cell induction from antigens present in the skin (Azukizawa *et al.*, 2011).

One of the best characterized DC subsets for iTreg cell induction is the CD103$^+$ gut-associated population (Belkaid and Oldenhove, 2008), which may represent the basis for the high rate of peripheral conversion in the gut (Atarashi *et al.*, 2011; Lathrop *et al.*, 2011; Sun *et al.*, 2007). CD103$^+$ DCs from the small intestine lamina propria and mesenteric lymph nodes are substantially better than splenic DCs in converting naïve T cells to Foxp3$^+$ Tregs in the presence of exogenous TGF-β (Benson *et al.*, 2007; Coombes *et al.*, 2007; Hill *et al.*, 2008; Sun *et al.*, 2007). The primary mechanism is thought to be mediated by the production of RA, as discussed above. However, it has been argued that lamina propria CD103$^+$CD11b$^+$ DCs preferentially induced Th17, whereas CD11b$^+$F4/80$^+$CD11c$^-$ macrophages were the most efficient at mediating Foxp3 induction due to IL-10 and TGF-β production (Denning *et al.*, 2011, 2007). Future experiments will be required to reconcile these interpretations. In summary, the bulk of the current evidence favors an important role for subsets of DCs in iTreg cell differentiation. Their role will depend on the source of antigen, and the expression level of MHC class II, costimulatory molecules, TGF-β, and synthetic enzymes for RA.

3.3.2. Integration of signals from the environment

There is now a growing body of evidence that the different T cell lineages regulate each other in the periphery. One of the most striking examples of this cross talk is the reciprocal regulation of tolerogenic iTreg and inflammatory Th17 cells—both of which can be generated after TCR stimulation in the presence of TGF-β (Bettelli *et al.*, 2006; Mangan *et al.*, 2006). High levels of TGF-β in the absence of IL-6 favor iTreg cell induction because too much TGF-β can repress IL-23R expression, thus preventing the expansion of the Th17 cells (Zhou *et al.*, 2008). Further, IL-2, which is required for TGF-β-dependent induction of Foxp3 (Davidson *et al.*, 2007), induces STAT5, which competes with STAT3 for binding to the IL-17 locus and represses its activation (Yang *et al.*, 2011). Last, Foxp3 physically interacts and inhibits the function of RORγt, the transcription factor that directs the development and mediates the function of Th17 cells (Zhou *et al.*, 2008).

By contrast, Th17 cells develop in the presence of IL-6 (Veldhoen *et al.*, 2006) or IL-21 (Korn *et al.*, 2007) via activation of STAT3, which activates the IL-17 locus and also appears to prevent Foxp3 induction (Durant *et al.*, 2010). Although STAT3 binding to the Foxp3 locus was initially reported to be marginal based on immunoprecipitation assays (Yao *et al.*, 2007), another study using luciferase reporters suggested that STAT3 bound to CNS2 (Xu *et al.*, 2010). Interestingly, binding of STAT3 resulted in decreased luciferase levels, implying that it is a silencer of Foxp3 expression. This finding is consistent with the observation that IL-6 causes the methylation of CpG islands in CNS2, and repression of Foxp3 expression (Lal *et al.*, 2009). Although the deletion of CNS2 was associated with Foxp3 stability (Zheng *et al.*, 2010), the deletion is not specific to the STAT3 site and likely affects more than one transcriptional pathway. Thus, Th17 and iTreg cell differentiation appears to be cross-regulated at multiple levels in response to cytokines.

The balance between Th17-mediated inflammation and iTreg-enforced tolerance can also be informed by other environmental signals. One mechanism is mediated by the aryl hydrocarbon receptor (AHR), a highly conserved transcription factor that is normally bound as an inactive complex in the cytosol with a number of chaperones (Quintana *et al.*, 2008). AHR is a promiscuous receptor, in that it can bind a great number of endogenous and xenobiotic ligands. It is believed to mediate normal development by transducing signals from endogenous agonists, but it also serves a protective role by adaptively mediating the catabolism of harmful chemicals—in particular, activation of cytochrome P450 monoxygenase, which can dispose of the various poisons that bind AHR (Nguyen and Bradfield, 2008). Ligand binding causes conformational change that not only exposes a nuclear localization signal but also alters its association with its chaperones. After it migrates to the nucleus, AHR associates with the aryl hydrocarbon receptor nuclear translocation, and their heterodimerization increases their collective capacity to bind specific enhancer elements.

Genomic analysis identified three nonconserved AHR-binding sites in the Foxp3 promoter and one conserved AHR-binding site around exon 11. AHR binding of the synthetic ligand 2,3,7,8-tetrachlorodibenzo-*p*-dioxin (TCDD) or the tryptophan-derived endogenous ligand 2-(1′H-indole-3′-carbonyl)-thiazole-4-carboxylic acid methyl ester (ITE) activates AHR so it binds the Foxp3 promoter and drives iTreg development (Quintana *et al.*, 2008). TCR stimulation along with activation of AHR by TCDD (Quintana *et al.*, 2008) or ITE (Quintana *et al.*, 2010) in the absence of exogenous TGF-β is sufficient to convert naïve T cells into Foxp3$^+$ iTreg cells that are suppressive *in vitro* and protective against experimental autoimmune encephalomyelitis *in vivo*. In humans, AHR stimulation by TCDD was required in addition to TGF-β for the *in vitro* generation of iTreg cell that can suppress effector T cell proliferation via the action of the ectonucleosidase, CD39 (Gandhi *et al.*, 2010). Thus, AHR can play an important role on iTreg cell selection.

Interestingly, other AHR ligands such as 6-formylindolo[3,2-b]carbazole, the product of tryptophan breakdown due to UV-B exposure (Nguyen and Bradfield, 2008), can augment the generation of Th17 cells in the presence of IL-6 and TGF-β (Quintana *et al.*, 2008; Veldhoen *et al.*, 2008). Thus, the nature of the AHR ligand may participate in Th17 versus iTreg cell selection.

Finally, Th17-inducing inflammatory cytokines are not the only ones that inhibit iTreg cell differentiation. For instance, chromatin immunoprecipitation experiments indicate that IL-12, a Th1-promoting cytokine, signals via STAT4 which prevents STAT5 binding to the Foxp3 promoter and CNS2, thereby inhibiting iTreg cell development (O'Malley *et al.*, 2009). Another mechanism by which Th1-polarizing conditions prevent Foxp3 induction may be via the direct binding of T-bet to the Foxp3 locus (Wei *et al.*, 2007). Signals favoring Th2 development such as IL-4 also appear to block iTreg cell development via GATA-3 binding to the Foxp3 promoter (Mantel *et al.*, 2007). Foxp3 expression also appears to be blocked by STAT6 binding to a second IL-4 sensitive region between CNS1 and CNS2 in the Foxp3 locus (Takaki *et al.*, 2008). Thus, the process of effector T helper differentiation is, in general, antagonistic to the generation of iTreg cells (Fig. 2.4).

4. SUMMARY

A substantial body of work has been generated regarding the thymic and peripheral selection of Treg cells. The generation of Treg cell in the thymus and periphery upon antigen encounter is likely related to the specificity of the Treg cells for self- and foreign antigens. Similarities between these processes include the notion that the TCR ligands tend to not be abundant and in the agonist affinity range. In the thymus, common ligands likely result in negative selection, whereas high antigen load in the periphery may favor an effector response, as it may indicate rapid proliferation of pathogens. Uncommon tissue-specific antigens appear to be the primary ligands of thymic Treg cell differentiation, whereas low-level foreign antigens such as those from commensal bacteria or food may

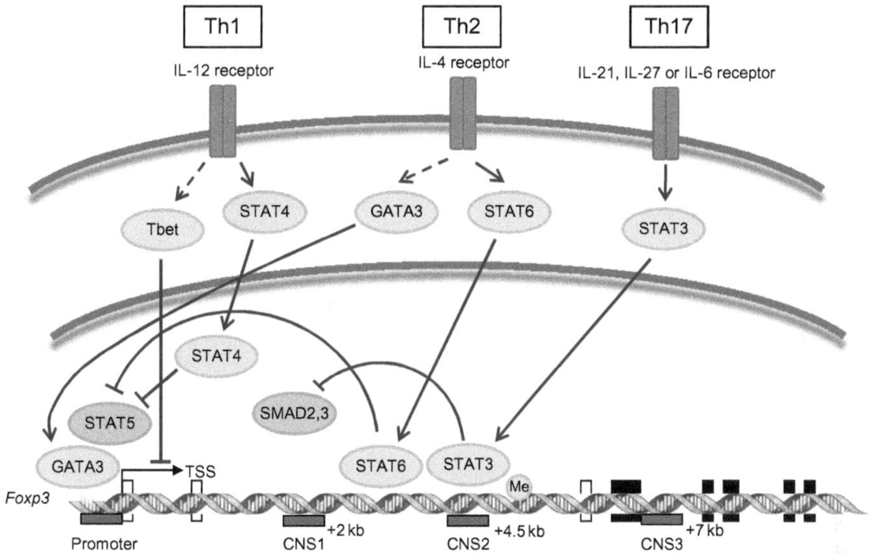

FIGURE 2.4 Negative regulation of peripheral Foxp3 induction. Cytokines associated with other T cell lineages can suppress the induction of Foxp3. Binding of IL-6, IL-21, or IL-27 to their respective type I cytokine receptors can activate STAT3. Binding of STAT3 to CNS2 can block SMAD2,3 binding to CNS1, thereby silencing this enhancer. However, IL-4 binding to its receptor causes the activation of STAT6 and the upregulation of GATA3. This pathway blocks Foxp3 induction via STAT6 binding to an intronic segment between CNS1 and CNS2. GATA3 is also presumed to bind the Foxp3 promoter, thereby contributing to the silencing of this gene. Finally, IL-12 binding to its receptor activates STAT4, which prevents the binding of STAT5 to both the promoter and the CNS2. IL-12 signaling also causes the upregulation of T-bet which can bind a currently unidentified segment of the Foxp3 locus to prevent its expression.

be predicted to facilitate iTreg cell differentiation. While cytokines such as IL-2 are involved in both thymic and peripheral Treg cell differentiation, it is likely that cytokines play the dominant role in the periphery in determining iTreg cell versus other T helper cell fate. Future studies will be required to validate or disprove these notions regarding thymic and peripheral Treg cell differentiation.

REFERENCES

Anderson, M. S., Venanzi, E. S., Klein, L., Chen, Z., Berzins, S. P., Turley, S. J., von Boehmer, H., Bronson, R., Dierich, A., Benoist, C., and Mathis, D. (2002). Projection of an immunological self shadow within the thymus by the aire protein. *Science* **298,** 1395–1401.

Anderson, M. S., Venanzi, E. S., Chen, Z., Berzins, S. P., Benoist, C., and Mathis, D. (2005). The cellular mechanism of Aire control of T cell tolerance. *Immunity* **23,** 227–239.

Apostolou, I., and von Boehmer, H. (2004). In vivo instruction of suppressor commitment in naive T cells. *J. Exp. Med.* **199,** 1401–1408.

Apostolou, I., Sarukhan, A., Klein, L., and von Boehmer, H. (2002). Origin of regulatory T cells with known specificity for antigen. *Nat. Immunol.* **3,** 756–763.

Appleby, M. W., and Ramsdell, F. (2008). Scurfy, the Foxp3 locus, and the molecular basis of peripheral tolerance. *Curr. Top. Microbiol. Immunol.* **321,** 151–168.

Asano, M., Toda, M., Sakaguchi, N., and Sakaguchi, S. (1996). Autoimmune disease as a consequence of developmental abnormality of a T cell subpopulation. *J. Exp. Med.* **184,** 387–396.

Aschenbrenner, K., D'Cruz, L. M., Vollmann, E. H., Hinterberger, M., Emmerich, J., Swee, L. K., Rolink, A., and Klein, L. (2007). Selection of Foxp3+ regulatory T cells specific for self antigen expressed and presented by Aire+ medullary thymic epithelial cells. *Nat. Immunol.* **8,** 351–358.

Atarashi, K., Tanoue, T., Shima, T., Imaoka, A., Kuwahara, T., Momose, Y., Cheng, G., Yamasaki, S., Saito, T., Ohba, Y., Taniguchi, T., Takeda, K., *et al.* (2011). Induction of colonic regulatory T cells by indigenous Clostridium species. *Science* **331,** 337–341.

Atibalentja, D. F., Byersdorfer, C. A., and Unanue, E. R. (2009). Thymus-blood protein interactions are highly effective in negative selection and regulatory T cell induction. *J. Immunol.* **183,** 7909–7918.

Azukizawa, H., Dohler, A., Kanazawa, N., Nayak, A., Lipp, M., Malissen, B., Autenrieth, I., Katayama, I., Riemann, M., Weih, F., Berberich-Siebelt, F., and Lutz, M. B. (2011). Steady state migratory RelB+ langerin+ dermal dendritic cells mediate peripheral induction of antigen-specific CD4+ CD25+ Foxp3+ regulatory T cells. *Eur. J. Immunol.* **41,** 1420–1434.

Barnes, M. J., and Powrie, F. (2009). Regulatory T cells reinforce intestinal homeostasis. *Immunity* **31,** 401–411.

Barnes, M. J., Krebs, P., Harris, N., Eidenschenk, C., Gonzalez-Quintial, R., Arnold, C. N., Crozat, K., Sovath, S., Moresco, E. M., Theofilopoulos, A. N., Beutler, B., and Hoebe, K. (2009). Commitment to the regulatory T cell lineage requires CARMA1 in the thymus but not in the periphery. *PLoS Biol.* **7,** e51.

Battaglia, M., Stabilini, A., and Roncarolo, M. G. (2005). Rapamycin selectively expands CD4+CD25+FoxP3+ regulatory T cells. *Blood* **105,** 4743–4748.

Bautista, J. L., Lio, C. W., Lathrop, S. K., Forbush, K., Liang, Y., Luo, J., Rudensky, A. Y., and Hsieh, C. S. (2009). Intraclonal competition limits the fate determination of regulatory T cells in the thymus. *Nat. Immunol.* **10,** 610–617.

Bayer, A. L., Lee, J. Y., de la Barrera, A., Surh, C. D., and Malek, T. R. (2008). A function for IL-7R for CD4+CD25+Foxp3+ T regulatory cells. *J. Immunol.* **181,** 225–234.

Belkaid, Y., and Oldenhove, G. (2008). Tuning microenvironments: Induction of regulatory T cells by dendritic cells. *Immunity* **29,** 362–371.

Belkaid, Y., and Tarbell, K. (2009). Regulatory T cells in the control of host-microorganism interactions (*). *Annu. Rev. Immunol.* **27,** 551–589.

Bensinger, S. J., Bandeira, A., Jordan, M. S., Caton, A. J., and Laufer, T. M. (2001). Major histocompatibility complex class II-positive cortical epithelium mediates the selection of CD4(+)25(+) immunoregulatory T cells. *J. Exp. Med.* **194,** 427–438.

Benson, M. J., Pino-Lagos, K., Rosemblatt, M., and Noelle, R. J. (2007). All-trans retinoic acid mediates enhanced T reg cell growth, differentiation, and gut homing in the face of high levels of co-stimulation. *J. Exp. Med.* **204,** 1765–1774.

Bettelli, E., Carrier, Y., Gao, W., Korn, T., Strom, T. B., Oukka, M., Weiner, H. L., and Kuchroo, V. K. (2006). Reciprocal developmental pathways for the generation of pathogenic effector TH17 and regulatory T cells. *Nature* **441,** 235–238.

Bosco, N., Agenes, F., Rolink, A. G., and Ceredig, R. (2006). Peripheral T cell lymphopenia and concomitant enrichment in naturally arising regulatory T cells: The case of the pre-Talpha gene-deleted mouse. *J. Immunol.* **177,** 5014–5023.

Bouneaud, C., Kourilsky, P., and Bousso, P. (2000). Impact of negative selection on the T cell repertoire reactive to a self-peptide: A large fraction of T cell clones escapes clonal deletion. *Immunity* **13,** 829–840.

Burchill, M. A., Yang, J., Vang, K. B., Moon, J. J., Chu, H. H., Lio, C. W., Vegoe, A. L., Hsieh, C. S., Jenkins, M. K., and Farrar, M. A. (2008). Linked T cell receptor and cytokine signaling govern the development of the regulatory T cell repertoire. *Immunity* **28,** 112–121.

Cai, Q., Dierich, A., Oulad-Abdelghani, M., Chan, S., and Kastner, P. (2009). Helios deficiency has minimal impact on T cell development and function. *J. Immunol.* **183,** 2303–2311.

Canelles, M., Park, M. L., Schwartz, O. M., and Fowlkes, B. J. (2003). The influence of the thymic environment on the CD4-versus-CD8 T lineage decision. *Nat. Immunol.* **4,** 756–764.

Chan, A. C., Iwashima, M., Turck, C. W., and Weiss, A. (1992). ZAP-70: A 70 kDa protein-tyrosine kinase that associates with the TCR zeta chain. *Cell* **71,** 649–662.

Chen, W., Jin, W., Hardegen, N., Lei, K. J., Li, L., Marinos, N., McGrady, G., and Wahl, S. M. (2003). Conversion of peripheral CD4+CD25− naive T cells to CD4+CD25+ regulatory T cells by TGF-beta induction of transcription factor Foxp3. *J. Exp. Med.* **198,** 1875–1886.

Chen, Q., Kim, Y. C., Laurence, A., Punkosdy, G. A., and Shevach, E. M. (2011). IL-2 controls the stability of Foxp3 expression in TGF-beta-induced Foxp3+ T cells in vivo. *J. Immunol.* **186,** 6329–6337.

Coombes, J. L., Siddiqui, K. R., Arancibia-Carcamo, C. V., Hall, J., Sun, C. M., Belkaid, Y., and Powrie, F. (2007). A functionally specialized population of mucosal CD103+ DCs induces Foxp3+ regulatory T cells via a TGF-beta and retinoic acid-dependent mechanism. *J. Exp. Med.* **204,** 1757–1764.

Curotto de Lafaille, M. A., Lino, A. C., Kutchukhidze, N., and Lafaille, J. J. (2004). CD25− T cells generate CD25+Foxp3+ regulatory T cells by peripheral expansion. *J. Immunol.* **173,** 7259–7268.

Daniels, M. A., Teixeiro, E., Gill, J., Hausmann, B., Roubaty, D., Holmberg, K., Werlen, G., Hollander, G. A., Gascoigne, N. R., and Palmer, E. (2006). Thymic selection threshold defined by compartmentalization of Ras/MAPK signalling. *Nature* **444,** 724–729.

Daniely, D., Kern, J., Cebula, A., and Ignatowicz, L. (2010). Diversity of TCRs on natural Foxp3+ T cells in mice lacking Aire expression. *J. Immunol.* **184,** 6865–6873.

Davidson, T. S., DiPaolo, R. J., Andersson, J., and Shevach, E. M. (2007). Cutting edge: IL-2 is essential for TGF-beta-mediated induction of Foxp3+ T regulatory cells. *J. Immunol.* **178,** 4022–4026.

D'Cruz, L. M., and Klein, L. (2005). Development and function of agonist-induced CD25+Foxp3+ regulatory T cells in the absence of interleukin 2 signaling. *Nat. Immunol.* **6,** 1152–1159.

Delgoffe, G. M., Kole, T. P., Zheng, Y., Zarek, P. E., Matthews, K. L., Xiao, B., Worley, P. F., Kozma, S. C., and Powell, J. D. (2009). The mTOR kinase differentially regulates effector and regulatory T cell lineage commitment. *Immunity* **30,** 832–844.

Denning, T. L., Wang, Y. C., Patel, S. R., Williams, I. R., and Pulendran, B. (2007). Lamina propria macrophages and dendritic cells differentially induce regulatory and interleukin 17-producing T cell responses. *Nat. Immunol.* **8,** 1086–1094.

Denning, T. L., Norris, B. A., Medina-Contreras, O., Manicassamy, S., Geem, D., Madan, R., Karp, C. L., and Pulendran, B. (2011). Functional specializations of intestinal dendritic cell and macrophage subsets that control Th17 and regulatory T cell responses are dependent on the T cell/APC ratio, source of mouse strain, and regional localization. *J. Immunol.* **187,** 733–747.

d'Hennezel, E., Ben-Shoshan, M., Ochs, H. D., Torgerson, T. R., Russell, L. J., Lejtenyi, C., Noya, F. J., Jabado, N., Mazer, B., and Piccirillo, C. A. (2009). FOXP3 forkhead domain mutation and regulatory T cells in the IPEX syndrome. *N. Engl. J. Med.* **361,** 1710–1713.

DiPaolo, R. J., and Shevach, E. M. (2009). CD4+ T-cell development in a mouse expressing a transgenic TCR derived from a Treg. *Eur. J. Immunol.* **39,** 234–240.

Dooley, J., Erickson, M., and Farr, A. G. (2008). Alterations of the medullary epithelial compartment in the Aire-deficient thymus: Implications for programs of thymic epithelial differentiation. *J. Immunol.* **181,** 5225–5232.

Durant, L., Watford, W. T., Ramos, H. L., Laurence, A., Vahedi, G., Wei, L., Takahashi, H., Sun, H. W., Kanno, Y., Powrie, F., and O'Shea, J. J. (2010). Diverse targets of the transcription factor STAT3 contribute to T cell pathogenicity and homeostasis. *Immunity* **32,** 605–615.

Fang, D., and Liu, Y. C. (2001). Proteolysis-independent regulation of PI3K by Cbl-b-mediated ubiquitination in T cells. *Nat. Immunol.* **2,** 870–875.

Fantini, M. C., Becker, C., Monteleone, G., Pallone, F., Galle, P. R., and Neurath, M. F. (2004). Cutting edge: TGF-beta induces a regulatory phenotype in CD4+CD25− T cells through Foxp3 induction and down-regulation of Smad7. *J. Immunol.* **172,** 5149–5153.

Feuerer, M., Jiang, W., Holler, P. D., Satpathy, A., Campbell, C., Bogue, M., Mathis, D., and Benoist, C. (2007). Enhanced thymic selection of FoxP3+ regulatory T cells in the NOD mouse model of autoimmune diabetes. *Proc. Natl. Acad. Sci. USA* **104,** 18181–18186.

Feuerer, M., Hill, J. A., Kretschmer, K., von Boehmer, H., Mathis, D., and Benoist, C. (2010). Genomic definition of multiple ex vivo regulatory T cell subphenotypes. *Proc. Natl. Acad. Sci. USA* **107,** 5919–5924.

Floess, S., Freyer, J., Siewert, C., Baron, U., Olek, S., Polansky, J., Schlawe, K., Chang, H. D., Bopp, T., Schmitt, E., Klein-Hessling, S., Serfling, E., *et al.* (2007). Epigenetic control of the foxp3 locus in regulatory T cells. *PLoS Biol.* **5,** e38.

Fontenot, J. D., Dooley, J. L., Farr, A. G., and Rudensky, A. Y. (2005a). Developmental regulation of Foxp3 expression during ontogeny. *J. Exp. Med.* **202,** 901–906.

Fontenot, J. D., Rasmussen, J. P., Gavin, M. A., and Rudensky, A. Y. (2005b). A function for interleukin 2 in Foxp3-expressing regulatory T cells. *Nat. Immunol.* **6,** 1142–1151.

Francisco, L. M., Salinas, V. H., Brown, K. E., Vanguri, V. K., Freeman, G. J., Kuchroo, V. K., and Sharpe, A. H. (2009). PD-L1 regulates the development, maintenance, and function of induced regulatory T cells. *J. Exp. Med.* **206,** 3015–3029.

Furtado, G. C., Curotto de Lafaille, M. A., Kutchukhidze, N., and Lafaille, J. J. (2002). Interleukin 2 signaling is required for CD4(+) regulatory T cell function. *J. Exp. Med.* **196,** 851–857.

Gandhi, R., Kumar, D., Burns, E. J., Nadeau, M., Dake, B., Laroni, A., Kozoriz, D., Weiner, H. L., and Quintana, F. J. (2010). Activation of the aryl hydrocarbon receptor induces human type 1 regulatory T cell-like and Foxp3(+) regulatory T cells. *Nat. Immunol.* **11,** 846–853.

Garza, K. M., Agersborg, S. S., Baker, E., and Tung, K. S. (2000). Persistence of physiological self antigen is required for the regulation of self tolerance. *J. Immunol.* **164,** 3982–3989.

Getnet, D., Grosso, J. F., Goldberg, M. V., Harris, T. J., Yen, H. R., Bruno, T. C., Durham, N. M., Hipkiss, E. L., Pyle, K. J., Wada, S., Pan, F., Pardoll, D. M., *et al.* (2010). A role for the transcription factor Helios in human CD4(+)CD25(+) regulatory T cells. *Mol. Immunol.* **47,** 1595–1600.

Gillard, G. O., Dooley, J., Erickson, M., Peltonen, L., and Farr, A. G. (2007). Aire-dependent alterations in medullary thymic epithelium indicate a role for Aire in thymic epithelial differentiation. *J. Immunol.* **178,** 3007–3015.

Gorelik, L., Constant, S., and Flavell, R. A. (2002). Mechanism of transforming growth factor beta-induced inhibition of T helper type 1 differentiation. *J. Exp. Med.* **195,** 1499–1505.

Gottschalk, R. A., Corse, E., and Allison, J. P. (2010). TCR ligand density and affinity determine peripheral induction of Foxp3 in vivo. *J. Exp. Med.* **207,** 1701–1711.

Granucci, F., Vizzardelli, C., Pavelka, N., Feau, S., Persico, M., Virzi, E., Rescigno, M., Moro, G., and Ricciardi-Castagnoli, P. (2001). Inducible IL-2 production by dendritic cells revealed by global gene expression analysis. *Nat. Immunol.* **2,** 882–888.

Guo, F., Iclozan, C., Suh, W. K., Anasetti, C., and Yu, X. Z. (2008). CD28 controls differentiation of regulatory T cells from naive CD4 T cells. *J. Immunol.* **181,** 2285–2291.

Gupta, S., Manicassamy, S., Vasu, C., Kumar, A., Shang, W., and Sun, Z. (2008). Differential requirement of PKC-theta in the development and function of natural regulatory T cells. *Mol. Immunol.* **46,** 213–224.

Hall, J. A., Cannons, J. L., Grainger, J. R., Dos Santos, L. M., Hand, T. W., Naik, S., Wohlfert, E. A., Chou, D. B., Oldenhove, G., Robinson, M., Grigg, M. E., Kastenmayer, R., *et al.* (2011). Essential role for retinoic acid in the promotion of CD4 (+) T cell effector responses via retinoic acid receptor alpha. *Immunity* **34,** 435–447.

Harada, Y., Harada, Y., Elly, C., Ying, G., Paik, J. H., DePinho, R. A., and Liu, Y. C. (2010). Transcription factors Foxo3a and Foxo1 couple the E3 ligase Cbl-b to the induction of Foxp3 expression in induced regulatory T cells. *J. Exp. Med.* **207,** 1381–1391.

Haribhai, D., Williams, J. B., Jia, S., Nickerson, D., Schmitt, E. G., Edwards, B., Ziegelbauer, J., Yassai, M., Li, S. H., Relland, L. M., Wise, P. M., Chen, A., *et al.* (2011). A requisite role for induced regulatory T cells in tolerance based on expanding antigen receptor diversity. *Immunity* **35,** 109–122.

Hawiger, D., Inaba, K., Dorsett, Y., Guo, M., Mahnke, K., Rivera, M., Ravetch, J. V., Steinman, R. M., and Nussenzweig, M. C. (2001). Dendritic cells induce peripheral T cell unresponsiveness under steady state conditions in vivo. *J. Exp. Med.* **194,** 769–779.

Haxhinasto, S., Mathis, D., and Benoist, C. (2008). The AKT-mTOR axis regulates de novo differentiation of CD4+Foxp3+ cells. *J. Exp. Med.* **205,** 565–574.

Hill, J. A., Hall, J. A., Sun, C. M., Cai, Q., Ghyselinck, N., Chambon, P., Belkaid, Y., Mathis, D., and Benoist, C. (2008). Retinoic acid enhances Foxp3 induction indirectly by relieving inhibition from CD4+CD44hi Cells. *Immunity* **29,** 758–770.

Hinterberger, M., Aichinger, M., da Costa, O. P., Voehringer, D., Hoffmann, R., and Klein, L. (2010). Autonomous role of medullary thymic epithelial cells in central CD4(+) T cell tolerance. *Nat. Immunol.* **11,** 512–519.

Hinterberger, M., Wirnsberger, G., and Klein, L. (2011). B7/CD28 in central tolerance: Costimulation promotes maturation of regulatory T cell precursors and prevents their clonal deletion. *Front.Immunol.* **2,** 1–12.

Hori, S. (2010). c-Rel: A pioneer in directing regulatory T-cell lineage commitment? *Eur. J. Immunol.* **40,** 664–667.

Hsieh, C. S., Liang, Y., Tyznik, A. J., Self, S. G., Liggitt, D., and Rudensky, A. Y. (2004). Recognition of the peripheral self by naturally arising CD25+ CD4+ T cell receptors. *Immunity* **21,** 267–277.

Hsieh, C. S., Zheng, Y., Liang, Y., Fontenot, J. D., and Rudensky, A. Y. (2006). An intersection between the self-reactive regulatory and nonregulatory T cell receptor repertoires. *Nat. Immunol.* **7,** 401–410.

Hsu, L. Y., Tan, Y. X., Xiao, Z., Malissen, M., and Weiss, A. (2009). A hypomorphic allele of ZAP-70 reveals a distinct thymic threshold for autoimmune disease versus autoimmune reactivity. *J. Exp. Med.* **206,** 2527–2541.

Huesmann, M., Scott, B., Kisielow, P., and von Boehmer, H. (1991). Kinetics and efficacy of positive selection in the thymus of normal and T cell receptor transgenic mice. *Cell* **66,** 533–540.

Isomura, I., Palmer, S., Grumont, R. J., Bunting, K., Hoyne, G., Wilkinson, N., Banerjee, A., Proietto, A., Gugasyan, R., Wu, L., McNally, A., Steptoe, R. J., *et al.* (2009). c-Rel is required for the development of thymic Foxp3+ CD4 regulatory T cells. *J. Exp. Med.* **206,** 3001–3014.

Itoh, M., Takahashi, T., Sakaguchi, N., Kuniyasu, Y., Shimizu, J., Otsuka, F., and Sakaguchi, S. (1999). Thymus and autoimmunity: Production of CD25+CD4+ naturally anergic and suppressive T cells as a key function of the thymus in maintaining immunologic self-tolerance. *J. Immunol.* **162,** 5317–5326.

Jiang, W., Anderson, M. S., Bronson, R., Mathis, D., and Benoist, C. (2005). Modifier loci condition autoimmunity provoked by Aire deficiency. *J. Exp. Med.* **202,** 805–815.

Jordan, M. S., Boesteanu, A., Reed, A. J., Petrone, A. L., Holenbeck, A. E., Lerman, M. A., Naji, A., and Caton, A. J. (2001). Thymic selection of CD4+CD25+ regulatory T cells induced by an agonist self-peptide. *Nat. Immunol.* **2,** 301–306.

Josefowicz, S. Z., and Rudensky, A. (2009). Control of regulatory T cell lineage commitment and maintenance. *Immunity* **30,** 616–625.

Josefowicz, S. Z., Wilson, C. B., and Rudensky, A. Y. (2009). Cutting edge: TCR stimulation is sufficient for induction of Foxp3 expression in the absence of DNA methyltransferase 1. *J. Immunol.* **182,** 6648–6652.

Kawahata, K., Misaki, Y., Yamauchi, M., Tsunekawa, S., Setoguchi, K., Miyazaki, J., and Yamamoto, K. (2002). Generation of CD4(+)CD25(+) regulatory T cells from autoreactive T cells simultaneously with their negative selection in the thymus and from nonautoreactive T cells by endogenous TCR expression. *J. Immunol.* **168,** 4399–4405.

Kerdiles, Y. M., Stone, E. L., Beisner, D. R., McGargill, M. A., Ch'en, I. L., Stockmann, C., Katayama, C. D., and Hedrick, S. M. (2010). Foxo transcription factors control regulatory T cell development and function. *Immunity* **33,** 890–904.

Killebrew, J. R., Perdue, N., Kwan, A., Thornton, A. M., Shevach, E. M., and Campbell, D. J. (2011). A self-reactive TCR drives the development of Foxp3+ regulatory T cells that prevent autoimmune disease. *J. Immunol.* **187,** 861–869.

Kim, H. P., and Leonard, W. J. (2007). CREB/ATF-dependent T cell receptor-induced FoxP3 gene expression: A role for DNA methylation. *J. Exp. Med.* **204,** 1543–1551.

Kim, J. M., and Rudensky, A. (2006). The role of the transcription factor Foxp3 in the development of regulatory T cells. *Immunol. Rev.* **212,** 86–98.

Kim, J. M., Rasmussen, J. P., and Rudensky, A. Y. (2007). Regulatory T cells prevent catastrophic autoimmunity throughout the lifespan of mice. *Nat. Immunol.* **8,** 191–197.

Knoechel, B., Lohr, J., Kahn, E., Bluestone, J. A., and Abbas, A. K. (2005). Sequential development of interleukin 2-dependent effector and regulatory T cells in response to endogenous systemic antigen. *J. Exp. Med.* **202,** 1375–1386.

Koonpaew, S., Shen, S., Flowers, L., and Zhang, W. (2006). LAT-mediated signaling in CD4+CD25+ regulatory T cell development. *J. Exp. Med.* **203,** 119–129.

Korn, T., Bettelli, E., Gao, W., Awasthi, A., Jager, A., Strom, T. B., Oukka, M., and Kuchroo, V. K. (2007). IL-21 initiates an alternative pathway to induce proinflammatory T(H)17 cells. *Nature* **448,** 484–487.

Kramer, S., Schimpl, A., and Hunig, T. (1995). Immunopathology of interleukin (IL) 2-deficient mice: Thymus dependence and suppression by thymus-dependent cells with an intact IL-2 gene. *J. Exp. Med.* **182,** 1769–1776.

Kretschmer, K., Apostolou, I., Hawiger, D., Khazaie, K., Nussenzweig, M. C., and von Boehmer, H. (2005). Inducing and expanding regulatory T cell populations by foreign antigen. *Nat. Immunol.* **6,** 1219–1227.

Kronin, V., Wu, L., Gong, S., Nussenzweig, M. C., and Shortman, K. (2000). DEC-205 as a marker of dendritic cells with regulatory effects on CD8 T cell responses. *Int. Immunol.* **12,** 731–735.

Kuroda, N., Mitani, T., Takeda, N., Ishimaru, N., Arakaki, R., Hayashi, Y., Bando, Y., Izumi, K., Takahashi, T., Nomura, T., Sakaguchi, S., Ueno, T., *et al.* (2005). Development of autoimmunity against transcriptionally unrepressed target antigen in the thymus of Aire-deficient mice. *J. Immunol.* **174,** 1862–1870.

Lahl, K., Loddenkemper, C., Drouin, C., Freyer, J., Arnason, J., Eberl, G., Hamann, A., Wagner, H., Huehn, J., and Sparwasser, T. (2007). Selective depletion of Foxp3+ regulatory T cells induces a scurfy-like disease. *J. Exp. Med.* **204,** 57–63.

Lal, G., Zhang, N., van der Touw, W., Ding, Y., Ju, W., Bottinger, E. P., Reid, S. P., Levy, D. E., and Bromberg, J. S. (2009). Epigenetic regulation of Foxp3 expression in regulatory T cells by DNA methylation. *J. Immunol.* **182,** 259–273.

Lathrop, S. K., Santacruz, N. A., Pham, D., Luo, J., and Hsieh, C. S. (2008). Antigen-specific peripheral shaping of the natural regulatory T cell population. *J. Exp. Med.* **205,** 3105–3117.

Lathrop, S. K., Bloom, S. M., Rao, S. M., Nutsch, K., Lio, C.-W., Santacruz, N., Peterson, D. A., Stappenbeck, T. S., and Hsieh, C. S. (2011). Peripheral education of the immune system by colonic commensal microbiota. *Nature* doi:10.1038/nature10434.

Lee, H. M., and Hsieh, C. S. (2009). Rare development of Foxp3+ thymocytes in the CD4+CD8+ subset. *J. Immunol.* **183,** 2261–2266.

Lei, Y., Ripen, A. M., Ishimaru, N., Ohigashi, I., Nagasawa, T., Jeker, L. T., Bosl, M. R., Hollander, G. A., Hayashi, Y., Malefyt Rde, W., Nitta, T., and Takahama, Y. (2011). Aire-dependent production of XCL1 mediates medullary accumulation of thymic dendritic cells and contributes to regulatory T cell development. *J. Exp. Med.* **208,** 383–394.

Leung, M. W., Shen, S., and Lafaille, J. J. (2009). TCR-dependent differentiation of thymic Foxp3+ cells is limited to small clonal sizes. *J. Exp. Med.* **206,** 2121–2130.

Li, M. O., Sanjabi, S., and Flavell, R. A. (2006). Transforming growth factor-beta controls development, homeostasis, and tolerance of T cells by regulatory T cell-dependent and -independent mechanisms. *Immunity* **25,** 455–471.

Liang, S., Alard, P., Zhao, Y., Parnell, S., Clark, S. L., and Kosiewicz, M. M. (2005). Conversion of CD4+ CD25− cells into CD4+ CD25+ regulatory T cells in vivo requires B7 costimulation, but not the thymus. *J. Exp. Med.* **201,** 127–137.

Lio, C. W., and Hsieh, C. S. (2008). A two-step process for thymic regulatory T cell development. *Immunity* **28,** 100–111.

Lio, C. W., Dodson, L. F., Deppong, C. M., Hsieh, C. S., and Green, J. M. (2010). CD28 facilitates the generation of Foxp3(−) cytokine responsive regulatory T cell precursors. *J. Immunol.* **184,** 6007–6013.

Liston, A., Lesage, S., Wilson, J., Peltonen, L., and Goodnow, C. C. (2003). Aire regulates negative selection of organ-specific T cells. *Nat. Immunol.* **4,** 350–354.

Liston, A., Lesage, S., Gray, D. H., O'Reilly, L. A., Strasser, A., Fahrer, A. M., Boyd, R. L., Wilson, J., Baxter, A. G., Gallo, E. M., Crabtree, G. R., Peng, K., *et al.* (2004). Generalized resistance to thymic deletion in the NOD mouse; a polygenic trait characterized by defective induction of Bim. *Immunity* **21,** 817–830.

Liston, A., Nutsch, K. M., Farr, A. G., Lund, J. M., Rasmussen, J. P., Koni, P. A., and Rudensky, A. Y. (2008). Differentiation of regulatory Foxp3+ T cells in the thymic cortex. *Proc. Natl. Acad. Sci. USA* **105,** 11903–11908.

Liu, Y., Zhang, P., Li, J., Kulkarni, A. B., Perruche, S., and Chen, W. (2008). A critical function for TGF-beta signaling in the development of natural CD4+CD25+Foxp3+ regulatory T cells. *Nat. Immunol.* **9,** 632–640.

Lohr, J., Knoechel, B., Jiang, S., Sharpe, A. H., and Abbas, A. K. (2003). The inhibitory function of B7 costimulators in T cell responses to foreign and self-antigens. *Nat. Immunol.* **4,** 664–669.

Long, M., Park, S. G., Strickland, I., Hayden, M. S., and Ghosh, S. (2009). Nuclear factor-kappaB modulates regulatory T cell development by directly regulating expression of Foxp3 transcription factor. *Immunity* **31,** 921–931.

Maloy, K. J., and Powrie, F. (2001). Regulatory T cells in the control of immune pathology. *Nat. Immunol.* **2,** 816–822.

Mangan, P. R., Harrington, L. E., O'Quinn, D. B., Helms, W. S., Bullard, D. C., Elson, C. O., Hatton, R. D., Wahl, S. M., Schoeb, T. R., and Weaver, C. T. (2006). Transforming growth factor-beta induces development of the T(H)17 lineage. *Nature* **441,** 231–234.

Mantel, P. Y., Kuipers, H., Boyman, O., Rhyner, C., Ouaked, N., Ruckert, B., Karagiannidis, C., Lambrecht, B. N., Hendriks, R. W., Crameri, R., Akdis, C. A., Blaser, K., *et al.* (2007). GATA3-driven Th2 responses inhibit TGF-beta1-induced FOXP3 expression and the formation of regulatory T cells. *PLoS Biol.* **5,** e329.

Marie, J. C., Letterio, J. J., Gavin, M., and Rudensky, A. Y. (2005). TGF-beta1 maintains suppressor function and Foxp3 expression in CD4+CD25+ regulatory T cells. *J. Exp. Med.* **201,** 1061–1067.

Marie, J. C., Liggitt, D., and Rudensky, A. Y. (2006). Cellular mechanisms of fatal early-onset autoimmunity in mice with the T cell-specific targeting of transforming growth factor-beta receptor. *Immunity* **25,** 441–454.

Maruyama, T., Li, J., Vaque, J. P., Konkel, J. E., Wang, W., Zhang, B., Zhang, P., Zamarron, B. F., Yu, D., Wu, Y., Zhuang, Y., Gutkind, J. S., *et al.* (2011). Control of the differentiation of regulatory T cells and T(H)17 cells by the DNA-binding inhibitor Id3. *Nat. Immunol.* **12,** 86–95.

McCaughtry, T. M., and Hogquist, K. A. (2008). Central tolerance: What have we learned from mice? *Semin. Immunopathol.* **30,** 399–409.

McCaughtry, T. M., Wilken, M. S., and Hogquist, K. A. (2007). Thymic emigration revisited. *J. Exp. Med.* **204,** 2513–2520.

Medoff, B. D., Sandall, B. P., Landry, A., Nagahama, K., Mizoguchi, A., Luster, A. D., and Xavier, R. J. (2009). Differential requirement for CARMA1 in agonist-selected T-cell development. *Eur. J. Immunol.* **39,** 78–84.

Molinero, L. L., Yang, J., Gajewski, T., Abraham, C., Farrar, M. A., and Alegre, M. L. (2009). CARMA1 controls an early checkpoint in the thymic development of FoxP3+ regulatory T cells. *J. Immunol.* **182,** 6736–6743.

Moon, J. J., Chu, H. H., Pepper, M., McSorley, S. J., Jameson, S. C., Kedl, R. M., and Jenkins, M. K. (2007). Naive CD4(+) T cell frequency varies for different epitopes and predicts repertoire diversity and response magnitude. *Immunity* **27,** 203–213.

Moran, A. E., Holzapfel, K. L., Xing, Y., Cunningham, N. R., Maltzman, J. S., Punt, J., and Hogquist, K. A. (2011). T cell receptor signal strength in Treg and iNKT cell development demonstrated by a novel fluorescent reporter mouse. *J. Exp. Med.* **208,** 1279–1289.

Mouly, E., Chemin, K., Nguyen, H. V., Chopin, M., Mesnard, L., Leite-de-Moraes, M., Burlen-defranoux, O., Bandeira, A., and Bories, J. C. (2010). The Ets-1 transcription factor controls the development and function of natural regulatory T cells. *J. Exp. Med.* **207,** 2113–2125.

Mucida, D., Kutchukhidze, N., Erazo, A., Russo, M., Lafaille, J. J., and Curotto de Lafaille, M. A. (2005). Oral tolerance in the absence of naturally occurring Tregs. *J. Clin. Invest.* **115,** 1923–1933.

Naeher, D., Daniels, M. A., Hausmann, B., Guillaume, P., Luescher, I., and Palmer, E. (2007). A constant affinity threshold for T cell tolerance. *J. Exp. Med.* **204,** 2553–2559.

Nakagawa, T., Roth, W., Wong, P., Nelson, A., Farr, A., Deussing, J., Villadangos, J. A., Ploegh, H., Peters, C., and Rudensky, A. Y. (1998). Cathepsin L: Critical role in Ii degradation and CD4 T cell selection in the thymus. *Science* **280,** 450–453.

Nguyen, L. P., and Bradfield, C. A. (2008). The search for endogenous activators of the aryl hydrocarbon receptor. *Chem. Res. Toxicol.* **21,** 102–116.

Nishio, J., Feuerer, M., Wong, J., Mathis, D., and Benoist, C. (2010). Anti-CD3 therapy permits regulatory T cells to surmount T cell receptor-specified peripheral niche constraints. *J. Exp. Med.* **207,** 1879–1889.

Nishizuka, Y., and Sakakura, T. (1969). Thymus and reproduction: Sex-linked dysgenesia of the gonad after neonatal thymectomy in mice. *Science* **166,** 753–755.

Ochando, J. C., Homma, C., Yang, Y., Hidalgo, A., Garin, A., Tacke, F., Angeli, V., Li, Y., Boros, P., Ding, Y., Jessberger, R., Trinchieri, G., *et al.* (2006). Alloantigen-presenting plasmacytoid dendritic cells mediate tolerance to vascularized grafts. *Nat. Immunol.* **7,** 652–662.

Olivares-Villagomez, D., Wang, Y., and Lafaille, J. J. (1998). Regulatory CD4(+) T cells expressing endogenous T cell receptor chains protect myelin basic protein-specific transgenic mice from spontaneous autoimmune encephalomyelitis. *J. Exp. Med.* **188,** 1883–1894.

Olivares-Villagomez, D., Wensky, A. K., Wang, Y., and Lafaille, J. J. (2000). Repertoire requirements of CD4+ T cells that prevent spontaneous autoimmune encephalomyelitis. *J. Immunol.* **164,** 5499–5507.

O'Malley, J. T., Sehra, S., Thieu, V. T., Yu, Q., Chang, H. C., Stritesky, G. L., Nguyen, E. T., Mathur, A. N., Levy, D. E., and Kaplan, M. H. (2009). Signal transducer and activator of transcription 4 limits the development of adaptive regulatory T cells. *Immunology* **127,** 587–595.

Oppenheim, J. J. (2007). IL-2: More than a T cell growth factor. *J. Immunol.* **179,** 1413–1414.

Ouyang, W., Beckett, O., Ma, Q., and Li, M. O. (2010a). Transforming growth factor-beta signaling curbs thymic negative selection promoting regulatory T cell development. *Immunity* **32,** 642–653.

Ouyang, W., Beckett, O., Ma, Q., Paik, J. H., DePinho, R. A., and Li, M. O. (2010b). Foxo proteins cooperatively control the differentiation of Foxp3+ regulatory T cells. *Nat. Immunol.* **11,** 618–627.

Pacholczyk, R., Ignatowicz, H., Kraj, P., and Ignatowicz, L. (2006). Origin and T cell receptor diversity of Foxp3+CD4+CD25+ T cells. *Immunity* **25,** 249–259.

Pacholczyk, R., Kern, J., Singh, N., Iwashima, M., Kraj, P., and Ignatowicz, L. (2007). Nonself-antigens are the cognate specificities of Foxp3+ regulatory T cells. *Immunity* **27,** 493–504.

Palmer, E., and Naeher, D. (2009). Affinity threshold for thymic selection through a T-cell receptor-co-receptor zipper. *Nat. Rev. Immunol.* **9,** 207–213.

Papiernik, M., de Moraes, M. L., Pontoux, C., Vasseur, F., and Penit, C. (1998). Regulatory CD4 T cells: Expression of IL-2R alpha chain, resistance to clonal deletion and IL-2 dependency. *Int. Immunol.* **10,** 371–378.

Pennington, D. J., Silva-Santos, B., Silberzahn, T., Escorcio-Correia, M., Woodward, M. J., Roberts, S. J., Smith, A. L., Dyson, P. J., and Hayday, A. C. (2006). Early events in the thymus affect the balance of effector and regulatory T cells. *Nature* **444,** 1073–1077.

Picca, C. C., Oh, S., Panarey, L., Aitken, M., Basehoar, A., and Caton, A. J. (2009). Thymocyte deletion can bias Treg formation toward low-abundance self-peptide. *Eur. J. Immunol.* **39,** 3301–3306.

Polansky, J. K., Schreiber, L., Thelemann, C., Ludwig, L., Kruger, M., Baumgrass, R., Cording, S., Floess, S., Hamann, A., and Huehn, J. (2010). Methylation matters: Binding of Ets-1 to the demethylated Foxp3 gene contributes to the stabilization of Foxp3 expression in regulatory T cells. *J. Mol. Med.* **88,** 1029–1040.

Proietto, A. I., van Dommelen, S., Zhou, P., Rizzitelli, A., D'Amico, A., Steptoe, R. J., Naik, S. H., Lahoud, M. H., Liu, Y., Zheng, P., Shortman, K., and Wu, L. (2008). Dendritic cells in the thymus contribute to T-regulatory cell induction. *Proc. Natl. Acad. Sci. USA* **105,** 19869–19874.

Quintana, F. J., Basso, A. S., Iglesias, A. H., Korn, T., Farez, M. F., Bettelli, E., Caccamo, M., Oukka, M., and Weiner, H. L. (2008). Control of T(reg) and T(H)17 cell differentiation by the aryl hydrocarbon receptor. *Nature* **453,** 65–71.

Quintana, F. J., Murugaiyan, G., Farez, M. F., Mitsdoerffer, M., Tukpah, A. M., Burns, E. J., and Weiner, H. L. (2010). An endogenous aryl hydrocarbon receptor ligand acts on dendritic cells and T cells to suppress experimental autoimmune encephalomyelitis. *Proc. Natl. Acad. Sci. USA* **107,** 20768–20773.

Rao, S., Gerondakis, S., Woltring, D., and Shannon, M. F. (2003). c-Rel is required for chromatin remodeling across the IL-2 gene promoter. *J. Immunol.* **170,** 3724–3731.
Relland, L. M., Mishra, M. K., Haribhai, D., Edwards, B., Ziegelbauer, J., and Williams, C. B. (2009). Affinity-based selection of regulatory T cells occurs independent of agonist-mediated induction of Foxp3 expression. *J. Immunol.* **182,** 1341–1350.
Ribot, J., Enault, G., Pilipenko, S., Huchenq, A., Calise, M., Hudrisier, D., Romagnoli, P., and van Meerwijk, J. P. (2007). Shaping of the autoreactive regulatory T cell repertoire by thymic cortical positive selection. *J. Immunol.* **179,** 6741–6748.
Roman, E., Shino, H., Qin, F. X., and Liu, Y. J. (2010). Cutting edge: Hematopoietic-derived APCs select regulatory T cells in thymus. *J. Immunol.* **185,** 3819–3823.
Ruan, Q., Kameswaran, V., Tone, Y., Li, L., Liou, H. C., Greene, M. I., Tone, M., and Chen, Y. H. (2009). Development of Foxp3(+) regulatory t cells is driven by the c-Rel enhanceosome. *Immunity* **31,** 932–940.
Rubtsov, Y. P., and Rudensky, A. Y. (2007). TGFbeta signalling in control of T-cell-mediated self-reactivity. *Nat. Rev. Immunol.* **7,** 443–453.
Rudra, D., Egawa, T., Chong, M. M., Treuting, P., Littman, D. R., and Rudensky, A. Y. (2009). Runx-CBFbeta complexes control expression of the transcription factor Foxp3 in regulatory T cells. *Nat. Immunol.* **10,** 1170–1177.
Sakaguchi, S., Sakaguchi, N., Asano, M., Itoh, M., and Toda, M. (1995). Immunologic self-tolerance maintained by activated T cells expressing IL-2 receptor alpha-chains (CD25). Breakdown of a single mechanism of self-tolerance causes various autoimmune diseases. *J. Immunol.* **155,** 1151–1164.
Salomon, B., Lenschow, D. J., Rhee, L., Ashourian, N., Singh, B., Sharpe, A., and Bluestone, J. A. (2000). B7/CD28 costimulation is essential for the homeostasis of the CD4+CD25+ immunoregulatory T cells that control autoimmune diabetes. *Immunity* **12,** 431–440.
Saoudi, A., Seddon, B., Heath, V., Fowell, D., and Mason, D. (1996). The physiological role of regulatory T cells in the prevention of autoimmunity: The function of the thymus in the generation of the regulatory T cell subset. *Immunol. Rev.* **149,** 195–216.
Sasaki, T., Irie-Sasaki, J., Jones, R. G., Oliveira-dos-Santos, A. J., Stanford, W. L., Bolon, B., Wakeham, A., Itie, A., Bouchard, D., Kozieradzki, I., Joza, N., Mak, T. W., *et al.* (2000). Function of PI3Kgamma in thymocyte development, T cell activation, and neutrophil migration. *Science* **287,** 1040–1046.
Sauer, S., Bruno, L., Hertweck, A., Finlay, D., Leleu, M., Spivakov, M., Knight, Z. A., Cobb, B. S., Cantrell, D., O'Connor, E., Shokat, K. M., Fisher, A. G., *et al.* (2008). T cell receptor signaling controls Foxp3 expression via PI3K, Akt, and mTOR. *Proc. Natl. Acad. Sci. USA* **105,** 7797–7802.
Schallenberg, S., Tsai, P. Y., Riewaldt, J., and Kretschmer, K. (2010). Identification of an immediate Foxp3(−) precursor to Foxp3(+) regulatory T cells in peripheral lymphoid organs of nonmanipulated mice. *J. Exp. Med.* **207,** 1393–1407.
Schmidt-Supprian, M., Courtois, G., Tian, J., Coyle, A. J., Israel, A., Rajewsky, K., and Pasparakis, M. (2003). Mature T cells depend on signaling through the IKK complex. *Immunity* **19,** 377–389.
Schmidt-Supprian, M., Tian, J., Grant, E. P., Pasparakis, M., Maehr, R., Ovaa, H., Ploegh, H. L., Coyle, A. J., and Rajewsky, K. (2004). Differential dependence of CD4+CD25+ regulatory and natural killer-like T cells on signals leading to NF-kappaB activation. *Proc. Natl. Acad. Sci. USA* **101,** 4566–4571.
Schorle, H., Holtschke, T., Hunig, T., Schimpl, A., and Horak, I. (1991). Development and function of T cells in mice rendered interleukin-2 deficient by gene targeting. *Nature* **352,** 621–624.
Seddon, B., and Mason, D. (1999). Peripheral autoantigen induces regulatory T cells that prevent autoimmunity. *J. Exp. Med.* **189,** 877–882.

Semple, K., Nguyen, A., Yu, Y., Wang, H., Anasetti, C., and Yu, X. Z. (2011). Strong CD28 costimulation suppresses induction of regulatory T cells from naive precursors through Lck signaling. *Blood* **117,** 3096–3103.

Siggs, O. M., Makaroff, L. E., and Liston, A. (2006). The why and how of thymocyte negative selection. *Curr. Opin. Immunol.* **18,** 175–183.

Siggs, O. M., Miosge, L. A., Yates, A. L., Kucharska, E. M., Sheahan, D., Brdicka, T., Weiss, A., Liston, A., and Goodnow, C. C. (2007). Opposing functions of the T cell receptor kinase ZAP-70 in immunity and tolerance differentially titrate in response to nucleotide substitutions. *Immunity* **27,** 912–926.

Spence, P. J., and Green, E. A. (2008). Foxp3+ regulatory T cells promiscuously accept thymic signals critical for their development. *Proc. Natl. Acad. Sci. USA* **105,** 973–978.

Starr, T. K., Jameson, S. C., and Hogquist, K. A. (2003). Positive and negative selection of T cells. *Annu. Rev. Immunol.* **21,** 139–176.

Sun, C. M., Hall, J. A., Blank, R. B., Bouladoux, N., Oukka, M., Mora, J. R., and Belkaid, Y. (2007). Small intestine lamina propria dendritic cells promote de novo generation of Foxp3 T reg cells via retinoic acid. *J. Exp. Med.* **204,** 1775–1785.

Suzuki, H., Kundig, T. M., Furlonger, C., Wakeham, A., Timms, E., Matsuyama, T., Schmits, R., Simard, J. J., Ohashi, P. S., Griesser, H., *et al.* (1995). Deregulated T cell activation and autoimmunity in mice lacking interleukin-2 receptor beta. *Science* **268,** 1472–1476.

Suzuki, H., Zhou, Y. W., Kato, M., Mak, T. W., and Nakashima, I. (1999). Normal regulatory alpha/beta T cells effectively eliminate abnormally activated T cells lacking the interleukin 2 receptor beta in vivo. *J. Exp. Med.* **190,** 1561–1572.

Taguchi, O., Kontani, K., Ikeda, H., Kezuka, T., Takeuchi, M., Takahashi, T., and Takahashi, T. (1994). Tissue-specific suppressor T cells involved in self-tolerance are activated extrathymically by self-antigens. *Immunology* **82,** 365–369.

Tai, X., Cowan, M., Feigenbaum, L., and Singer, A. (2005). CD28 costimulation of developing thymocytes induces Foxp3 expression and regulatory T cell differentiation independently of interleukin 2. *Nat. Immunol.* **6,** 152–162.

Takaki, H., Ichiyama, K., Koga, K., Chinen, T., Takaesu, G., Sugiyama, Y., Kato, S., Yoshimura, A., and Kobayashi, T. (2008). STAT6 Inhibits TGF-beta1-mediated Foxp3 induction through direct binding to the Foxp3 promoter, which is reverted by retinoic acid receptor. *J. Biol. Chem.* **283,** 14955–14962.

Thornton, A. M., Korty, P. E., Tran, D. Q., Wohlfert, E. A., Murray, P. E., Belkaid, Y., and Shevach, E. M. (2010). Expression of Helios, an Ikaros transcription factor family member, differentiates thymic-derived from peripherally induced Foxp3+ T regulatory cells. *J. Immunol.* **184,** 3433–3441.

Tone, Y., Furuuchi, K., Kojima, Y., Tykocinski, M. L., Greene, M. I., and Tone, M. (2008). Smad3 and NFAT cooperate to induce Foxp3 expression through its enhancer. *Nat. Immunol.* **9,** 194–202.

van Santen, H. M., Benoist, C., and Mathis, D. (2004). Number of T reg cells that differentiate does not increase upon encounter of agonist ligand on thymic epithelial cells. *J. Exp. Med.* **200,** 1221–1230.

Vang, K. B., Yang, J., Mahmud, S. A., Burchill, M. A., Vegoe, A. L., and Farrar, M. A. (2008). IL-2, -7, and -15, but not thymic stromal lymphopoeitin, redundantly govern CD4+Foxp3+ regulatory T cell development. *J. Immunol.* **181,** 3285–3290.

Vang, K. B., Yang, J., Pagan, A. J., Li, L. X., Wang, J., Green, J. M., Beg, A. A., and Farrar, M. A. (2010). Cutting edge: CD28 and c-Rel-dependent pathways initiate regulatory T cell development. *J. Immunol.* **184,** 4074–4077.

Veldhoen, M., Hocking, R. J., Atkins, C. J., Locksley, R. M., and Stockinger, B. (2006). TGFbeta in the context of an inflammatory cytokine milieu supports de novo differentiation of IL-17-producing T cells. *Immunity* **24,** 179–189.

Veldhoen, M., Hirota, K., Westendorf, A. M., Buer, J., Dumoutier, L., Renauld, J. C., and Stockinger, B. (2008). The aryl hydrocarbon receptor links TH17-cell-mediated autoimmunity to environmental toxins. *Nature* **453,** 106–109.

Venuprasad, K., Huang, H., Harada, Y., Elly, C., Subramaniam, M., Spelsberg, T., Su, J., and Liu, Y. C. (2008). The E3 ubiquitin ligase Itch regulates expression of transcription factor Foxp3 and airway inflammation by enhancing the function of transcription factor TIEG1. *Nat. Immunol.* **9,** 245–253.

Verginis, P., McLaughlin, K. A., Wucherpfennig, K. W., von Boehmer, H., and Apostolou, I. (2008). Induction of antigen-specific regulatory T cells in wild-type mice: Visualization and targets of suppression. *Proc. Natl. Acad. Sci. USA* **105,** 3479–3484.

Verhagen, J., Gabrysova, L., Minaee, S., Sabatos, C. A., Anderson, G., Sharpe, A. H., and Wraith, D. C. (2009). Enhanced selection of FoxP3+ T-regulatory cells protects CTLA-4-deficient mice from CNS autoimmune disease. *Proc. Natl. Acad. Sci. USA* **106,** 3306–3311.

Walsh, P. T., Buckler, J. L., Zhang, J., Gelman, A. E., Dalton, N. M., Taylor, D. K., Bensinger, S. J., Hancock, W. W., and Turka, L. A. (2006). PTEN inhibits IL-2 receptor-mediated expansion of CD4+ CD25+ Tregs. *J. Clin. Invest.* **116,** 2521–2531.

Wan, Y. Y., Chi, H., Xie, M., Schneider, M. D., and Flavell, R. A. (2006). The kinase TAK1 integrates antigen and cytokine receptor signaling for T cell development, survival and function. *Nat. Immunol.* **7,** 851–858.

Weaver, C. T., and Hatton, R. D. (2009). Interplay between the TH17 and TReg cell lineages: A (co-)evolutionary perspective. *Nat. Rev. Immunol.* **9,** 883–889.

Webster, K. E., Walters, S., Kohler, R. E., Mrkvan, T., Boyman, O., Surh, C. D., Grey, S. T., and Sprent, J. (2009). In vivo expansion of T reg cells with IL-2-mAb complexes: Induction of resistance to EAE and long-term acceptance of islet allografts without immunosuppression. *J. Exp. Med.* **206,** 751–760.

Wei, J., Duramad, O., Perng, O. A., Reiner, S. L., Liu, Y. J., and Qin, F. X. (2007). Antagonistic nature of T helper 1/2 developmental programs in opposing peripheral induction of Foxp3+ regulatory T cells. *Proc. Natl. Acad. Sci. USA* **104,** 18169–18174.

Wilson, C. B., Makar, K. W., Shnyreva, M., and Fitzpatrick, D. R. (2005). DNA methylation and the expanding epigenetics of T cell lineage commitment. *Semin. Immunol.* **17,** 105–119.

Wing, K., and Sakaguchi, S. (2010). Regulatory T cells exert checks and balances on self tolerance and autoimmunity. *Nat. Immunol.* **11,** 7–13.

Wirnsberger, G., Mair, F., and Klein, L. (2009). Regulatory T cell differentiation of thymocytes does not require a dedicated antigen-presenting cell but is under T cell-intrinsic developmental control. *Proc. Natl. Acad. Sci. USA* **106,** 10278–10283.

Wohlfert, E. A., Gorelik, L., Mittler, R., Flavell, R. A., and Clark, R. B. (2006). Cutting edge: Deficiency in the E3 ubiquitin ligase Cbl-b results in a multifunctional defect in T cell TGF-beta sensitivity in vitro and in vivo. *J. Immunol.* **176,** 1316–1320.

Wong, J., Mathis, D., and Benoist, C. (2007a). TCR-based lineage tracing: No evidence for conversion of conventional into regulatory T cells in response to a natural self-antigen in pancreatic islets. *J. Exp. Med.* **204,** 2039–2045.

Wong, J., Obst, R., Correia-Neves, M., Losyev, G., Mathis, D., and Benoist, C. (2007b). Adaptation of TCR repertoires to self-peptides in regulatory and nonregulatory CD4+ T cells. *J. Immunol.* **178,** 7032–7041.

Xiao, S., Jin, H., Korn, T., Liu, S. M., Oukka, M., Lim, B., and Kuchroo, V. K. (2008). Retinoic acid increases Foxp3+ regulatory T cells and inhibits development of Th17 cells by enhancing TGF-beta-driven Smad3 signaling and inhibiting IL-6 and IL-23 receptor expression. *J. Immunol.* **181,** 2277–2284.

Xu, L., Kitani, A., Stuelten, C., McGrady, G., Fuss, I., and Strober, W. (2010). Positive and negative transcriptional regulation of the Foxp3 gene is mediated by access and binding of the Smad3 protein to enhancer I. *Immunity* **33,** 313–325.

Yamaguchi, T., and Sakaguchi, S. (2006). Regulatory T cells in immune surveillance and treatment of cancer. *Semin. Cancer Biol.* **16,** 115–123.

Yamazaki, S., Bonito, A. J., Spisek, R., Dhodapkar, M., Inaba, K., and Steinman, R. M. (2007). Dendritic cells are specialized accessory cells along with TGF- for the differentiation of Foxp3+ CD4+ regulatory T cells from peripheral Foxp3 precursors. *Blood* **110,** 4293–4302.

Yang, X. P., Ghoreschi, K., Steward-Tharp, S. M., Rodriguez-Canales, J., Zhu, J., Grainger, J. R., Hirahara, K., Sun, H. W., Wei, L., Vahedi, G., Kanno, Y., O'Shea, J. J., *et al.* (2011). Opposing regulation of the locus encoding IL-17 through direct, reciprocal actions of STAT3 and STAT5. *Nat. Immunol.* **12,** 247–254.

Yano, M., Kuroda, N., Han, H., Meguro-Horike, M., Nishikawa, Y., Kiyonari, H., Maemura, K., Yanagawa, Y., Obata, K., Takahashi, S., Ikawa, T., Satoh, R., *et al.* (2008). Aire controls the differentiation program of thymic epithelial cells in the medulla for the establishment of self-tolerance. *J. Exp. Med.* **205,** 2827–2838.

Yao, Z., Kanno, Y., Kerenyi, M., Stephens, G., Durant, L., Watford, W. T., Laurence, A., Robinson, G. W., Shevach, E. M., Moriggl, R., Hennighausen, L., Wu, C., *et al.* (2007). Nonredundant roles for Stat5a/b in directly regulating Foxp3. *Blood* **109,** 4368–4375.

Zehn, D., and Bevan, M. J. (2006). T cells with low avidity for a tissue-restricted antigen routinely evade central and peripheral tolerance and cause autoimmunity. *Immunity* **25,** 261–270.

Zheng, S. G., Wang, J. H., Stohl, W., Kim, K. S., Gray, J. D., and Horwitz, D. A. (2006). TGF-beta requires CTLA-4 early after T cell activation to induce FoxP3 and generate adaptive CD4+CD25+ regulatory cells. *J. Immunol.* **176,** 3321–3329.

Zheng, Y., Josefowicz, S., Chaudhry, A., Peng, X. P., Forbush, K., and Rudensky, A. Y. (2010). Role of conserved non-coding DNA elements in the Foxp3 gene in regulatory T-cell fate. *Nature* **463,** 808–812.

Zhou, L., Lopes, J. E., Chong, M. M., Ivanov, I. I., Min, R., Victora, G. D., Shen, Y., Du, J., Rubtsov, Y. P., Rudensky, A. Y., Ziegler, S. F., and Littman, D. R. (2008). TGF-beta-induced Foxp3 inhibits T(H)17 cell differentiation by antagonizing RORgammat function. *Nature* **453,** 236–240.

Zhou, X., Bailey-Bucktrout, S. L., Jeker, L. T., Penaranda, C., Martinez-Llordella, M., Ashby, M., Nakayama, M., Rosenthal, W., and Bluestone, J. A. (2009). Instability of the transcription factor Foxp3 leads to the generation of pathogenic memory T cells in vivo. *Nat. Immunol.* **10,** 1000–1007.

Ziegler, S. F. (2006). FOXP3: Of mice and men. *Annu. Rev. Immunol.* **24,** 209–226.

Zorn, E., Nelson, E. A., Mohseni, M., Porcheray, F., Kim, H., Litsa, D., Bellucci, R., Raderschall, E., Canning, C., Soiffer, R. J., Frank, D. A., and Ritz, J. (2006). IL-2 regulates FOXP3 expression in human CD4+CD25+ regulatory T cells through a STAT-dependent mechanism and induces the expansion of these cells in vivo. *Blood* **108,** 1571–1579.

CHAPTER 3

Regulatory T Cells in Infection

Rick M. Maizels and **Katherine A. Smith**

Contents

Abstract

Infectious agents have intimately co-evolved with the host immune system, acquiring a portfolio of highly sophisticated mechanisms to modulate immunity. Among the common strategies developed by viruses, bacteria, protozoa, helminths, and fungi is the manipulation of the regulatory T cell network in order to favor pathogen survival and transmission. Treg activity also benefits the host in many

Institute of Immunology and Infection Research, University of Edinburgh, Edinburgh, United Kingdom

Advances in Immunology, Volume 112
ISSN 0065-2776, DOI: 10.1016/B978-0-12-387827-4.00003-6

circumstances by controlling immunopathogenic reactions to infection. Interestingly, some pathogens are able to directly induce the conversion of naive T cells into suppressive Foxp3-expressing Tregs, while others activate pre-existing natural Tregs, in both cases repressing pathogen-specific effector responses. However, Tregs can also act to promote immunity in certain settings, such as in initial stages of infection when effector cells must access the site of infection, and subsequently in ensuring generation of effector memory. Notably, there is little current information on whether infections selectively drive pathogen-specific Tregs, and if so whether these cells are also reactive to self-antigens. Further analysis of specificity, together with a clearer picture of the relative dynamics of Treg subsets over the course of disease, should lead to rational strategies for immune intervention to optimize immunity and eliminate infection.

1. INTRODUCTION

Regulatory T cells are now recognized as an absolute requirement for healthy function of the mammalian immune system to forestall autoimmune pathology by self-reactive lymphocytes and to prevent deleterious reactions against extrinsic commensal and dietary antigens (Kim *et al.*, 2007; Sakaguchi *et al.*, 2008). Tregs also control immune responsiveness to infective pathogens (Belkaid and Tarbell, 2009), and in this context, their influence is not always benign. In this review, we survey our current knowledge of the role of Tregs in a wide range of infection settings and highlight the examples in which these cells are of critical importance in conferring susceptibility, dampening pathogenesis, and maintaining functional immunity.

The significance of Treg involvement in infectious episodes is not limited simply to how the host handles a particular pathogen; there is abundant evidence from infections of all types that adaptation or modulation of immune capacity resulting from infection can profoundly impact on bystander immune responses, in particular allergies, autoimmune diseases, and gastrointestinal disorders (Bach, 2002; Maizels, 2005). While this interaction is popularly termed the "Hygiene Hypothesis" (see Section 6 below), it is clearly highly context-dependent, and we also review the emerging evidence that some infectious agents in particular are able to alter the immune status of their host through the regulatory T cell compartment.

2. REGULATORY T CELLS

Regulatory T cells encompass several distinct phenotypes of immune system cells able to block or suppress immune reactivity *in vivo* and *in vitro*. While the most well-characterized Treg subset is a dedicated lineage selected in the thymus, there is peripheral conversion of potential effector cells into the Treg compartment, and evidence that additional T cell types engage transiently in regulatory activity or combine production of regulatory cytokines with more conventional effector products. The predominant Treg types are $CD4^+$ and express either or both the surface IL-2Rα chain CD25 and the forkhead box transcription factor Foxp3 (Gavin *et al.*, 2007). The $CD4^+CD25^+Foxp3^+$ phenotype also commonly expresses the inhibitory receptor CTLA-4 (Alegre *et al.*, 2001) as well as the GITR receptor which can activate both regulatory and non-regulatory T cells (Shimizu *et al.*, 2002). In addition to Foxp3-expressing Tregs are other functional regulatory cells, which may produce IL-10 (termed Tr1) and IL-35 (Tr35) (Collison *et al.*, 2010). In the case of Tr1, as IL-10 can also be produced by both Th1 ($IFN\gamma^+$ (Jankovic *et al.*, 2007)) and Th2 ($IL\text{-}4^+$) effectors, definition of this subset is relatively fluid. Moreover, T cells producing TGF-β, originally named Th3, can also act in a regulatory capacity. Finally, while Tregs are generally found to be $CD4^+$, $CD8^+$ T cells can express Foxp3 (Nakagawa *et al.*, 2010) and produce the same suppressive cytokines.

$Foxp3^+$ Treg cells exist in two categories, which arise in fundamentally different ways. Thymic, or "natural," regulatory T cells recognize self-antigens in the thymus and are imprinted with regulatory function before being released into the periphery. Thymic Tregs are a $CD4^+CD25^+Foxp3^+$ phenotype, and deficiency of these cells results in fatal autoimmune inflammation; this subset remains committed to this function over time (Rubtsov *et al.*, 2010). In addition, naive peripheral $CD4^+$ T cells can be induced to adopt a regulatory function by initiating expression of Foxp3 (or indeed, TGF-β, IL-10, or IL-35). In particular, induced or "adaptive" Tregs can convert from $CD4^+Foxp3^-$ to $CD4^+Foxp3^+$, as discussed further below, thereby expanding the range of Treg specificities to exogenous antigens (Bluestone and Abbas, 2003). However, induced Tregs are not irrevocably programmed and may revert to an effector phenotype, losing expression of CD25 and Foxp3 under certain conditions.

Early experiments with Tregs used CD25 as an accessible surface marker for cell transfer and *in vivo* antibody-mediated depletion with the antibody PC61 (developed by Lowenthal *et al.* (1985)). While landmark studies were achieved with these methods, it was recognized that effector populations can also be compromised by anti-CD25 antibody, and that a substantial minority of $Foxp3^+$ Tregs do not, at any one time, express CD25. In recent years, it has been possible to better target $Foxp3^+$ Tregs, through the construction of transgenic mice expressing the diphtheria toxin receptor (DTR) under the control of the Treg-specific Foxp3 promoter. Constructs described to date include the DEREG (Lahl *et al.*, 2007) and Foxp3LuciDTR (Suffner *et al.*, 2010) mice, both of which have transgenic BACs randomly inserted, and the $Foxp3^{DTR}$ mouse (Kim *et al.*,

2007; Lund *et al.*, 2008) containing a construct knocked-in to the Foxp3 locus on the X chromosome. Administration of diphtheria toxin (DTx) selectively depletes Tregs, although transiently and sparing cells which downregulate Foxp3 or (in the case of the BAC transgenics) suppress the randomly integrated locus.

3. INFECTIONS

Infectious agents have developed, over long evolutionary time, effective and often exquisite means of surviving in the host sufficiently long to assure transmission, and very often to establish long-term residence. Among the many strategies infectious organisms employ, exploiting the regulatory T cell compartment is doubtless one of the most effective. However, in the twists and turns of evolving host-pathogen co-adaption, some surprising interactions have developed: thus Tregs do not simply suppress immunity and forestall pathology, but also facilitate appropriate effector mechanisms and maintain long-term memory.

Some thought needs to be given to how infectious diseases of humans are appropriately modeled in laboratory rodents. Acute microbial infections in mice are not reflective of long-term human chronic infections, in which immunoregulation and re-setting of immune homeostasis are the norm. However, mouse models are invaluable for understanding early events, in particular the induction and activation of regulatory networks, and elegant transgenic constructs are available which permit tracking, manipulation, and gene-deletion within individual cell phenotypes.

While the application of these more sophisticated tools to infectious disease systems is still in its infancy, a substantial body of information has accumulated on the expression of Treg subsets, their functions *in vitro,* and the role of either $CD25^+$ or $Foxp3^+$ Tregs in the course of infection *in vivo*. We summarize in the following sections the data from the major infection systems, in humans and experimental models, while presenting more a complete annotation of infections under study in Tables 3.1–3.5 for each of the taxonomic groupings.

3.1. Viruses

Since the early 1900s, virus infections have been associated with immune suppression, variously attributed to functional impairment of lymphocytes, compromised function of antigen presenting cells, and the triggering of a suppressive T cell subset (Rouse and Horohov, 1986). Over recent years, a wide variety of viral infections has been examined for Treg activity (as detailed in Table 3.1), supporting the contention that

TABLE 3.1 Tregs in viral infections

Retroviruses		
FIV	Tregs promote progression and restrain anti-viral responses	Vahlenkamp *et al*. (2004); Mikkelsen *et al*. (2010)
Friend retrovirus	Expansion of Tregs *in vivo* and loss of tumor immunity	Iwashiro *et al*. (2001); Zelinskyy *et al*. (2006)
	Tregs suppress CD8 antiviral immunity	Dittmer *et al*. (2004); Robertson *et al*. (2006); Zelinskyy *et al*. (2009a,b)
	Nonantigen-specific Tregs control pathology in RAG model	Antunes *et al*. (2008)
HIV	Treg numbers correlate with viral load but decline in persistent viremia	Andersson *et al*. (2005); Baker *et al*. (2007); Nilsson *et al*. (2006); Tsunemi *et al*. (2005)
	$CD25^+$ Tregs maintain suppressive capacity in infection	Kinter *et al*. (2007a,b)
	Tregs reduce activation of and inhibit infection of effector T cells	Chase *et al*. (2008); Moreno-Fernandez *et al*. (2011)
	gp120 binding to CD4 may activate Tregs	Becker *et al*. (2009)
	T cells recognising protective HLA allele specificities not suppressed by Tregs	Elahi *et al*. (2011)
LCMV	Blocks diabetes through Tregs	Diana *et al*. (2011)
	Superantigen-mediated expansion of Tregs	Punkosdy *et al*. (2011)
MAIDS	Tregs promote infection, ablation blocks	Beilharz *et al*. (2004)
MMTV	Tregs reduce viral load at outset, increase later. Superantigen-specific Tregs	Cabrera *et al*. (2008)
SIV	Foxp3 in both $CD25^-$ and $CD25^+$ T cells correlates with high viremia	Boasso *et al*. (2007)

(*continued*)

TABLE 3.1 *(continued)*

RNA viruses		
Hepatitis C	Elevated functional Foxp3^{+} Tregs	Ebinuma *et al.* (2008)
	Disease resolution associated with reduced Treg activity	Smyk-Pearson *et al.* (2008)
Influenza A	CD25^{+} depletion raises CD8^{+} response	Haeryfar *et al.* (2005)
MHV coronavirus	Treg depletion can be fatal	Anghelina *et al.* (2009)
	Tregs reduce demyelination	Trandem *et al.* (2010)
	Viral epitope-specific Tregs	Zhao *et al.* (2011)
Rhinovirus in humans	Induce IL-35^{+}Foxp3^{-} Tregs via DCs	Seyerl *et al.* (2010)
RSV	Tregs dampen response and limit pathology but depletion does not change viral load	Fulton *et al.* (2010); Lee *et al.* (2010)
DNA viruses		
CMV	CD25^{+} Tregs suppress CMV response	Aandahl *et al.* (2004)
EBV	CD8^{+} Tregs in active infection	Popescu *et al.* (2007)
Hepatitis B	CD4^{+}FoxP3^{+} Treg numbers correlate with viral load and serum TGF-β	Barboza *et al.* (2007); Yang *et al.* (2007a)
HSV-1	Tregs expand in infection, restraining responsiveness and pathology	Suvas *et al.* (2003, 2004)
	Ocular pathology controlled by Tregs (and IL-10), including *in vitro*-generated viral-specific Tregs	Sarangi *et al.* (2008); Sehrawat *et al.* (2008)
HSV-2	Treg ablation results in loss of immunity through reduced effectors at site of infection	Lund *et al.* (2008)
Human papillomavirus 16	CD25^{+} Tregs correlate with persistent infection	Molling *et al.* (2007)
Vaccinia	CD25^{+} depletion raises CD8^{+} response	Haeryfar *et al.* (2005)

suppression of antiviral effector cells may allow the establishment and maintenance of chronic viral infection (Li *et al.*, 2008; Mills, 2004).

3.1.1. Retroviral infections and suppression of $CD8^+$ effector function

Retroviral infections in both mice and humans can be influenced by Treg populations (Li *et al.*, 2008; Rouse *et al.*, 2006). In mice, infection with the chronic Friend retrovirus (FV) (Iwashiro *et al.*, 2001) or the LP-BM5 murine leukemia virus mixture (which causes mouse AIDS) stimulated expansion of $CD4^+$ Treg cells co-expressing CD25 or CD38, cell surface markers associated with regulatory cells (Antunes *et al.*, 2008; Beilharz *et al.*, 2004; Robertson *et al.*, 2006; Zelinskyy *et al.*, 2006). Treg expansion was associated with the detection of virus-specific $CD8^+$ T cells that displayed an exhausted phenotype with low levels of effector cytokines and cytotoxic molecules (Dittmer *et al.*, 2004; Zelinskyy *et al.*, 2005).

The protective $CD8^+$ T cell response could be experimentally inhibited by the transfer of virus-specific $CD4^+$ T cells from naive or persistently infected mice to an acutely infected host (Dittmer *et al.*, 2004; He *et al.*, 2004); while both $CD25^+$ and $CD25^-$ populations could mediate this effect, contact-dependent *in vitro* suppression of $CD8^+$ T cell function was associated with the $CD4^+CD25^+$ subset (Robertson *et al.*, 2006). Depletion of $CD25^+$ T cells from persistently infected mice did not consistently improve the ability of $CD8^+$ T cells to control virus loads, but treatment of FV-infected mice with anti-GITR resulted in rescue of $CD8^+$ T cell dysfunction and reversal of retrovirus-induced immunosuppression (Dittmer *et al.*, 2004; He *et al.*, 2004). Similarly, LP-BM5 viral progression was retarded in mice treated with combined anti-CD25, -CTLA-4, and GITR antibodies (Beilharz *et al.*, 2004). These results illustrate the important principle that in ongoing infection, immunity may not be restored simply by removing Tregs, if the resident effector population is anergized or exhausted. Rather, intervention to restimulate effector cells is also required.

The FV system also clearly displays a broader bystander suppression by Tregs, with $CD8^+$ T cell antitumor responsiveness inhibited in mice receiving $CD4^+$ cells from FV-infected mice, with TGF-β and CTLA-4 shown to act *in vitro* (Iwashiro *et al.*, 2001). Normally, FV infection does not elicit immune-related pathology, except in the bone marrow of $RAG1^{-/-}$ mice receiving virus-specific $CD4^+$ T cells. In this setting, pathology is suppressed in an antigen-noncognate manner by both polyclonal TCRβ-transgenic and wild-type $CD4^+CD25^+Foxp3^+$ T cells (Antunes *et al.*, 2008).

When DEREG mice (expressing a BAC-inserted Foxp3-promoter DTR construct) were depleted during the early, acute phase of Friend virus infection, they showed stronger and more multifunctional virus-specific $CD8^+$ T cells, and $>$10-fold lower viral loads, without evident cost to the host in terms of immunopathology (Zelinskyy *et al.*, 2009a,b).

Interestingly, the transient Treg depletion that is achieved in the DEREG mouse model not only rescued $CD4^+$ effector cells from functional exhaustion, but also had a lasting effect in reducing chronic virus loads (Dietze *et al.*, 2011). Hence, transient depletion of Tregs could be a safe therapy for chronic viral infection.

An important retroviral pathogen is feline immunodeficiency virus (FIV), which expands functionally suppressive $CD25^+$ T cells in chronically-infected cats (Vahlenkamp *et al.*, 2004). In vivo depletion of $CD25^+$ T cells in infected cats resulted in transient increases in both anti-viral and bystander responses (Mikkelsen *et al.*, 2010), although depletion prior to infection did not alter the course of disease (Mikkelsen *et al.*, 2011). In human retrovirus infection, Treg activity is largely inferred from phenotypic analysis of *ex vivo* lymphocytes, with reports showing a positive correlation between viral load and $FOXP3^+$ Treg numbers (Andersson *et al.*, 2005; Nilsson *et al.*, 2006; Tsunemi *et al.*, 2005). However, Tregs may impact on the course of HIV infection not only by impeding protective immunity but conversely by minimizing the pool of activated effector cells which are susceptible to virus infection (Eggena *et al.*, 2005).

In T cells from HIV patients, both HIV-specific and bystander (CMV) *in vitro* responses are enhanced by depletion of $CD4^+CD25^+$ Tregs (Aandahl *et al.*, 2004), and Tregs *in vitro* can suppress cytolytic capacity and cytokine secretion by HIV-specific $CD4^+$ (Weiss *et al.*, 2004) and $CD8^+$ (Kinter *et al.*, 2007a,b) effector T cells *in vitro*. The possibility that Tregs can also act to ameliorate infection was also raised by data showing that Treg inhibition correlated with lower levels of viremia (Kinter *et al.*, 2004), and patients with low Treg numbers had greater peripheral T cell activation, a poor prognostic indicator for disease (Eggena *et al.*, 2005). Later studies showed that molecules of the B7:CD28 family, programmed death-1 (PD-1) and CTLA-4, may maintain virus-specific T cell exhaustion typical of HIV infection (Kaufmann and Walker, 2009), as PD-1 expression correlated with viral load and disease progression in cohorts of HIV^+ untreated patients (Day *et al.*, 2006; Trautmann *et al.*, 2006).

3.1.2. Expansion of Tregs following LCMV infection

Lymphocytic choriomeningitis virus (LCMV) is a natural murine RNA virus transmitted directly from mother to offspring, and different isolates cause infections of varying duration. Chronic, but not acute, infection of mice with LCMV results in a marked expansion of TGF-β-producing $CD4^+CD25^+$ Tregs (Filippi *et al.*, 2009). More precisely, expansion occurs within a TCR Vβ5 $CD4^+Foxp3^+$ population, which derived from pre-existing "natural" Tregs, as there was no conversion of GFP^-Foxp3^- T cells transferred to mice immediately prior to infection (Punkosdy *et al.*, 2011). This TCR reacts with the endogenous mouse mammary tumor provirus (MMTV) Mtv9 superantigen, one of several MMTVs which have segregated during inbreeding of laboratory mouse strains (Cohen and Varmus, 1979), and MMTV superantigen-specific Foxp3$^+$ Tregs have been previously reported (Cabrera *et al.*, 2008). The expansion of regulatory T cell mediated by endogenous retroviral superantigens provides a unique mechanism of immune-evasion following chronic LCMV infection.

Interestingly, LCMV is a potent inhibitor of type 1 autoimmune diabetes in mice (Filippi *et al.*, 2009), an effect associated with the activity of Tregs (Diana *et al.*, 2011; Filippi *et al.*, 2011), as discussed below in Section 6.

3.1.3. Tregs and protective immunity to herpes simplex virus

Herpes simplex virus (HSV) is an acute cytolytic virus, immunity to which depends upon a protective CD8$^+$ T cell response. However, HSV-1 infection heightened the suppressive function of CD4$^+$CD25$^+$ Tregs in mice (Suvas *et al.*, 2003) and PC61-mediated depletion of CD25$^+$ cells prior to infection amplified the virus-specific CD8$^+$ response, whereas CD25$^+$ Treg transfer had the opposite effect. Moreover, anti-CD25-mediated Treg depletion enhanced memory responses and protective immunity following primary infection with HSV or re-exposure to viral antigen following HSV antigen immunization or primary infection (Toka *et al.*, 2004).

Depletion of Tregs using Foxp3^{DTR} knock-in mice in a local HSV-2 infection resulted in increased viral loads in the mucosa and nervous system and fatal infection. Treg depletion, however, attenuated cellular trafficking to the site of infection and reduced inflammatory cytokine levels to a degree that significantly compromised protective immunity (Lund *et al.*, 2008). In this instance, and in supporting studies ablating Tregs during LCMV infection of nonlymphoid tissue (Wherry *et al.*, 2003), Tregs play an important role in controlling viral load as well as intensifying the cytokine milieu in secondary lymphoid organs.

3.1.4. Tregs benefit both host and pathogen in hepatitis C virus infection

The outcome of human hepatitis C virus (HCV) infection can range from complete control to viral persistence and associated liver disease (Rehermann, 2009); however, the development of therapeutic strategies for treatment have been hampered by difficulties in establishing *in vitro* and *in vivo* models of viral replication, so that currently all data pertain to infection in primate systems.

In HCV-infected patients, frequencies of peripheral CD4$^+$CD25$^+$ and Foxp3$^+$ Tregs are elevated (Cabrera *et al.*, 2004; Ebinuma *et al.*, 2008; Sugimoto *et al.*, 2003) and *in vitro* analysis indicated that CD4$^+$CD25$^+$ cells suppress virus-specific CD4$^+$ and CD8$^+$ responses through IL-10 and TGF-β secretion (Cabrera *et al.*, 2004). Although other authors confirmed the suppressive activity of HCV patient-derived $CD4^+CD25^+$ cells, antibodies to these same mediators did not block the suppression of purified $CD8^+$ effectors, suggesting action through a $CD4^+$ intermediary (Boettler *et al.*, 2005; Rushbrook *et al.*, 2005). Significantly, HCV-associated Tregs were able to suppress influenza virus-specific $CD8^+$ T cell function (Boettler *et al.*, 2005). Comparison with a cohort recovering from acute HCV infection suggested a decline in $CD4^+CD25^+$ regulatory function (Boettler *et al.*, 2005), while longitudinal studies in individual patients more convincingly concluded that spontaneous recovery from HCV infection is associated with the temporal loss of $Foxp3^+$ Treg function (Smyk-Pearson *et al.*, 2008).

Clear evidence has also been provided that HCV-antigen-specific Tregs evolve during infection. $Foxp3^+$ Tregs can be isolated from HCV-positive PBMC stimulated with HCV peptides, with different peptides proving optimal for different patients (Li *et al.*, 2007b), and epigenetic analysis of the Foxp3 locus indicating stable rather than transient commitment to the Treg phenotype (Li *et al.*, 2009). Similarly, HCV peptides were used to expand Tregs from infected patients, some $Foxp3^+$ Tregs reacting with HCV-specific class II-peptide tetramers (Ebinuma *et al.*, 2008).

Although Tregs may impair immunity to HCV, they may also protect the patient from excessive pathology. Thus, liver inflammation is inversely correlated to $CD4^+CD25^+$ T cell numbers in chronic HCV infection (Cabrera *et al.*, 2004). Moreover, $CD25^+$ Tregs from patients with low pathological scores exerted more suppressive effects on HCV-specific $CD4^+$ T cell responses than Tregs from patients with advanced clinical disease (Bolacchi *et al.*, 2006), demonstrating that loss of Treg function can be correlated with organ-specific viral-induced inflammation and pathology.

3.2. Bacteria

Historically, most bacterial immunology focused on the acutely pathogenic species representing the most pressing threat to human health; more recently, research has also encompassed the commensal microbiome, particularly in the gut in which intense interactions occur with the immune system. It is appropriate to consider both pathogens and commensals in terms of Treg activity, in part because they form a biological continuum (with many commensals being opportunistic pathogens), and also because of the shared signaling pathways and specific receptors that are involved in their recognition. We summarize below the data from some of the principal bacterial systems, with additional details listed in Table 3.2.

TABLE 3.2 Tregs in bacterial infections

Mycobacteria		
Mycobacterium bovis BCG in mice	Elevated pulmonary Foxp3$^+$ cells, bacterial load unchanged by anti-CD25 depletion	Quinn *et al.* (2006)
Mycobacterium tuberculosis in humans	Elevated FOXP3$^+$ cells, inversely correlating with immunity, reduced after treatment	Chen *et al.* (2007b); Guyot-Revol *et al.* (2006); Li *et al.* (2007a); Qin *et al.* (2008)
Mycobacterium tuberculosis in mice	Tregs expand, increase bacterial load	Kursar *et al.* (2007); Ordway *et al.* (2007, 2011)
	Tregs delay priming and migration of effectors	Scott-Browne *et al.* (2007)
	Mtb-specific Tregs activated	Shafiani *et al.* (2010)
Mycobacterium vaccae	Induce Tregs, block allergy	Zuany-Amorim *et al.* (2002)
Other intracellular		
Brucella abortus (Gram –)	Tregs block protective immunity	Pasquali *et al.* (2010)
Chlamydia trachomatis	Tregs stimulated but no correlation with disease	Gall *et al.* (2011)
Listeria monocytogenes (Gram +)	Tregs suppress memory CD8$^+$ T cells	Kursar *et al.* (2002)
	No antigen-specific Tregs *in vivo*	Fontenot *et al.* (2005)
Respiratory		
Bordetella pertussis (Gram –)	Tr1 generation through filamentous hemagglutinin	McGuirk *et al.* (2002)
Gastrointestinal pathogens		
Haemophilus ducreyi (Gram –)	Tregs enriched in lesions	Li *et al.* (2010)
Helicobacter pylori (Gram –)	Treg expansion in the mucosa, CD25 depletion reduces bacterial load but generates pathology	Lundgren *et al.* (2005); Rad *et al.* (2006); Raghavan *et al.* (2003)
	Infection-related Tregs suppress airway allergy	Arnold *et al.* (2011)

(continued)

TABLE 3.2 *(continued)*

Salmonella enterica (Typhimurium; Gram −)	Treg depletion or anti-CTLA-4 boosts clearance and memory	Johanns *et al*. (2010)
Commensal bacteria		
Bacteroides fragilis (Gram −)	Drives Treg expansion, through PSA binding to TLR2	Round and Mazmanian, (2010); Round *et al*. (2011)
Bifidobacterium infantis (Gram +)	Induction of Tregs, bystander suppression of inflammation following mucosal *S. typhimurium* infection	O'Mahony *et al*. (2008)
Clostridium species (Gram +)	Mediates Treg induction through TGF-β, protects against DSS colitis	Atarashi *et al*. (2011)
Helicobacter hepaticus (Gram −)	Tr1-like IL-10-producing cells block gut inflammation	Kullberg *et al*. (2002)
Streptococcus pneumoniae (Gram +)	$CD8^+CD28^+$ suppressive Tregs producing IL-10 and TGF-β	Mertens *et al*. (2009)

3.2.1. Mycobacteria

Mycobacterium tuberculosis (Mtb) is present in two billion individuals worldwide and remains a major cause of morbidity and mortality around the world (Dye, 2006). Most infectious episodes are effectively resolved, but where elimination of the bacteria does not occur, Th1 immunity is impaired (Jo *et al.*, 2003; Lienhardt *et al.*, 2002).

In mice, Tregs expand in the lung and associated lymph nodes following Mtb infection (Kursar *et al.*, 2007), and bacterial loads are 10-fold lower following depletion of Thy1.1^{+}Foxp3^{+} cells in Thy1.1-wild-type: Thy1.2-Foxp3$^{-/-}$ mixed bone marrow chimeras (Scott-Browne *et al.*, 2007). Conversely, co-transfer of CD4^{+}CD25^{+} Tregs neutralizes immunity to infection mediated by effector CD4^{+}CD25^{-} T cells in RAG-1$^{-/-}$ mice (Kursar *et al.*, 2007). Treg expansion was particularly rapid in mice infected with a hypervirulent strain of Mtb with the emergence of a CD4^{+}CD25^{+}CD223^{+}Foxp3^{+}IL-10^{+} regulatory T cell population in the lung (Ordway *et al.*, 2007). The expansion of Mtb-specific Tregs was followed *in vivo*, using an Mtb-specific TCR transgenic mouse, P25; in a RAG-sufficient background, a subset of P25 Mtb-reactive T cells express Foxp3, possibly representing natural Tregs with dual specificity for this pathogen (Shafiani *et al.*, 2010). The Foxp3^{+} pathogen-specific Tregs proliferated faster than the effector populations in the lung, delayed the infiltration of CD4^{+} and CD8^{+} populations, and caused a significant rise in bacterial titer (Shafiani *et al.*, 2010).

The saprophytic species *Mycobacterium vaccae* may also have Treg-stimulating activity. While interest in this organism was initially focused as a possible immunogen against Mtb, it was also found to be beneficial in downregulating human atopic dermatitis (Arkwright and David, 2001). Treatment of mice with a heat-killed *M. vaccae* suspension prior to ovalbumin sensitization gave rise to a population of Ova-specific CD4^{+}CD45RB^{lo} regulatory T cells, which mediated inhibition of airway allergy through IL-10 and TGF-β (Zuany-Amorim *et al.*, 2002).

In humans, the frequency of FOXP3^{+} cells is substantially higher in TB patients (Guyot-Revol *et al.*, 2006; Li *et al.*, 2007a) and declines following successful chemotherapy of infection (Chen *et al.*, 2007b). In patients, GITR expression is also significantly raised in CD4^{+}CD25^{+} T cells, which functionally suppress effector responses (Li *et al.*, 2007a), while depletion of CD4^{+}CD25^{+} Tregs restores *in vitro* responsiveness of peripheral T cells (Ribeiro-Rodrigues *et al.*, 2006). Moreover, the number of CD4^{+}CD25^{+}Foxp3^{+} cells present in patients' pleural fluid (PF) inversely correlates with the ability of PF CD4^{+}CD25^{-} T cells to mount a IFN-γ response to Mtb antigens (Chen *et al.*, 2007b). CD4^{+}CD25^{+}Foxp3^{+} cells from healthy carriers multiplied *in vitro* in response to heat-killed Mtb, and the active principle shown to be the 19-kDa *M. tuberculosis* lipoprotein

ManLAM (mannose-capped lipoarabinomannan), acting via the mannose receptor of human monocytes (Garg *et al.*, 2008).

An important issue is how Treg activity will be influenced by vaccination. Following infection of BCG-vaccinated mice with naturally virulent strains of *M. tuberculosis,* initial effector responses declined while Treg activity increased, with pathology accentuating over time (Ordway *et al.*, 2011). Whether regulatory T cell populations can be specifically depleted or modified to favor the outcome of mycobacterium vaccination remains to be determined.

3.2.2. *Listeria monocytogenes*

Regulatory T cells can also control the magnitude of a secondary response in infections where $CD8^+$ T cells are important in controlling protective immunity against pathogens such as *Listeria monocytogenes*. In this instance, depletion of $CD4^+$ T cells significantly enhanced the formation of a memory $CD8^+$ T cell response following secondary infection or immunization. Anti-CD25 depletion and transfer experiments demonstrated that this suppressive activity was enriched within the $CD4^+CD25^+$ T cell population from naive or *L. monocytogenes*-infected mice (Kursar *et al.*, 2002). Through the use of transgenic mice where all the T cells recognize Ova presented in the context of H-$2A^b$ (OT-II), or where Foxp3 was coupled to a GFP reporter (Foxp3–GFP), it was possible to demonstrate that acute infection with *L. monocytogenes* expressing OVA was not associated with the induction of antigen-specific regulatory T cells (Fontenot *et al.*, 2005), suggesting that downstream suppression of immunity was more likely via an interaction with $CD8^+$ T cells than a direct antigen-specific regulation of $CD4^+$ T cell function.

$Foxp3^+$ Tregs also inhibit *Listeria*-specific $CD8^+$ T cell responses *in vivo*; however, $Foxp3^+$ Tregs were found to be less potent at suppressing effector responsiveness, and specific depletion of the Treg population in Foxp3-DTR mice did not alter bacterial clearance or the expansion and activation of virus-specific $CD8^+$ T cells following infection of mice with *L. monocytogenes* (Ertelt *et al.*, 2011). These findings highlight the importance of Tregs in controlling inflammatory responses in the steady state and raise the possibilities of this function being overcome following infection.

Infection with *L. monocytogenes* is more common, and more hazardous, in pregnancy. In this context, a recent study in mice reported that the physiological increase in $Foxp3^+$ Tregs during allogeneic pregnancy was associated with greater susceptibility to *Listeria* (and *Salmonella*) infections, an effect attributable to IL-10 production by these cells (Rowe *et al.*, 2011). Moreover, Treg depletion in pregnant $Foxp3^{DTR}$ mice restored normal levels of resistance to infection while reducing live births by 70%;

hence, the extraordinary balance between infection and reproduction is managed to optimal effect by Foxp3$^+$ Tregs.

3.2.3. *Helicobacter pylori*

Human *Helicobacter pylori* infection correlates with a higher number of regulatory T cells in the gastric mucosa (Lundgren *et al.*, 2005; Rad *et al.*, 2006), which are also found in *H. pylori*-induced gastric adenocarcinoma (Enarsson *et al.*, 2006). The inability of the host to eradicate *H. pylori* infection can therefore be linked to Treg suppression of *H. pylori*-specific effector T cell responses in humans and mice (Lundgren *et al.*, 2003; Raghavan *et al.*, 2003). Accordingly, depletion of CD25$^+$ T cells increased the gastric inflammatory response and reduced bacterial burden in infected mice, but also resulted in development of severe gastritis (Rad *et al.*, 2006; Raghavan *et al.*, 2003), although another laboratory reported no effect of depletion (Kaparakis *et al.*, 2006). Tregs purified from gastric tumors were able to suppress *H. pylori*-specific effector responses *in vitro*, suggesting that antigen-specific regulatory T cells might contribute to tumor progression through bystander suppression, as noted above in Friend virus infection (Enarsson *et al.*, 2006).

Studies on induction of Tregs during *H. pylori* infection have shown that gastric epithelial cells (GECs) exposed to this organism upregulate the PD1 ligand B7-H1, and that increased conversion of naive cells into Tregs is inhibited by anti-B7-H1 antibody (Beswick *et al.*, 2007). In addition, GEC production of TGF-β both acts to induce Foxp3$^+$ Tregs and to inhibit effector T cell responses *in vitro* (Beswick *et al.*, 2011). The systemic impact of *H. pylori* on the generation of Tregs is so strong that infected mice are protected from airway allergic inflammation induced by ovalbumin, and CD4$^+$CD25$^+$ Tregs from infected mice can confer this protection on uninfected, allergen-sensitized animals (Arnold *et al.*, 2011).

3.2.4. *Bordetella pertussis*

Bordetella pertussis infection is associated with a severe and protracted disease, which is often fatal in young children. Although the development of antigen-specific Th1 cells promotes recovery from infection and clearance of bacteria from the respiratory tract, these responses are suppressed in acute infections (McGuirk *et al.*, 1998). One virulence factor implicated in this is filamentous hemagglutinin (FHA), and FHA-specific Tr1 clones have been generated from infected mice, expressing high levels of IL-10 but little IFN-γ (McGuirk *et al.*, 2002).

The major virulence factor of *B. pertussis* is its toxin (PTx), which is widely used to enhance the incidence and severity of disease in murine experimental autoimmune encephalomyelitis (EAE). A single injection is reported to inhibit Tregs and promote Th17 responsiveness

(Chen *et al.*, 2007a). Most recently, however, it has been reported that weekly PTx administration causes expansion and persistence of peripheral $CD4^+CD25^+Foxp3^+$ regulatory T cells and elevations in serum IL-10 and TGF-β (Weber *et al.*, 2010). It will be interesting to ascertain if, in active infection, sustained release of pertussis toxin in fact promotes suppressive Tregs rather than proinflammatory effector responses.

3.2.5. Commensal microbes

The development of germ-free (GF) mice has allowed us to analyze the specific impact of certain commensal bacteria species on the immune system. GF mice appear to have site-specific differences in the phenotype and suppressive capacity of their $CD4^+CD25^+$ regulatory T cell population (Ostman *et al.*, 2006). In particular, GF mice lack $Foxp3^+$ Tregs in the colonic lamina propria, which are induced as a predominantly $Helios^-$ population when animals are colonized with defined commensals (Geuking *et al.*, 2011). However, the generation of $Foxp3^{DTR}$ mice in specific-pathogen free (SPF) and GF conditions demonstrated that the suppressive activity of splenic and lymph node $CD4^+Foxp3^+$ Tregs was equivalent in both mice (Chinen *et al.*, 2010). Treg depletion in either mice also resulted in equivalent systemic inflammatory responses; however, inflammation was much more severe in the small intestine of Treg-depleted SPF mice, reflecting the substantial load of nonself antigen represented by the commensal microbiota, and the critical role of Tregs in subduing reactivity to gut flora.

Earlier work which had established this principle includes the transfer of naive ($CD45RB^{high}$) T cells into T cell-deficient mice, provoking massive gut inflammation, and its suppression by co-transfer of Tregs through IL-10, TGF-β, and CTLA-4 (Maloy *et al.*, 2003). Similarly, transfer of naive T cells from an IL-10-deficient $RAG^{-/-}$ mouse enhanced inflammation induced following *Helicobacter hepaticus* infection, whereas co-transfer of IL-10-sufficient $CD45RB^{low}CD4^+$ T cells, of either $CD25^+$ or $CD25^-$ phenotype, from *H. hepaticus*-infected but not uninfected mice was most able to prevent disease (Kullberg *et al.*, 2002). These studies were key steps toward the concept that regulatory IL-10-producing T cells are essential to prevent bacteria-induced colitis.

Normally asymptomatically resident within the colon, *Bacteroides fragilis* is a Gram-negative bacteria that has been detected within abscesses formed throughout the peritoneal cavity as a result of bowel perforation (Polk and Kasper, 1977). *B. fragilis* was found to protect animals from experimental colitis induced by *H. hepaticus* via a single microbial molecule (polysaccharide A, PSA) (Mazmanian *et al.*, 2008). This molecule induces IL-10 production from T cells, suppresses proinflammatory IL-17 production, and was further shown to promote the differentiation of $CD4^+Foxp3^+$ regulatory T cells through TLR2 during protection from experimental colitis, as further discussed in Section 4.4 below (Round and Mazmanian, 2010). This microbial polysaccharide and *B. fragilis* were also shown protect against pathology in a mouse model of experimentally induced EAE, where both stimulated $Foxp3^+$ regulatory T cell expansion *in vivo* (Ochoa-Reparaz *et al.*, 2010a,b).

A defined mix of 46 spore-forming *Clostridium* species, prominent and indigenous to the murine gastrointestinal tract, was also found to enhance TGF-β production and expand $IL\text{-}10^+Foxp3^+Helios^-$ regulatory T cells in the intestine of previously GF mice (Atarashi *et al.*, 2011). Clostridial enrichment of the neonatal gut flora resulted in resistance to DSS-mediated colitis and reduced polyclonal IgE responsiveness to OVA–alum. A parallel probiotic effect has been found with *Bifidobacterium infantis*, which increases $Foxp3^+$ Tregs and counters inflammation following *Salmonella typhimurium* infection (O'Mahony *et al.*, 2008), while *Lactobacillus reuteri* evokes a similar $Foxp3^+$ Treg expansion and mediates suppression of airway allergy in mice (Karimi *et al.*, 2009).

3.3. Protozoa

Protozoa are single-celled organisms which include parasites of both extracellular and intracellular niches; the major global health problems from protozoal pathogens are caused by *Plasmodium* (malaria) and *Leishmania* species, along with human trypanosomes in South America and trypanosomes of livestock in Africa. These species are highlighted below, with further details given of Tregs in protozoal infections in Table 3.3.

3.3.1. *Leishmania*

One of the founding paradigms of T cell immunology emerged from research into infections of mice with *Leishmania major*, in which the progressive disease in BALB/c mice compared to C57BL/6 is linked to their dominant Th2 response to this parasite (Reiner and Locksley, 1995). C57BL/6 mice resolve infection after several weeks unless their Th1 response is compromised. However, IL-4R-deficient BALB/c are not resistant to all strains of *L. major* (Noben-Trauth *et al.*, 1999), and other *Leishmania* species (which cause cutaneous or visceral forms of disease) elicit little immunity in any strain of mouse. In *Leishmania tropica*, a cutaneous leishmaniasis agent which is equally infective to BALB/c and C57BL/10, only the combined neutralization of TGF-β and IL-10 signaling was able to induce immune clearance of parasites (Anderson *et al.*, 2008). The prominence of these two cytokines is repeated in other species, including the cutaneous *L. major* (Belkaid *et al.*, 2001), and the visceral species *Leishmania donovani* (Murphy *et al.*, 2001; Rodrigues *et al.*, 1998)

TABLE 3.3 Tregs in protozoal infections

Human malaria		
Plasmodium falciparum	Elevated $CD25^+$ and $FOXP3^+$ in infection, correlate with parasite load, and in cord blood of newborns to infected mothers	Brustoski *et al.* (2006); Mackroth *et al.* (2011); Walther *et al.* (2005, 2009)
	$FOXP3^+$ numbers expand in severe malaria, decline following treatment	Minigo *et al.* (2009)
	Human patients have high $FOXP3^+$ Tregs	Goncalves *et al.* (2010) but see Finney *et al.* (2009)
	Bystander FOXP3 (hi) induction in human T cells	Scholzen *et al.* (2009)
	Human placenta Treg induction	Bisseye *et al.* (2009)
P. vivax	Elevated $FOXP3^+$ in infection	Bueno *et al.* (2010); Goncalves *et al.* (2010); Jangpatarapongsa *et al.* (2008)
Murine malaria		
P. berghei	CD25 depletion alleviates cerebral malaria (CM)	Amante *et al.* (2007); Vigario *et al.* (2007); Wu *et al.* (2010)
	$Foxp3^+$ depletion does not alter CM while expansion through IL-2/IL-2R complexes protects from CM	Haque *et al.* (2010); Steeg *et al.* (2009)
P. chabaudi	Foxp3 overexpression compromises protection; Tregs are anti-inflammatory	Berretta *et al.* (2011); Cambos *et al.* (2008)
P. yoelii	Anti-CD25 prevents malaria immune evasion through TLR9 signaling	Hisaeda *et al.* (2004, 2005, 2008)
	IL-10 and anti-CD25 in malaria	Chen *et al.* (2009a); Couper *et al.* (2007, 2008b)
	Early $CD25^+$ Treg expansion in susceptible strain	Wu *et al.* (2007)

Leishmaniasis		
L. braziliensis (cutaneous)	Human lesions have $FOXP3^+$	Campanelli *et al.* (2006)
L. donovani	CD40-low DCs induce Tregs, exacerbate infection	Martin *et al.* (2010)
	Human lesions have high $FOXP3^+$, abated with treatment	Ganguly *et al.* (2010)
L. guyanensis (cutaneous)	Human lesions have high $FOXP3^+$	Bourreau *et al.* (2009a,b)
L. infantum (visceral)	Elevated $Foxp3^+CD103^+$ in infection	Rodrigues *et al.* (2009)
L. major	Tregs maintain low-level infection and protective immunity, require CD103 to access infection site and suppress	Belkaid *et al.* (2002); Suffia *et al.* (2005)
	Treg depletion raises Th2 response and susceptibility; Tregs reactivate infection	Aseffa *et al.* (2002); Mendez *et al.* (2004)
Toxoplasmosis		
Toxoplasma gondii	Tregs reduce parasite-induced abortion in pregnant mice	Ge *et al.* (2008)
Trypanosomiasis		
T. congolense	$Foxp3^+$ Tregs suppress protective $CD8^+$ NKT cells	Wei and Tabel (2008)
T. congolense	Natural Tregs suppress $CD4^+$, $CD8^+$, and macrophage inflammation	Guilliams *et al.* (2007)
T. cruzi	Increased $FOXP3^+$ in human infection, and $CD25^+$ Tregs prolong survival in mice	de Araujo *et al.* (2011); Mariano *et al.* (2008)

and *Leishmania infantum* (Rodrigues *et al.*, 2009). In the latter, most recent, study, infection was associated with elevated levels of $CD4^+$ $Foxp3^+$ $CD103^+$ Tregs, which contributed toward a high IL-10 profile (Rodrigues *et al.*, 2009).

The role of IL-10 is particularly well documented in *L. major* infection, as for example in the C57BL/6 mouse, in which sterile immunity only takes effect if IL-10 is neutralized (Belkaid *et al.*, 2001). Similarly, in IL-4R$\alpha^{-/-}$ BALB/c mice, IL-10R blockade is required for complete parasite elimination (Nagase *et al.*, 2007). Importantly, IL-10 is derived primarily from $CD25^-$ T cells, including some co-expressing IFN-γ (Anderson *et al.*, 2007), explaining why anti-CD25 depletion is less effective than anti-IL-10R in conferring immunity in IL-4R$\alpha^{-/-}$ mice (Nagase *et al.*, 2007).

The modulatory effect of Tregs therefore depends critically on the genetic and immunological status of the host. Thus, anti-CD25 depletion of BALB/c mice resulted in enhanced Th2 responsiveness and greater susceptibility rather than resistance (Aseffa *et al.*, 2002). The ability of $CD4^+CD25^+$ T cells to suppress both Th1 and Th2 responses to *L. major* was then shown in co-transfer experiments (Xu *et al.*, 2003). Although equal suppression of both Th subsets would have no net effect on protection, it is interesting that BALB/c mice lacking CD103 are resistant to infection (Suffia *et al.*, 2005), presumably because T regs cannot access or be retained at the infection site.

In wild-type C57BL/6 mice, $CD25^+$ Tregs down-modulate immunity sufficiently to allow low-level persistence of parasites in the dermal site. In the absence of Tregs, parasites are eliminated but mice also lose their long-term immunity to reinfection (Belkaid *et al.*, 2002). Because parasite persistence and reactivation of infection in humans are major issues, it is relevant to note that high-dose reinfection in mice can expand $CD4^+CD25^+$ Tregs thereby allowing latent *L. major* at a distal site to reactivate (Mendez *et al.*, 2004). By co-transfer of allotype-marked $CD4^+CD25^+$ and $CD4^+CD25^-$ T cells from naive mice into RAG-$2^{-/-}$ recipients prior to *L. major* infection, it was also established that the infection only stimulates pre-existing "natural" Tregs, with little conversion observed from $CD4^+CD25^-$ to $Foxp3^+$ Tregs (Suffia *et al.*, 2006); these authors also showed that the $Foxp3^+$ natural Treg population was reactive to *L. major* antigens, and indeed, they were able to propagate parasite-specific Treg clones that maintained this specificity for months *in vitro*.

Each of these factors appears to be at play in human Leishmaniasis. Cutaneous lesions caused by *Leishmania braziliensis* show elevated $Foxp3^+$ Tregs which co-express CTLA-4 and GITR while producing both TGF-β and IL-10 (Campanelli *et al.*, 2006), while an independent study on this infection found that IL-10 production (by both Tregs and monocytes) strongly correlated with lesion activity (Salhi *et al.*, 2008). Similarly, in a related cutaneous species (*Leishmania guyanensis*), high IL-10 and Foxp3 expression were reported in patients with long-standing lesions who were unresponsive to chemotherapy (Bourreau *et al.*, 2009a,b).

3.3.2. Malaria

Malaria, caused by *Plasmodium* species, is one of the world's most prevalent lethal diseases, causing anemia (due to parasitism of erythrocytes) and cerebral inflammation (due to trapping of infected red cells in the vasculature). The complexity of both immunity and inflammation with the parasite is reflected in the dual roles of Tregs as protectors, in different settings, of both the host and the parasite (Finney *et al.*, 2010; Scholzen *et al.*, 2010), although as with *Leishmania*, IL-10 (Couper *et al.*, 2008a,b) and TGF-β (Omer *et al.*, 2003) are the critical regulators in malaria. Consistent with the latter study, a human malaria vaccine trial with healthy European volunteers found elevated serum TGF-β in individuals who did not respond to vaccination with inflammatory cytokines (Walther *et al.*, 2005).

In endemic humans, many studies have reported elevated $CD25^+FOXP3^+$ cell numbers in *Plasmodium falciparum* malaria (Finney *et al.*, 2009; Goncalves *et al.*, 2010; Minigo *et al.*, 2009; Walther *et al.*, 2009) as well as *Plasmodium vivax* (Bueno *et al.*, 2010; Goncalves *et al.*, 2010; Jangpatarapongsa *et al.*, 2008). However, although the frequencies of Tregs can vary significantly between individuals, the ratios of Treg:Th1 may not differ (Finney *et al.*, 2009). Of further note is the suggestion that Tregs in humans repress development of malaria-specific T cell memory rather than act on inflammation itself (Walther *et al.*, 2009); this study also implicated Tr1 (IL-10^+IFN-γ^+) regulatory cells which do not express FOXP3. Hence, there is currently little compelling evidence that $FOXP3^+$ Tregs suppress immunity to malaria in endemic populations.

A prominent aspect of human malaria is its effect on infants born to infected mothers. In two recent studies, it has been reported that following delivery from infected mothers, cord blood lymphocytes show high IL-10 and low Th1 responsiveness to malaria antigens, which can be reversed by $CD25^+$ T cell depletion (Bisseye *et al.*, 2009; Brustoski *et al.*, 2006). Prenatal exposure to *P. falciparum* antigens also correlated with greater frequency of $CD4^+CD25^{hi}$ or $CD25^+$CD127lo Tregs in newborns' cord blood, able to suppress malaria antigen-specific IFN-γ production *in vitro* (Mackroth *et al.*, 2011; Walther *et al.*, 2009). It is interesting to consider whether these Tregs may persist and so determine the susceptibility of the child to malaria infection and disease.

Murine models of malaria infection reflect a major, but not exclusive, role for Tregs in determining infection outcome. The rodent malaria species *Plasmodium yoelii* is frequently studied in both susceptible (BALB/c) and resistant (DBA/2) mice. Within 3–4 days of infection, the susceptible BALB/c mice raise $CD4^+CD25^+$ Treg frequency and overall IL-10 production, suggesting a functional link with their poor protective

Th1 response (Wu *et al.*, 2007), and supporting an early finding that anti-CD25 depletion generated protective immunity to this parasite (Hisaeda *et al.*, 2004). However, a subsequent study in C57BL/6 mice compared lethal (Py17XL) and nonlethal (Py17X, NL) strains of *P. yoelii* and found that both elicited similar, modest, rises in Foxp3$^+$ Treg numbers and that in neither case did CD4$^+$CD25$^+$ cell depletion alter the course of infection (Couper *et al.*, 2008b). In contrast, IL-10 from CD4$^+$CD25$^-$Foxp3$^-$ T cells following *P. yoelii* infection was the critical factor in impeding parasite clearance and ameliorating liver pathology following infection, with IL-10$^{-/-}$ mice surviving the otherwise lethal Py17XL infection. Nevertheless, a demonstration that Treg activation can suppress immunity to *P. yoelii* comes from mice co-infected with the helminth *Heligmosomoides polygyrus* (see below, Section 3.4.2); co-infected mice developed more severe malaria infections which were rescued by anti-CD25 antibody treatment (Tetsutani *et al.*, 2009).

The best available mouse model for *P. falciparum*-mediated cerebral malaria (CM) is another rodent species, *Plasmodium berghei* in the C57BL/6 mouse, associated with parasite vascular adhesion and overproduction of Th1 inflammatory mediators within the brain. Perhaps counter-intuitively, anti-CD25 Treg depletion protects mice from CM, reducing parasite sequestration and also CD8$^+$ T cell infiltration (Amante *et al.*, 2007; Randall *et al.*, 2008; Wu *et al.*, 2010). Interestingly, the effect of depletion is time dependent (Vigario *et al.*, 2007) suggesting that the action of Tregs could be to facilitate entry of effector cells into the CNS, as described above (see Section 3.1.3) in HSV-2 infections (Lund *et al.*, 2008). An alternative explanation is that key effector populations for CM express CD25 after infection and are co-depleted by antibody treatment. In support of this, Treg ablation in DEREG mice showed a substantial population of CD25$^+$Foxp3$^-$ T cells developing after infection, and no amelioration of CM disease (Haque *et al.*, 2010; Steeg *et al.*, 2009). However, when Treg numbers are experimentally boosted with IL-2/anti-IL-2 complexes, mice were fully protected from CM (Haque *et al.*, 2010), arguing again that the action of Tregs depends critically on their proportions and activation state in vivo.

Other mouse strains are more resistant to *P. berghei*-induced CM, but in the BALB/c, anti-CD25-depletion had the opposite effect and accentuated CM symptoms (Nie *et al.*, 2007). However, as in *P. yoelii*, Treg depletion had little effect on overall parasitemias or progression to death from fulminant infections (Wu *et al.*, 2010). Clearly, regulation of the immune response to both human and murine malaria involves multiple cellular components, particularly at the level of tissue infiltration, and is greatly dependent upon dynamic and kinetic factors that have yet to be defined; while Tregs may not be uniquely responsible for susceptibility to infection, clearly it is essential to strike the appropriate balance with effector mechanisms for a health outcome to this potentially devastating infection (Hansen and Schofield, 2010; Scholzen *et al.*, 2010).

3.3.3. Trypanosomes

The trypanosomes encompass two very different groups of parasites, as the African species (e.g., *Trypanosoma brucei*) are extracellular pathogens, which can cause disease in humans and livestock, while the South American species (*Trypanosoma cruzi*) has an intracellular niche in human phagocytes and smooth muscle cells. Immunosuppression has long been a prominent feature in African trypanosomiasis, and active suppressor cell populations were described in mice by the late George Roelants and colleagues (Roelants *et al.*, 1979).

C57BL/6 mice, which escape lethality to *Trypanosoma congolense* infection through limitation of an early IFN-γ response, show expansion of IL-10 producing Foxp3$^+$ Tregs. In this "trypanotolerant" strain, Tregs were able to downregulate classical activation of macrophages and limit tissue pathology resulting from the inflammatory immune response (Guilliams *et al.*, 2007). This role of Tregs in limiting pathology, but allowing increased resistance following trypanosome infection, was also demonstrated following *T. brucei* infection. Treg expansion with the CD28 superagonist resulted in downregulation of inflammatory type 1 cytokines and the development of macrophages into the alternatively activated phenotype (Guilliams *et al.*, 2008). Later studies demonstrated that anti-CD25 antibody treatment and effective depletion of natural Foxp3$^+$ Tregs before *T. congolense* infection protects BALB/c mice against this normally lethal disease. Protection was reversed in CD25-depleted mice by administration of a specific inhibitor of inducible nitric oxide synthase (Wei and Tabel, 2008).

T. cruzi is the causative agent of Chagas' disease in South America, and infection has again been associated with immunosuppression of humoral and cell-mediated immunity, in part attributable to the action of IL-10 and TGF-β which disable iNOS-mediated killing by infected macrophages (Gazzinelli *et al.*, 1992). In patients, greater CD4$^+$CD25$^+$FOXP3$^+$ Treg numbers are found in both asymptomatic carriers and those developing pathology due to parasites in the myocardium; however, only in healthy patients did Tregs produce IL-10, indicating that cardiomyopathy may results from insufficient production of this cytokine (de Araujo *et al.*, 2011). In infected mice, similar phenotype Tregs were found to migrate to the heart, but depleting interventions with anti-CD25 increased mortality, while administration of anti-GITR antibody additionally increased myocarditis and tissue parasitism (Mariano *et al.*, 2008).

3.4. Helminths

Helminths are multicellular worms comprised of three broad taxa, the Nematodes (round worms, including the model organism *Caenorhabditis elegans*), Trematodes (flukes), and Cestodes (tapeworms), each separated by approximately 500 million years of evolution. While taxonomically distant, the parasitic species share many immunological features which are likely to have co-evolved under similar selective pressure from the immune system of the host (Allen and Maizels, 2011).

Most helminth infections, in man and livestock, are long-term chronic infestations which are maintained in the population by repeated cycles of reinfection; hence protective immunity is slow to develop, and indeed, most helminth species are associated to some degree with a state of immune suppression. Classic studies demonstrated that peripheral blood T cells from Schistosome and filariasis-infected patients showed parasite antigen-specific hyporesponsiveness, as detailed below, which could be reversed by chemotherapeutic removal of the parasite burden (Cooper *et al.*, 2000; Greene *et al.*, 1985; Sartono *et al.*, 1995).

A marked contrast from infections with microbial agents is seen for the role of IL-10 in helminth infections; while in viral, bacterial, and protozoal infections, IL-10 generally impairs resistance (Couper *et al.*, 2008a; Moore *et al.*, 2001), the role of IL-10 in Th2-dominated helminth infections is both complex and double-edged (Hoffmann *et al.*, 2000a). For example, IL-10 is essential to protect against potentially fatal immunopathology in chronic schistosome infection, but it is equally necessary in the initial stages of infection to establish dominant (and generally protective) Th2 responses by suppressing competing Th1/Th17 activity. Similarly, IL-10 is required for Th2-mediated expulsion of adult *Trichinella spiralis* nematodes from the intestine, and yet acts to block immunity to their offspring, larvae which encyst in tissue musculature (Beiting *et al.*, 2007; Helmby and Grencis, 2003). In human helminth infections, IL-10 acts more unequivocally as an immunoregulatory player, perhaps because patients are studied in the chronic, homeostatic phase rather than during the initial events of priming and Th subset selection.

3.4.1. Filarial nematodes

Human filarial nematodes include the causative agents of lymphatic filariasis (*Brugia malayi*, *Brugia timori*, and *Wuchereria bancrofti*) and onchocerciasis or river blindness (*Onchocerca volvulus*). In these long-lived infections, many infected patients are asymptomatic but carry large numbers of transmission stages (microfilariae, MF) in the blood (for lymphatic filariasis) or skin (in onchocerciasis). Typically, peripheral blood T cells from these patients fail to respond to parasite antigen challenge *in vitro*, and are hence termed hyporesponsive (Piessens *et al.*, 1980;

TABLE 3.4 Tregs in helminth infections

Filarial nematodes		
Brugia malayi	Induces Foxp3 expression, including in DO11.10 T cells	McSorley *et al.* (2008)
Brugia pahangi	CD25 depletion raises Th2 response	Gillan and Devaney (2005)
Litomosoides sigmodontis	Tregs maintain infection through CTLA-4 and inhibit allergy	Dittrich *et al.* (2008); Taylor *et al.* (2005, 2007)
Onchocerca volvulus	TGF-β-producing clones from human infection site	Doetze *et al.* (2000)
Wuchereria bancrofti	Raised $FOXP3^+$ T cells in infected patients	Babu *et al.* (2006)
Intestinal nematodes		
Enterobius vermicularis	High FOXP3 expression in uninflamed mucosa of UC patient	Büning *et al.* (2008)
Heligmosomoides polygyrus	*De novo* induction of Tregs; Tregs reduce intestinal pathology, suppress Th2 response and bystander airway allergy	Finney *et al.* (2007); Grainger *et al.* (2010); Rausch *et al.* (2008, 2009); Wilson *et al.* (2005)
Strongyloides ratti	Treg depletion reduces worm burden	Blankenhaus *et al.* (2011)
Strongyloides stercoralis	In HTLV-1 co-infection, excessive $FOXP3^+$ Tregs, suppression of IL-5 and high worm burdens	Montes *et al.* (2009)
Toxocara canis	Tissue-migrating larvae induce Foxp3 in mice	Othman *et al.* (2010)
Trichinella spiralis	$IL\text{-}10^-$ Tregs restrain Th2 responses	Beiting *et al.* (2007)
Trichuris muris	$IL\text{-}10^-$ Tregs restrain Th2 responses	D'Elia *et al.* (2009)

(*continued*)

TABLE 3.4 *(continued)*

Trematodes (flatworms)		
Fasciola hepatica	Infection induces IL-10 and TGF-β from Tr1-like Tregs	Walsh *et al.* (2007)
Schistosoma haematobium	FOXP3^{+} Tregs correlate with infection intensity in children	Nausch *et al.* (2011)
Schistosoma japonicum	Egg antigens stimulate CD25^{+} suppression of airway allergy	Yang *et al.* (2007b)
	Treg induction via TLR2 ligation to HSP60 peptide	Wang *et al.* (2009)
	Anti-CD25 treatment reduces worm load	Tang *et al.* (2011)
Schistosoma mansoni	IL-10^{-} Tregs elevated CD103^{+}, dampen IL-4 responses to eggs	Baumgart *et al.* (2006)
	IL-10^{+} CD25^{+} Tregs control pathology, dampen Th1 allowing Th2 to expand	Hesse *et al.* (2004); McKee and Pearce (2004)
	CD25^{+} Tregs expand through TLR2 to control pathology, upregulating CD103, CTLA4, and many other genes	Layland *et al.* (2007, 2010)
	Tregs induced by eggs, inhibit Th1	Taylor *et al.* (2006)
	Foxp3 expression decreases following chemotherapeutic cure	Watanabe *et al.* (2007)
	Pathology patients have fewer CD25^{high} Tregs	Teixeira-Carvalho *et al.* (2008)
Cestodes (tapeworms)		
Echinococcus multilocularis	Peritoneal T cells express high Foxp3	Mejri *et al.* (2011)

Yazdanbakhsh *et al.*, 1993). Treg activity was presaged in this system by Piessens' report on suppressor T cells in hyporesponsive MF+ patients (Piessens *et al.*, 1982), and by later work showing that the hyporesponsiveness can be reversed, *in vitro*, with anti-IL-10 and TGF-β antibodies (King *et al.*, 1993). Most recently, the link between Tregs and the human filarial infection has been firmly established with elevations of both natural and adaptive Treg numbers (Metenou *et al.*, 2010). Moreover, in individuals who are more reactive to parasite infections, with low or zero circulating MF and immunopathological symptoms such as lymphoedema and elephantiasis, Treg activity is deficient (Babu *et al.*, 2009b).

Additional evidence for Treg-like cells in human onchocerciasis came from analysis of T cells in the subcutaneous granulomas surrounding adult *O. volvulus* (Doetze *et al.*, 2000), with $CD4^+$ T cell clones from this tissue expressing IL-10 and TGF-β (Satoguina *et al.*, 2002). At this time, their FOXP3 status was not determined. In lymphatic filariasis, asymptomatic carriers were found to express higher CTLA-4 levels (Steel and Nutman, 2003), with anti-CTLA-4 antibody also raising the cytokine responses of patients' cells *in vitro*. Interestingly, CTLA-4 may act with PD-1 in filariasis patients to block protective Th1 and Th17 responses to tuberculosis (Babu *et al.*, 2009a).

A particularly striking feature of human filarial infections is the extremely high levels of IgG4 antibodies, both parasite-specific and total, that are rapidly lost once parasites are removed by chemotherapy (Atmadja *et al.*, 1995). Hyporesponsive patients show the maximal IgG4 levels alongside depressed IgE responses (Yazdanbakhsh *et al.*, 1993), a relationship which can now be explained by the action of Tregs, as *in vitro* switching of B cells to the IgG4 isotype is promoted by IL-10 (Satoguina *et al.*, 2005) as well as TGF-β and GITR ligation, although not CTLA-4 (Satoguina *et al.*, 2008). Hence, circulating IgG4 levels in humans could be a marker not only for helminth infection but also for elevated Treg activity.

The conclusion that human filariasis activates Tregs is well supported by studies in animal models; although the mosquito-borne infective larvae of *B. malayi* are tolerated for less than 14 days in mice, the parasites induce a short-lived expansion in $Foxp3^+$ Tregs, as occurs more strongly in mice transplanted with adult worms of the same species (McSorley *et al.*, 2008). Dead parasites of either stage did not elicit this response, indicating that the presence or products of live filarial worms were responsible for stimulating Tregs. Moreover, bystander-specificity T cells (carrying the DO11.10 ovalbumin-specific TCR) were induced to express Foxp3 when transferred into BALB/c mice carrying either larval or adult *B. malayi* (McSorley *et al.*, 2008).

Because human filariae cannot complete their infection cycle in mice, it is necessary to study related, rodent-compatible, species to ascertain the functional importance of Tregs in the natural context. As such, the model system of *Litomosoides sigmodontis* (Hoffmann *et al.*, 2000b) has proven exceptionally informative. Very soon after infection, there is expansion of natural Tregs, as determined by BrdU uptake *in vivo*, followed by a second wave of inducible Tregs (Taylor *et al.*, 2009), with the initial wave at least essential for parasite establishment. Transfer of cells from infected mice protected allergic recipients from allergic airway hypersensitivity, in a manner inhibited by blockade of TGF-β or anti-CD25 Treg depletion (Dittrich *et al.*, 2008). One consequence of regulatory expansion is silencing of effector cell responses (an interesting parallel to hyporesponsiveness in humans), and the emergence of a $Foxp3^{-}GITR^{+}CTLA4^{+}$ unresponsive $CD4^{+}$ population (Taylor *et al.*, 2005). Most significantly, intervention with depleting antibodies, using anti-CD25 in combination with either anti-GITR (Taylor *et al.*, 2005), or anti-CTLA4 (Taylor *et al.*, 2007) boosted responsiveness and elicited immune killing of worms. These studies were the first to demonstrate that interfering with Treg function (and re-stimulating hyporesponsive effectors through GITR ligation) can reverse susceptibility to a helminth infection.

3.4.2. Intestinal nematodes

Intestinal nematode infections (also termed geohelminths reflecting their fecal-oral transmission) are extraordinarily prevalent in humans in tropical countries, with approximately two billion cases in the world today (Hotez *et al.*, 2008). Studies have not found significant increases in systemic $Foxp3^{+}$ Treg frequencies, but qualitative changes are apparent, which may well be immunologically significant.

In areas hyperendemic for the intestinal helminth infections *Ascaris lumbricoides* and *Trichuris suum*, lymphocytes from infected children constitutively express high levels of IL-10 and TGF-β, while antigen-specific responses are inversely depressed (Turner *et al.*, 2008). For example, $CD4^{+}CTLA4^{+}$ T cells are more numerous in children with intestinal helminths than uninfected subjects (García-Hernández *et al.*, 2009). Functionally, peripheral T cells from geohelminth-infected children show depressed *in vitro* immune responses to malarial and mycobacterial antigens that are rescued by removal of the $CD25^{high}$ cells (Wammes *et al.*, 2010).

In a less common infection, *Strongyloides stercoralis*, patients co-infected with HTLV-1 show exaggerated levels of circulating $Foxp3^{+}$ T cells, reaching $\sim$18% of the total $CD4^{+}$ T cell population (Montes *et al.*, 2009), together with higher worm burdens, while IL-5 and eosinophilia were suppressed.

Research into mouse gastrointestinal parasites has employed several model systems, with the most information to date obtained from *H. polygyrus*, a relative of the human hookworms, which spends its entire parasitic phase within the gastrointestinal tract (Monroy and Enriquez, 1992). This species is particularly associated with immunosuppression, down-modulating responses to allergens, autoantigens, and other infectious organisms (reviewed by Maizels *et al.* (2011)). Early in infection, both the proportion and absolute numbers of Foxp3^{+} Treg cells expand in the mesenteric lymph nodes (Finney *et al.*, 2007; Rausch *et al.*, 2008), while within the Foxp3^{+} population, there is also increased expression of CD103, considered to be a marker of Treg activation (Huehn *et al.*, 2004). CD25^{+} Tregs from *H. polygyrus*-infected mice are suppressive when transferred to uninfected recipients, as shown by inhibition of airway allergic inflammation (Wilson *et al.*, 2005). CD8^{+} Tregs are also found to expand in the lamina propria (Metwali *et al.*, 2006).

Remarkably, *H. polygyrus* attenuates colitis in IL-10-deficient mice (Elliott *et al.*, 2004), although IL-10 is necessary for this helminth to protect normal mice from chemically induced colitis (Setiawan *et al.*, 2007). In contrast, infection cannot block colitis in mice expressing a T cell-specific kinase-dead TGF-β receptor II (Ince *et al.*, 2009), demonstrating that both IL-10 and TGF-β can be invoked by the regulatory pathways activated by the infection. The importance of TGF-β is emphasized, however, both by the finding that *H. polygyrus* secretes a functional mimic of this cytokine (see Section 4.1 below) and by the successful boosting of immunity to adult worms by *in vivo* administration of an inhibitor of TGF-β receptor kinase I (Grainger *et al.*, 2010).

A recent study reported on *H. polygyrus* infection in DEREG mice, expressing DTR under a BAC transgene (Lahl *et al.*, 2007); in this report, Foxp3-depleted mice showed heightened Th2 responses but similar infection levels (Rausch *et al.*, 2009). It should be noted, however, that intestinal worms were enumerated at an early time point before genetically resistant mice expel most worms (Maizels *et al.*, 2011), and this system will need further investigation.

As with *H. polygyrus*, many (but not all) mouse intestinal nematode infections cause an expansion of CD4^{+}Foxp3^{+} Tregs; in the case of *T. spiralis* infections, this occurs in both mice (Beiting *et al.*, 2007) and rats (Gruden-Movsesijan *et al.*, 2010). In mice, anti-CD25 antibody-mediated Treg depletion does not reduce worm numbers, although treatment results in a heightened Th2 and intact IL-10 production by CD4^{+}CD25^{-} T cells (Beiting *et al.*, 2007). In a separate study, anti-CTLA-4 antibody administration to infected mice did reduce muscle larval numbers (Furze *et al.*, 2006), indicating that perhaps a CD25^{-}CTLA-4^{+}IL-10^{+} Tr1-like population is in play.

Trichuris muris is, like *H. polygyrus*, a well-studied intestinal model and inhabits the cecum of mice (Cliffe and Grencis, 2004). Interestingly, different strains of *T. muris* survive for varying times *in vivo*, and the longest-lived isolate elicits the strongest Foxp3$^+$ Treg response (D'Elia *et al.*, 2009). This parasite also elicits a population of intestinal Foxp3$^-$ IL-35-producing suppressive T cells (Tr35 cells), which *in vitro* differentiate under the influence of IL-10 and IL-35 (Collison *et al.*, 2010). An important role in limiting pathology has also been established in these infections, as anti-CD25- and anti-GITR-treated mice develop aggravated pathology, as well as lower worm numbers in the case of anti-GITR treatment (D'Elia *et al.*, 2009).

3.4.3. Schistosomes

Schistosomes are trematode worms causing schistosomiasis (Bilharzia) in some 220 million people worldwide. Like other helminths, they form long-lived, chronic infections which are associated with a degree of parasite-specific immune suppression, deviation (e.g., to IgG4 in humans), and susceptibility to repeated reinfection. As with human filariasis (see Section 3.4.1 above), peripheral T cells from infected patients often fail to respond to parasite antigen challenge *in vitro* (Grogan *et al.*, 1998), and two reports have charted Treg activity in human schistosomiasis. In *Schistosoma mansoni* (in which adult worms live in the mesenteric vasculature), CD4$^+$CD25^{high} T cell frequencies were inversely proportional to effector phenotype (CD25^{mid}HLA-DR$^+$) cells, but not to parasite intensity; however, curative chemotherapy significantly reduced the frequency of Tregs using these markers (Watanabe *et al.*, 2007). More recently, analysis of the urogenital parasite *S. haematobium* found a significant positive correlation between CD4^{dim}CD25^{high}CD127^{low}Foxp3$^+$ T cells and parasite intensity in children at the age at which they are still susceptible to reinfection; interestingly in adults, the reverse was the case (Nausch *et al.*, 2011).

More information is available from mice, for which *S. mansoni* is infective. The infection follows two phases, a Th1-dominated maturation period, during which skin-penetrating cercaria migrate to the lung and then the hepatic portal vasculature; and a later Th2-dominated stage which is provoked by egg release from adult worms (Pearce and MacDonald, 2002). Because eggs become lodged in the liver, this later stage is accompanied by severe granulomatous immunopathology that is moderated by IL-10-producing T cells, both Th2 and Treg, which become numerous at this time (Hesse *et al.*, 2004; McKee and Pearce, 2004). While CD25$^+$Foxp3$^+$ Tregs are not the major contributor of IL-10, they dampen Th2 responses, with anti-CD25 depletion resulting in significantly enhanced IL-4 production (Baumgart *et al.*, 2006). Moreover, anti-CD25 treatment increased egg destruction and aggravated liver pathology around the eggs, demonstrating a beneficial role for Tregs at this stage of the infection (Layland *et al.*, 2007). Subsequent studies confirmed the Foxp3$^+$ phenotype of Tregs surrounding the site of inflammation (Layland *et al.*, 2010), and the interaction was more formally demonstrated by retroviral expression of Foxp3 in mice resulting in the suppression of liver granuloma formation (Singh *et al.*, 2005).

Schistosome eggs produce a number of immunologically active substances, including an IL-4-inducing protein (IPSE or α-1 (Schramm *et al.*, 2007)) and a ribonuclease, ω-1 (Everts *et al.*, 2009; Steinfelder *et al.*, 2009). While both can drive Th2 responses *in vivo* and *in vitro*, only ω-1 can induce Foxp3 in T cells, requiring the presence of DCs, TGF-β, and retinoic acid (Zaccone *et al.*, 2011). Unrelated to this protein, an HSP60-derived peptide SJMHE1 from *Schistosoma japonicum* was shown to expand $CD4^{+}CD25^{+}Foxp3^{+}$ T cell populations *in vivo* and *in vitro*, with such cells able to inhibit delayed-type hypersensitivity on transfer to mice 1 day prior to allergen sensitization (Wang *et al.*, 2009).

3.5. Fungi

Fungal pathogens are found in a variety of niches and in different developmental forms; in addition, a number are commensals which can adapt opportunistically to immunodeficiency. One such example is oropharyngeal *Candida albicans* infection, immunity to which is compromised by the $CD4^{+}CD25^{+}$ IL10-producing Treg population that is deficient in $TLR2^{-/-}$ mice, and which when depleted *in vivo* with anti-CD25 antibody, results in improved resistance to infection (Netea *et al.*, 2004). Interestingly, as observed for *L. major* infection (Belkaid *et al.*, 2002), the $CD4^{+}CD25^{+}$ subset was also required to generate normal protective memory responses to infection, establishing a "protective tolerance" that restrains pathology while allowing a form of commensalism to persist (Montagnoli *et al.*, 2002). Recent work has elucidated a fascinating dynamic in which, when confronted with an acute infection, Tregs promote Th17 responses to *C. albicans*, while at later time points act to restrain the same effector population from mediating inflammatory bowel disease (Pandiyan *et al.*, 2011); thus early CD25 depletion resulted in diminished Th17 immunity and increased fungal burden, whereas transfer of $CD25^{+}$ Tregs could prevent colitis in infected RAG mice caused by *in vitro* polarized Th17 cells.

Paracoccidioides brasiliensis is regarded as the most prevalent primary fungal pathogen of Latin America and is the causative agent of a systemic granulomatous disease in the host. $CD4^{+}CD25^{+}CD103^{+}CTLA\text{-}4^{+}Foxp3^{+}GITR^{+}$ Tregs are found in the lesions of infected patients (Cavassani *et al.*, 2006). In a mouse model, adoptive transfer of $CD4^{+}CD25^{+}$ but not $CD4^{+}CD25^{-}$ T cells from infected mice increased the fungal load in recipients, except in a $CCR5^{-/-}$ setting (Moreira *et al.*, 2008). Mice lacking CCR5 had a reduced number of Tregs in the lungs, and did not exhibit suppressed T cell proliferation *ex vivo* following a more contained infection. $CCR5^{-/-}$ mice may have a generalized defect

TABLE 3.5 Tregs in fungal infections

Aspergillus fumigatus TLR	Inflammation controlled by $CD4^+CD25^+$ Tregs	Montagnoli *et al.* (2006)
Candida albicans	Early Th17 promoted by Tregs, but later immunity suppressed; Tregs neutralized by TLR2 ligation	Montagnoli *et al.* (2002); Netea *et al.* (2004); Pandiyan *et al.* (2011)
Cordyceps sinensis	Increased $Foxp3^+$ Tregs and reduced T1D in NOD mice	Shi *et al.* (2009)
Histoplasma capsulatum	Tregs suppress Th17 at site of infection	Kroetz and Deepe (2011)
Onychomycosis	Higher $CD4^+CD25^+$ cell numbers in patients	Kaya *et al.* (2009)
Paracoccidioides brasiliensis	Tregs control inflammation and limit fungal clearance; migration of $Foxp3^+$ Tregs to lesions	Cavassani *et al.* (2006); Loures *et al.* (2010); Moreira *et al.* (2008)
Pneumocystis carinii	$CD4^+CD25^+$ Tregs suppress inflammation	Hori *et al.* (2002); McKinley *et al.* (2006)

in egress of thymic Tregs, as also demonstrated by their greater resistance to another fungal pathogen, *Histoplasma capsulatum* (Kroetz and Deepe, 2011). The correlation between Treg activity and extent of fungal infection did not hold, however, in TLR4-deficient mice, which showed higher Foxp3^{+} Treg numbers and yet were able to control *P. brasiliensis* infection more efficiently (Loures *et al.*, 2010); whether this reflects an early stimulatory role by Tregs as observed for *C. albicans* has not yet been tested.

Aspergillus fumigatus is a further fungal pathogen and a causative agent of airway hypersensitivity and allergy. Exposure of mice to *Aspergillus conidia* resulted in the early expansion, activation, and recruitment of CD4^{+}CD25^{+} Tregs, which correlated with decreases in inflammation at this time point. Depletion of natural Tregs using cyclophosphamide or anti-CD25 reduced CD4^{+}CD25^{+} T cell numbers, exacerbated inflammation, and decreased the survival of infected wild-type mice (Montagnoli *et al.*, 2006). This work also highlighted the role of IDO, as well as IL-10 and CTLA-4, as a mediator feeding back to tolerize DCs and forestall hypersensitivity in the later stages of infection.

4. TREG ACTIVATION—A COMMON IMMUNE-EVASION STRATEGY ACHIEVED THROUGH DIVERSE ROUTES

The evidence from the many and diverse infectious agents reviewed above is that Tregs often suppress protective immunity: examples can be given from the retrovirus (FV) model, through malaria to the helminth worms. However, where pathogens are reliant on Treg activity, this offers a therapeutic route to eliminate infection, which has been reproduced in a number of these same models (Hisaeda *et al.*, 2004). Hence, identifying the pathway(s) for Treg activation (Figure 3.1) is crucial for future intervention strategies.

Treg activation can benefit both host and pathogen however. Most frequently, this is evident at the level of dampening pathology. T cell-mediated responses to HSV in the corneal stroma are a frequent cause of human blindness. Depletion of natural regulatory T cells was shown to enhance lesion formation and keratitis following HSV infection by impairing antiviral immunity and T cell migration to lesion sites (Suvas *et al.*, 2004). Similarly, Tregs restrain intestinal pathology in infections with *T. muris*; in this system, anti-GITR antibody results in lower worm burdens, but incurs more intense gut pathology (D'Elia *et al.*, 2009). In the long term (especially in chronic human infections), the key to a healthy status is the balance between controlling infection and limiting pathology—maintaining a recalibrated homeostasis in chronic infection.

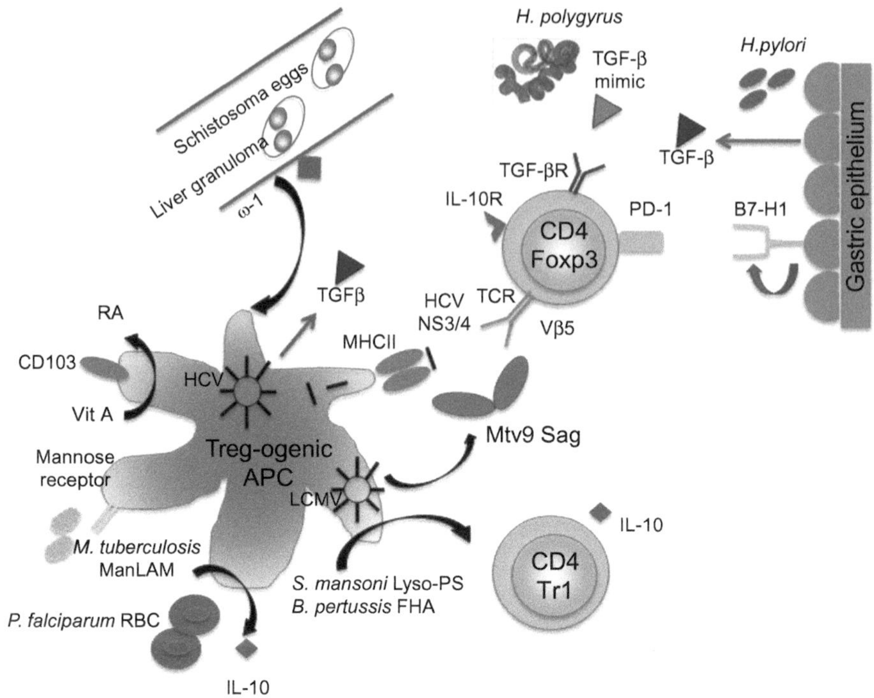

FIGURE 3.1 Pathways of Treg induction and activation in infection.

4.1. Direct conversion of T cells into Tregs

A few examples are now established in which pathogen products directly activate, promote, or induce Tregs; clearly these are important proof-of-principle that the expansion of Tregs *in vivo* is not purely a homeostatic response that accompanies every effector expansion. In these instances, at least, we can surmise that pathogens have evolved to stimulate and exploit the host's down-modulatory Treg populations.

A recent example is the induction of host Tregs by *H. polygyrus*, which secretes a TGF-β-like mimic that activates the TGF-β signaling pathway (Grainger *et al.*, 2010). *H. polygyrus* adult worm excretory-secretory products (HES), when added to naive Foxp3$^-$ murine T cells together with TCR ligation, induce *de novo* Foxp3. As this fails to occur in cells expressing the T cell-specific dominant negative TGF-β-RII, the interaction is directly between HES and the T cells in question. Foxp3$^+$ T cells, whether induced by HES or mammalian TGF-β, are equally able to suppress airway allergy in recipient mice (Grainger *et al.*, 2010). A similar activity was also found in secreted products of the related sheep parasite *Teladorsagia circumcincta*, indicating that some helminth parasites have evolved to exploit a key immunosuppressive pathway of their host. The helminth TGF-β mimic may be the biological equivalent of viral cytokine-like molecules, such as the EBV IL-10 homolog and other examples summarized elsewhere (Tortorella *et al.*, 2000).

A very different mechanism is employed by *Streptococcus pneumoniae* which elaborates a zwitterionic polysaccharide able to directly cross-link the TCR of $CD8^+$ T cells and switch them into a regulatory phenotype (Mertens *et al.*, 2009). As with the *H. polygyrus* HES, this is a process which can by-pass any requirement for an APC population.

4.2. Induction of Tregs via DCs

Treg cells, like all T cells, require cognate APC interactions for their activation, and DCs are the major cell type responsible; consequently, it is not surprising that in the majority of systems studied, Treg generation involves the DC population. This is most clearly demonstrated where individual molecular components from pathogens are able to drive Treg differentiation through a DC pathway: for example, the FHA from *B. pertussis* modifies DC interactions with naive antigen-specific T cells, polarizing them to an IL-10-producing phenotype (McGuirk *et al.*, 2002); similarly, the lysophosphatidylserine molecule from *S. mansoni* acts through DCs to induce $IL\text{-}10^+$ Tr1 cells (van der Kleij *et al.*, 2002). In the former case, the FHA binds TLR4 (Higgins *et al.*, 2003) while the lyso-PS ligates TLR2, highlighting the regulatory face of the Toll ligand family as discussed in a following section.

Most recently, the induction of Foxp3 in naive T cells by DCs exposed to the *S. mansoni* egg antigen ω-1 has been described (Zaccone *et al.*, 2011), providing a mechanistic pathway that may be followed *in vivo* when DCs from *S. japonicum*-infected mice promote $Foxp3^+$ and $IL\text{-}10^+$ Treg *in vitro*, inhibiting airway allergy in the process (Liu *et al.*, 2011). Likewise, the Ac-TMP-1 protein released by adult *Ancylostoma caninum* hookworm skews DCs to induce both $CD4^+$ and $CD8^+CD25^+Foxp3^+$ T cells; both subsets expressed IL-10 while the $CD4^+$ Tregs also produced TGF-β (Cuéllar *et al.*, 2009). In addition, *Anisakis simplex* (a nematode from marine mammals), elaborates a homologue of MIF (macrophage migration inhibitory factor) which when injected into mice elicits increased numbers of $Foxp3^+$ T cells (Park *et al.*, 2009).

Many further examples of Treg generation by DCs in infectious settings employ whole organisms, or their secreted products. For example, bone marrow DCs exposed to live *H. pylori* bacteria are able to drive *de novo* induction of Tregs *in vitro* (Zhang *et al.*, 2010), and expand $Foxp3^+$ Tregs when adoptively transferred to mice shortly before infection (Kao *et al.*, 2010). In a different system, secreted products of the nematode *H. polygyrus* were used to pulse DCs which preferentially induced functional $CD4^+CD25^+IL\text{-}10^+Foxp3^-$ Tregs (Segura *et al.*, 2007). *In vivo*, expansion of pro-regulatory DC subsets occurs, such as the predominant $CD11c^{lo}CD103^-$ DC subset in the mesenteric lymph nodes of *H. polygyrus*-infected mice which are potent inducers of Foxp3 in naive murine T cells (Smith *et al.*, 2011).

In each of these systems, the mechanisms by which DCs induce Tregs are similar to those established in model systems in the absence of infection: TGF-β, IL-10, and retinoic acid are all implicated. For example, the induction of $Foxp3^+$ Tregs by malaria-infected red blood cells is dependent on both TGF-β and IL-10 (Scholzen *et al.*, 2009). Similarly, the Treg-inducing ability of Schistosome egg antigen and its component ω-1 requires target DCs to produce TGF-β (Zaccone *et al.*, 2011).

4.3. Bystander induction by other cell types

Although not involving DCs, the strategy mounted by HCV to induce Tregs is similar to that described for several other pathogens above, except that infected human hepatocytes are induced to express TGF-β, which, using well-characterized pathways, is able to drive expression of Foxp3 and other regulatory markers (including CD25, CTLA-4, and LAP) in human $CD4^+$ T cells (Hall *et al.*, 2010). GECs are similarly stimulated by *H. pylori* to express TGF-β and B7H-1 to evoke $Foxp3^+$ Tregs (Beswick *et al.*, 2011).

In *S. mansoni* infection, regulatory B cells are able to induce $Foxp3^+$ Tregs to infiltrate airways, suppressing allergic inflammation (Amu *et al.*, 2010). B cells are also implicated in *Helicobacter felix* infection, in which TLR2 activation of B cells is required to induce IL-10-producing $CD4^+CD25^+$ regulatory T cells and control immunopathology (Sayi *et al.*, 2011).

4.4. TLRs in Treg activation

Most pathogens are initially recognized by the innate immune system through one or more Toll-like receptors (TLRs), ligating to archetypal molecular species characteristic of particular microbe classes, and expressed by DCs and other sentinels of immunity (Medzhitov, 2007). Adaptive immune lymphocytes, including T cells, may also express TLRs, raising the question of whether in infection Tregs are either stimulated or inhibited by interacting with TLR ligands.

In general, TLR stimulation of DCs and other APCs is considered to be strongly proinflammatory and likely to overcome homeostatic Treg control. However, some pathogen TLR ligands (such as the TLR2-binding *S. haematobium* lysophosphatidylserine (van der Kleij *et al.*, 2002)) drive human DCs to induce IL-10-secreting Tr1 cells, while *S. typhimurium* LPS-driven TLR4 ligation was reported to promote proliferation and suppressive activity of murine $CD4^+CD25^+$ T cells (Caramalho *et al.*, 2003).

Subsequently, numerous studies into pathogen-derived pro-regulatory TLR stimulation have been reported with some clear instances of direct effects on Tregs, rather than pathways routed through APC populations (Himmel *et al.*, 2008; van Maren *et al.*, 2008).

TLR2 emerges as a significant enhancer of Treg activity in the steady state (Sutmuller *et al.*, 2006) as well as in the context of several infections, as indicated by the greater resistance of $TLR2^{-/-}$ mice to *C. albicans* (Bellocchio *et al.*, 2004; Netea *et al.*, 2004) and *Yersinia enterocolitica* (Sing *et al.*, 2002). Likewise, in infections with the helminth *S. mansoni*, $TLR2^{-/-}$ mice, in which $CD4^+CD25^+$ Tregs do not expand, suffer aggravated liver pathology which can be rescued by transfer of wild-type schistosome-primed $CD4^+CD25^+$ T cells (Layland *et al.*, 2007). An increase in Tregs occurred in TLR4- but not in TLR2-deficient mice following administration of the HSP-60-derived *S. japonicum* peptide SJMHE1, as bone marrow-derived macrophage and dendritic cells from $TLR2^{-/-}$ mice, but not wild-type mice, primed *in vitro* or *in vivo* were unable to induce Tregs *in vitro* (Wang *et al.*, 2009).

Treg proliferation can be promoted by TCR signaling and TLR2 ligation with a synthetic agonist; however, in *C. albicans* infection, the effect of TLR2 ligation is to reverse $CD4^+CD25^+$ Treg suppression of anti-fungal responses and allow fungal outgrowth (Netea *et al.*, 2004). Hence, $TLR2^+$ Tregs transferred into $TLR2^{-/-}$ mice promote a 100-fold rise in *C. albicans* infection, which is prevented in the presence of the TLR2 ligand Pam-3-Cys (Sutmuller *et al.*, 2006). It is important to note that later studies indicate that the effects of TLR2 expression on Tregs are largely to promote proliferation independently of APCs (Chen *et al.*, 2009b), and that loss of suppressive activity may be relatively transient (Liu *et al.*, 2006) or circumscribed in effect (Oberg *et al.*, 2010; van Maren *et al.*, 2011). Moreover, other TLR2 ligands can enhance Treg function (Zanin-Zhorov *et al.*, 2006). The apparently contradictory role of TLR2 may be explained by its ability to heterodimerize with different partners (TLR1, TLR6, and TLR10), and the property of some pathogen-derived ligands to selectively stimulate an immunosuppressive, rather than an activating, signal (Depaolo *et al.*, 2008).

Recently, TLR2 activation of B cells has been shown to be critical for microbial Treg induction by the PSA of the commensal bacterium *B. fragilis* (Round *et al.*, 2011). These authors showed that wild-type bacteria, but not PSA-deficient organisms, stimulated IL-10-producing $Foxp3^+$ Tregs in wild-type, but not in TLR2-deficient mice. Since induction was intact in both TLR1- and TLR6-deficient hosts, *B. fragilis* signaling does not appear to be mediated by heterodimers with either of these components, while the TLR2-dependent expression of IL-10 by purified T cells in the absence of APCs argues that the bacterial product acts directly on the T cell without requiring an intermediary population.

Other TLRs show similar involvement in both enhancement and inhibition of Tregs, depending upon the setting. Bacterial flagellin binding through TLR5, for example, was reported to promote human Tregs (Crellin *et al.*, 2005). In other systems, the reversal of Treg suppression by potent synthetic ligands for TLR7 (Hackl *et al.*, 2011; Van *et al.*, 2011) and TLR8 (Peng *et al.*, 2005) does not necessarily reflect the potential for more subtle pathogen-derived molecules to activate Tregs in different ways, and this will clearly be a fertile area for future research.

5. ANTIGEN SPECIFICITY OF NATURAL AND ADAPTIVE TREGS IN INFECTION

Two central issues in regulatory T cell biology are the division between thymic ("natural") and induced ("adaptive") Tregs, and the nature of the antigen specificity of Tregs functional in any particular setting (Bluestone and Abbas, 2003; Rudensky, 2011). Although natural Tregs will have been selected in the thymus for self-reactivity, this does not preclude them recognizing exogenous ligands through cross-reactivity, mimicry, or dual specificity; conversely, induced Tregs in infection are not necessarily pathogen-specific, but may carry a third-party specificity having been activated as bystanders in a pro-regulatory cytokine environment. Indeed, molecular analysis of TCR usage among natural Tregs argues that they do not have a self-restricted repertoire (Pacholczyk *et al.*, 2007). In any event, specificity is not essential for suppressive function, as once a regulatory cell has been triggered, their production of downregulatory cytokines and ability to tolerize DCs allow them to modify systemic reactions to bystander antigens, whether of pathogenic or nonpathogenic (e.g., allergen) origin.

A number of studies have addressed whether Foxp3$^+$ T cell responses observed in infection represent stimulation of pre-existing natural/thymic Tregs or conversion of naive/effector T cells into adaptive Tregs in the periphery. *De novo* induction of Tregs did not occur among Foxp3-negative OVA-transgenic (OT-II) T cells adoptively transferred into mice prior to infection with Ova-expressing *L. monocytogenes* (Fontenot *et al.*, 2005). In the *L. major* system, co-transfer of allotype-marked CD25$^+$ and CD25$^-$ T cells demonstrated that Foxp3 expression remained entirely within the CD25$^+$ population, and conversion from naive/effector cells did not occur (Suffia *et al.*, 2006), while in *L. sigmodontis* infections, BrdU labeling showed early natural Treg proliferation in response to infection (Taylor *et al.*, 2009). However, in *H. polygyrus* infections, conversion into induced Tregs was demonstrated in Foxp3-negative T cells from a Foxp3–GFPxDO11.10 F1 mouse when transferred to an infected wild-type recipient. When mice were given oral ovalbumin, up to 50% switched on Foxp3 expression (Grainger *et al.*, 2010). This discrepancy may lie in the specific localization of the pathogen following infection in these models, as peripheral induction of Tregs has been shown to occur most efficiently in gut-associated lymphoid tissue (Sun *et al.*, 2007). This highly regulated site is populated in the steady state by tolerogenic dendritic cell populations producing high levels of TGF-β (Coombes *et al.*, 2007) or by pro-regulatory DCs in helminth infection (Smith *et al.*, 2011).

A closely related issue is whether pathogen persistence depends on either or both thymic and/or induced Tregs. In *L. sigmodontis* infection, prior depletion of existing (thymic and noncognate induced) Tregs (with anti-CD25 antibody) amplified Th2 responses and reduced worm survival (Taylor *et al.*, 2009). Remarkably, the effects of natural Treg depletion were not evident until some 60 days post-infection; this has been attributed to the early rapid proliferation of pre-existing Tregs on infection, which are able then to dominate the course of the ensuing response (Taylor *et al.*, 2009).

An alternative approach is to identify whether Tregs specific for exogenous, pathogen-derived epitopes are generated in infection. For example, it was noted that following chronic LCMV infection, there is selective expansion of a $V\beta5^+Foxp3^+$ Treg population (Punkosdy *et al.*, 2011). However, these Tregs evolved from a pre-existing pool and are found only in mouse strains carrying an endogenous Mtv9 superantigen-encoding provirus. Hence, LCMV represents a potentially unique case in which a viral Treg epitope is also encoded in the host genome, with specific Treg activation resulting in chronic infection.

A different transgenic TCR system involved the P25 TCR expressed by $CD4^+$ T cells specific for an immunodominant *M. tuberculosis* peptide $Ag85B_{240-254}$ presented by the class II molecule $I\text{-}A^b$. When P25 transgenic T cells (negative for Foxp3–GFP) were transferred to mice prior to infection, there was no conversion to Foxp3 expression despite an overall increase in the endogenous Treg population (Shafiani *et al.*, 2010); hence the predominant Treg type in this infection could be considered natural Tregs. Interestingly, when donor P25 TCR transgenic mice were analyzed in detail, a population of splenic $P25^+Foxp3^+$ Tregs was observed, representing "dual specificity" natural Tregs. Purified $P25^+$ Tregs from uninfected mice were then transferred to naive recipients and shown to delay the priming of effector T cells following Mtb infection (Shafiani *et al.*, 2010). Hence, immunity to infection is impaired by Tregs which are both naturally arising and pathogen-specific. In *L. major* infection, it was previously reported that natural Tregs respond specifically to parasite antigen, as shown by propagation through repeated antigen stimulation (Suffia *et al.*, 2006). As this was a nontransgenic, polyclonal TCR population, this observation implies that the "dual specificity" natural Tregs may be found more extensively than in the case of P25 alone.

Despite these illustrations of natural Treg involvement in microbial infections, there are few clear examples of the specificity of peripherally induced Tregs in infection setting. In RSV-infected mice, preferential binding of an MHC class II tetramer containing a defined epitope (M209) was observed to $CD4^+Foxp3^+$ Tregs (Liu *et al.*, 2009); these cells downregulate virus-induced pathology when transferred to infected recipients (Liu *et al.*, 2010), but as they emerge in a polyclonal environment, it is not established whether they represent dual specificity natural Tregs or have been selectively induced by infection. If Helios proves to be an authentic transcription factor for natural/thymic Tregs (Thornton *et al.*, 2010), staining for this molecule would be a valuable adjunct to these analyses.

In addition to specificity, there remains a major unresolved question of whether natural or induced Tregs are more important in the dampening of immunity and control of pathogenesis. While the answer to this question will depend on the infection setting, the tissue site, and the kinetics of the response, it may also underestimate the complexity of immune regulation *in vivo*: in most cases, both types of Treg are likely to be necessary. An interesting perspective has emerged from studies in the filarial nematode *L. sigmodontis* indicating that early natural Tregs act to limit responsiveness (to the detriment of long-term protective immunity), before a second wave of induced Tregs come into play (Taylor *et al.*, 2009).

6. TREGS AND THE HYGIENE HYPOTHESIS

One of the most significant implications of regulatory T cell activation by infectious agents may be the downregulation of immune responsiveness to other coincident antigens. The impact of this modulation may be either beneficial to the host, in suppressing responses to allergens, autoantigens, and commensals (Maizels, 2005), or detrimental, in compromising immunity to life-threatening infections such as malaria (Su *et al.*, 2005).

Robust experimental and epidemiological evidence that infections can protect against allergies and other immunological over-reactions has been established across the board for infectious organisms from helminths (Cooper *et al.*, 2003; Fleming and Cook, 2006; Maizels, 2005; Smits and Yazdanbakhsh, 2007), mycobacteria (Zuany-Amorim *et al.*, 2002), and viruses (Filippi *et al.*, 2009; Richer *et al.*, 2008), extending also to probiotic and commensal bacteria (Feleszko *et al.*, 2007; Karimi *et al.*, 2009; Repa *et al.*, 2003). It should be noted, however, that the same infections can result in poorer responses to childhood vaccination (Cooper *et al.*, 1998, 2001), as well as to co-infections with pathogens such as malaria (Hartgers *et al.*, 2009).

The close link between helminth infections, Tregs, and suppression is one of the most active research areas in this regard (Maizels and Wiedermann, 2009). In mouse models, *H. polygyrus* was originally shown to generate $CD4^+CD25^+$ Tregs which, on transfer to allergen-sensitized recipients, protected them from airway allergy (Wilson *et al.*, 2005). Similarly, *L. sigmodontis* nematode infection has been shown to inhibit airway allergy to a bystander antigen (Dittrich *et al.*, 2008) and block development of autoimmune disease in diabetes-prone NOD mice, alongside expansion of both Th2 and $CD4^+CD25^+Foxp3^+$ Treg cells (Hübner *et al.*, 2009). These findings with helminth infections are remarkably similar to reports of viral infections such as LCMV and Coxsackie virus, involving the expansion of $CD4^+CD25^+$ T cells producing TGF-β and suppression of allergy and diabetes (Diana *et al.*, 2011; Filippi *et al.*, 2011).

Currently, several avenues are being explored for the therapeutic treatment of immunopathological conditions with certain parasite species associated with regulatory effects, such as the porcine intestinal worm *Trichuris suis* which has been reported to benefit patients with inflammatory bowel diseases (Summers *et al.*, 2005, 2006). A further report on the remission of disease in MS patients with adventitious intestinal helminthiases (Correale and Farez, 2007) correlated these benefits of infection with enhanced Treg and TGF-β levels. These findings, together with an intriguing case report of an ulcerative colitis patient (Broadhurst *et al.*, 2010) and reduced diabetes incidence in filariasis patients (Aravindhan *et al.*, 2010), have heightened interest in helminth therapy of severe immunological dysfunction (Fleming and Fabry, 2007). However, not all trials utilizing this strategy have had positive outcomes (Bager *et al.*, 2010), and it remains to be demonstrated that any beneficial effects in humans of helminth therapy are mediated by Tregs rather than parallel regulatory pathways which are likely to be activated by the same infections.

7. CONCLUSION

Any overview of the impact of Tregs across a diverse range of infectious organisms will inevitably highlight the individual features of each system, with unique niches, dynamics, and molecular interactions. Nevertheless, it is clear that Tregs are involved in the outcome of nearly every infectious episode studied, not necessarily in the central role, but invariably modifying the scale and mode of immunity. Moreover, in many cases from viruses to worms, their intervention is pivotal in differentiating healthy from pathogenic outcomes; we can now learn from these examples, and understand the exceptions, to design new strategies for the control of diseases.

Targeting Tregs for the therapy of infectious diseases is an attractive option in settings where the host mounts an immune response of low pathogenic potential, but which normally is muted and cannot attain the level or intensity required to eliminate the pathogen. More caution may be required if Tregs shelter the host from pathogenesis or contribute to immunity in nonintuitive ways such as facilitating tissue access for effector cells or maintaining low-level persistent antigen for immunological memory. As fundamental understanding of Treg function deepens, more precise targeting may become feasible, for example, by preventing *de novo* induction of Tregs in infection, by blocking specific molecular interactions (such as CTLA-4), and by interfering with key molecules required for Treg migration, including CD103, CCR4, and CCR5 (Moreira *et al.*, 2008; Sather *et al.*, 2007).

A key issue that requires advancing in all the infection systems under consideration, is that of the breadth and antigen specificity of Treg populations in infection. If there is global activation of the natural Treg compartment (e.g., as a physiological response to major inflammatory reactions), then Treg targeting is likely to unleash an unacceptable level of immunopathology if not autoimmunity. However, if the key factor in infection is the activity of a selective subset of pathogen-specific Tregs (or a small subset of natural Tregs which also react to a pathogen determinant), these could be ablated in a more restricted fashion.

Vaccination will, of course, remain our primary strategy to eradicate infectious diseases. Two very interesting perspectives on vaccination emerge from our survey of Tregs in infection. Firstly, the efficacy of vaccines against the major microbial infections is compromised in children harboring common helminth infections, most likely due to the higher level of Tregs in those individuals; hence anthelmintic and/or Treg-reducing interventions may be necessary if current vaccines are to achieve further effects in populations within endemic areas. Secondly, the question is raised of whether vaccines should be fine-tuned to minimize Treg activation. While empirically we have developed adjuvants which stimulate effector immunity, less consideration has been given to whether new vaccines (particularly to parasitic organisms in which vaccines have been notoriously inefficient) should be purged of Treg-stimulating specificities, or even of epitopes which stimulate pathogenic rather than protective effector responses.

Finally, we continue to learn much from those most accomplished immunologists, the successful organisms which can establish themselves in the human body. The complexity of the regulatory network in most infections remains to be defined, but will surely reveal many critically important features of the sequence and hierarchy through which immune suppression is established *in vivo*. In addition, at a molecular level, there is every expectation that new "drugs from bugs" will be developed that specifically enhance Treg development and activity, which should prove invaluable in the treatment of noninfectious immunopathologies such as allergy and autoimmunity.

ACKNOWLEDGMENTS

The authors thank the Wellcome Trust for Programme Grant support.

REFERENCES

Aandahl, E. M., Michaelsson, J., Moretto, W. J., Hecht, F. M., and Nixon, D. F. (2004). Human $CD4^+$ $CD25^+$ regulatory T cells control T-cell responses to human immunodeficiency virus and cytomegalovirus antigens. *J. Virol.* **78,** 2454–2459.

Alegre, M. L., Frauwirth, K. A., and Thompson, C. B. (2001). T-cell regulation by CD28 and CTLA-4. *Nat. Rev. Immunol.* **1,** 220–228.

Allen, J. E., and Maizels, R. M. (2011). Diversity and dialogue in immunity to helminths. *Nat. Rev. Immunol.* **11,** 375–388.

Amante, F. H., Stanley, A. C., Randall, L. M., Zhou, Y., Haque, A., McSweeney, K., Waters, A. P., Janse, C. J., Good, M. F., Hill, G. R., and Engwerda, C. R. (2007). A role for natural regulatory T cells in the pathogenesis of experimental cerebral malaria. *Am. J. Pathol.* **171,** 548–559.

Amu, S., Saunders, S. P., Kronenberg, M., Mangan, N. E., Atzberger, A., and Fallon, P. G. (2010). Regulatory B cells prevent and reverse allergic airway inflammation via FoxP3-positive T regulatory cells in a murine model. *J. Allergy Clin. Immunol.* **125,** 1114–1124.

Anderson, C. F., Oukka, M., Kuchroo, V. J., and Sacks, D. (2007). $CD4^+CD25^-Foxp3^-$ Th1 cells are the source of IL-10-mediated immune suppression in chronic cutaneous leishmaniasis. *J. Exp. Med.* **204,** 285–297.

Anderson, C. F., Lira, R., Kamhawi, S., Belkaid, Y., Wynn, T. A., and Sacks, D. (2008). IL-10 and TGF-β control the establishment of persistent and transmissible infections produced by *Leishmania tropica* in C57BL/6 mice. *J. Immunol.* **180,** 4090–4097.

Andersson, J., Boasso, A., Nilsson, J., Zhang, R., Shire, N. J., Lindback, S., Shearer, G. M., and Chougnet, C. A. (2005). The prevalence of regulatory T cells in lymphoid tissue is correlated with viral load in HIV-infected patients. *J. Immunol.* **174,** 3143–3147.

Anghelina, D., Zhao, J., Trandem, K., and Perlman, S. (2009). Role of regulatory T cells in coronavirus-induced acute encephalitis. *Virology* **385,** 358–367.

Antunes, I., Tolaini, M., Kissenpfennig, A., Iwashiro, M., Kuribayashi, K., Malissen, B., Hasenkrug, K., and Kassiotis, G. (2008). Retrovirus-specificity of regulatory T cells is neither present nor required in preventing retrovirus-induced bone marrow immune pathology. *Immunity* **29,** 782–794.

Aravindhan, V., Mohan, V., Surendar, J., Rao, M. M., Ranjani, H., Kumaraswami, V., Nutman, T. B., and Babu, S. (2010). Decreased prevalence of lymphatic filariasis among subjects with type-1 diabetes. *Am. J. Trop. Med. Hyg.* **83,** 1336–1339.

Arkwright, P. D., and David, T. J. (2001). Intradermal administration of a killed *Mycobacterium vaccae* suspension (SRL 172) is associated with improvement in atopic dermatitis in children with moderate-to-severe disease. *J. Allergy Clin. Immunol.* **107,** 531–534.

Arnold, I. C., Dehzad, N., Reuter, S., Martin, H., Becher, B., Taube, C., and Müller, A. (2011). *Helicobacter pylori* infection prevents allergic asthma in mouse models through the induction of regulatory T cells. *J. Clin. Invest.* **121,** 3088–3093.

Aseffa, A., Gumy, A., Launois, P., MacDonald, H. R., Louis, J. A., and Tacchini-Cottier, F. (2002). The early IL-4 response to *Leishmania major* and the resulting Th2 cell maturation steering progressive disease in BALB/c mice are subject to the control of regulatory $CD4^+CD25^+$ T cells. *J. Immunol.* **169,** 3232–3241.

Atarashi, K., Tanoue, T., Shima, T., Imaoka, A., Kuwahara, T., Momose, Y., Cheng, G., Yamasaki, S., Saito, T., Ohba, Y., Taniguchi, T., Takeda, K., *et al.* (2011). Induction of colonic regulatory T cells by indigenous *Clostridium* species. *Science* **331,** 337–341.

Atmadja, A. K., Atkinson, R., Sartono, E., Partono, F., Yazdanbakhsh, M., and Maizels, R. M. (1995). Differential decline in filarial-specific IgG1, IgG4 and IgE antibodies following diethylcarbamazine chemotherapy of *Brugia malayi* infected patients. *J. Infect. Dis.* **172,** 1567–1572.

Babu, S., Blauvelt, C. P., Kumaraswami, V., and Nutman, T. B. (2006). Regulatory networks induced by live parasites impair both Th1 and Th2 pathways in patent lymphatic filariasis: Implications for parasite persistence. *J. Immunol.* **176,** 3248–3256.

Babu, S., Bhat, S. Q., Kumar, N. P., Jayantasri, S., Rukmani, S., Kumaran, P., Gopi, P. G., Kolappan, C., Kumaraswami, V., and Nutman, T. B. (2009a). Human type 1 and 17 responses in latent tuberculosis are modulated by coincident filarial infection through cytotoxic T lymphocyte antigen-4 and programmed death-1. *J. Infect. Dis.* **200,** 288–298.

Babu, S., Bhat, S. Q., Pavan Kumar, N., Lipira, A. B., Kumar, S., Karthik, C., Kumaraswami, V., and Nutman, T. B. (2009b). Filarial lymphedema is characterized by antigen-specific Th1 and Th17 proinflammatory responses and a lack of regulatory T cells. *PLoS Negl. Trop. Dis.* **3,** e420.

Bach, J. F. (2002). The effect of infections on susceptibility to autoimmune and allergic diseases. *N. Engl. J. Med.* **347,** 911–920.

Bager, P., Arnved, J., Rønborg, S., Wohlfahrt, J., Poulsen, L. K., Westergaard, T., Petersen, H. W., Kristensen, B., Thamsborg, S., Roepstorff, A., Kapel, C., and Melbye, M. (2010). *Trichuris suis* ova therapy for allergic rhinitis: A randomized, double-blind, placebo-controlled clinical trial. *J. Allergy Clin. Immunol.* **125,** 123–130.

Baker, C. A., Clark, R., Ventura, F., Jones, N. G., Guzman, D., Bangsberg, D. R., and Cao, H. (2007). Peripheral CD4 loss of regulatory T cells is associated with persistent viraemia in chronic HIV infection. *Clin. Exp. Immunol.* **147,** 533–539.

Barboza, L., Salmen, S., Goncalves, L., Colmenares, M., Peterson, D., Montes, H., Cartagirone, R., Gutierrez Mdel, C., and Berrueta, L. (2007). Antigen-induced regulatory T cells in HBV chronically infected patients. *Virology* **368,** 41–49.

Baumgart, M., Tomkins, F., Leng, J., and Hesse, M. (2006). Naturally-occurring $CD4^+Foxp3^+$ regulatory T cells are an essential, IL-10-independent part of the immunoregulatory network in *Schistosoma mansoni* egg-induced inflammation. *J. Immunol.* **176,** 5374–5387.

Becker, C., Taube, C., Bopp, T., Michel, K., Kubach, J., Reuter, S., Dehzad, N., Neurath, M. F., Reifenberg, K., Schneider, F. J., Schmitt, E., and Jonuleit, H. (2009). Protection from graft-versus-host disease by HIV-1 envelope protein gp120-mediated activation of human $CD4^+CD25^+$ regulatory T cells. *Blood* **114,** 1263–1269.

Beilharz, M. W., Sammels, L. M., Paun, A., Shaw, K., van Eeden, P., Watson, M. W., and Ashdown, M. L. (2004). Timed ablation of regulatory $CD4^+$ T cells can prevent murine AIDS progression. *J. Immunol.* **172,** 4917–4925.

Beiting, D. P., Gagliardo, L. F., Hesse, M., Bliss, S. K., Meskill, D., and Appleton, J. A. (2007). Coordinated control of immunity to muscle stage *Trichinella spiralis* by IL-10, regulatory T cells, and TGF-β. *J. Immunol.* **178,** 1039–1047.

Belkaid, Y., and Tarbell, K. (2009). Regulatory T cells in the control of host-microorganism interactions. *Annu. Rev. Immunol.* **27,** 551–589.

Belkaid, Y., Hoffmann, K. F., Mendez, S., Kamhawi, S., Udey, M. C., Wynn, T. A., and Sacks, D. L. (2001). The role of interleukin (IL)-10 in the persistence of *Leishmania major* in the skin after healing and the therapeutic potential of anti-IL-10 receptor antibody for sterile cure. *J. Exp. Med.* **194,** 1497–1506.

Belkaid, Y., Piccirillo, C. A., Mendez, S., Shevach, E. M., and Sacks, D. L. (2002). $CD4^{+}CD25^{+}$ regulatory T cells control *Leishmania major* persistence and immunity. *Nature* **420,** 502–507.

Bellocchio, S., Montagnoli, C., Bozza, S., Gaziano, R., Rossi, G., Mambula, S. S., Vecchi, A., Mantovani, A., Levitz, S. M., and Romani, L. (2004). The contribution of the Toll-like/IL-1 receptor superfamily to innate and adaptive immunity to fungal pathogens in vivo. *J. Immunol.* **172,** 3059–3069.

Berretta, F., St-Pierre, J., Piccirillo, C. A., and Stevenson, M. M. (2011). IL-2 contributes to maintaining a balance between $CD4^{+}Foxp3^{+}$ regulatory T cells and effector $CD4^{+}$ T cells required for immune control of blood-stage malaria infection. *J. Immunol.* **186,** 4862–4871.

Beswick, E. J., Pinchuk, I. V., Das, S., Powell, D. W., and Reyes, V. E. (2007). Expression of the programmed death ligand 1, B7-H1, on gastric epithelial cells after *Helicobacter pylori* exposure promotes development of $CD4^{+}CD25^{+}FoxP3^{+}$ regulatory T cells. *Infect. Immun.* **75,** 4334–4341.

Beswick, E. J., Pinchuk, I. V., Earley, R. B., Schmitt, D. A., and Reyes, V. E. (2011). Role of gastric epithelial cell-derived transforming growth factor beta in reduced CD4+ T cell proliferation and development of regulatory T cells during Helicobacter pylori infection. *Infect. Immun.* **79,** 2737–2745.

Bisseye, C., van der Sande, M., Morgan, W. D., Holder, A. A., Pinder, M., and Ismaili, J. (2009). *Plasmodium falciparum* infection of the placenta impacts on the T helper type 1 (Th1)/Th2 balance of neonatal T cells through $CD4^{+}CD25^{+}$forkhead box $P3^{+}$ regulatory T cells and interleukin-10. *Clin. Exp. Immunol.* **158,** 287–293.

Blankenhaus, B., Klemm, U., Eschbach, M. L., Sparwasser, T., Huehn, J., Kuhl, A. A., Loddenkemper, C., Jacobs, T., and Breloer, M. (2011). *Strongyloides ratti* infection induces expansion of $Foxp3^{+}$ regulatory T cells that interfere with immune response and parasite clearance in BALB/c mice. *J. Immunol.* **186,** 4295–4305.

Bluestone, J. A., and Abbas, A. K. (2003). Natural versus adaptive regulatory T cells. *Nat. Rev. Immunol.* **3,** 253–257.

Boasso, A., Vaccari, M., Hryniewicz, A., Fuchs, D., Nacsa, J., Cecchinato, V., Andersson, J., Franchini, G., Shearer, G. M., and Chougnet, C. (2007). Regulatory T-cell markers, indoleamine 2,3-dioxygenase, and virus levels in spleen and gut during progressive simian immunodeficiency virus infection. *J. Virol.* **81,** 11593–11603.

Boettler, T., Spangenberg, H. C., Neumann-Haefelin, C., Panther, E., Urbani, S., Ferrari, C., Blum, H. E., von Weizsacker, F., and Thimme, R. (2005). T cells with a $CD4^{+}CD25^{+}$ regulatory phenotype suppress in vitro proliferation of virus-specific $CD8^{+}$ T cells during chronic hepatitis C virus infection. *J. Virol.* **79,** 7860–7867.

Bolacchi, F., Sinistro, A., Ciaprini, C., Demin, F., Capozzi, M., Carducci, F. C., Drapeau, C. M., Rocchi, G., and Bergamini, A. (2006). Increased hepatitis C virus (HCV)-specific $CD4^{+}CD25^{+}$ regulatory T lymphocytes and reduced HCV-specific $CD4^{+}$ T cell response in HCV-infected patients with normal versus abnormal alanine aminotransferase levels. *Clin. Exp. Immunol.* **144,** 188–196.

Bourreau, E., Ronet, C., Darcissac, E., Lise, M. C., Sainte Marie, D., Clity, E., Tacchini-Cottier, F., Couppie, P., and Launois, P. (2009a). Intralesional regulatory T-cell suppressive function during human acute and chronic cutaneous leishmaniasis due to *Leishmania guyanensis*. *Infect. Immun.* **77,** 1465–1474.

Bourreau, E., Ronet, C., Darsissac, E., Lise, M. C., Marie, D. S., Clity, E., Tacchini-Cottier, F., Couppie, P., and Launois, P. (2009b). In leishmaniasis due to *Leishmania guyanensis* infection, distinct intralesional interleukin-10 and Foxp3 mRNA expression are associated with unresponsiveness to treatment. *J. Infect. Dis.* **199,** 576–579.

Broadhurst, M. J., Leung, J. M., Kashyap, V., McCune, J. M., Mahadevan, U., McKerrow, J. H., and Loke, P. (2010). IL-22$^+$ CD4$^+$ T cells are associated with therapeutic *Trichuris trichiura* infection in an ulcerative colitis patient. *Sci. Transl. Med.* **2,** 60ra88.

Brustoski, K., Moller, U., Kramer, M., Hartgers, F. C., Kremsner, P. G., Krzych, U., and Luty, A. J. (2006). Reduced cord blood immune effector-cell responsiveness mediated by CD4$^+$ cells induced in utero as a consequence of placental *Plasmodium falciparum* infection. *J. Infect. Dis.* **193,** 146–154.

Bueno, L. L., Morais, C. G., Araujo, F. F., Gomes, J. A., Correa-Oliveira, R., Soares, I. S., Lacerda, M. V., Fujiwara, R. T., and Braga, E. M. (2010). *Plasmodium vivax*: Induction of CD4$^+$CD25$^+$FoxP3$^+$ regulatory T cells during infection are directly associated with level of circulating parasites. *PLoS One* **5,** e9623.

Büning, J., Homann, N., von Smolinski, D., Borcherding, F., Noack, F., Stolte, M., Kohl, M., Lehnert, H., and Ludwig, D. (2008). Helminths as governors of inflammatory bowel disease. *Gut* **57,** 1182–1183.

Cabrera, R., Tu, Z., Xu, Y., Firpi, R. J., Rosen, H. R., Liu, C., and Nelson, D. R. (2004). An immunomodulatory role for CD4$^+$CD25$^+$ regulatory T lymphocytes in hepatitis C virus infection. *Hepatology* **40,** 1062–1071.

Cabrera, G., Burzyn, D., Mundinano, J., Courreges, M. C., Camicia, G., Lorenzo, D., Costa, H., Ross, S. R., Nepomnaschy, I., and Piazzon, I. (2008). Early increases in superantigen-specific Foxp3$^+$ regulatory T cells during mouse mammary tumor virus infection. *J. Virol.* **82,** 7422–7431.

Cambos, M., Bélanger, B., Jacques, A., Roulet, A., and Scorza, T. (2008). Natural regulatory (CD4$^+$CD25$^+$FOXP$^+$) T cells control the production of pro-inflammatory cytokines during *Plasmodium chabaudi adami* infection and do not contribute to immune evasion. *Int. J. Parasitol.* **38,** 229–238.

Campanelli, A. P., Roselino, A. M., Cavassani, K. A., Pereira, M. S., Mortara, R. A., Brodskyn, C. I., Goncalves, H. S., Belkaid, Y., Barral-Netto, M., Barral, A., and Silva, J. S. (2006). CD4+CD25+ T cells in skin lesions of patients with cutaneous leishmaniasis exhibit phenotypic and functional characteristics of natural regulatory T cells. *J. Infect. Dis.* **193,** 1313–1322.

Caramalho, I., Lopes-Carvalho, T., Ostler, D., Zelenay, S., Haury, M., and Demengeot, J. (2003). Regulatory T cells selectively express toll-like receptors and are activated by lipopolysaccharide. *J. Exp. Med.* **197,** 403–411.

Cavassani, K. A., Campanelli, A. P., Moreira, A. P., Vancim, J. O., Vitali, L. M., Mamede, R. C., Martinez, R., and Silva, J. S. (2006). Systemic and local characterization of regulatory T cells in a chronic fungal infection in humans. *J. Immunol.* **177,** 5811–5818.

Chase, A. J., Yang, H. C., Zhang, H., Blankson, J. N., and Siliciano, R. F. (2008). Preservation of FoxP3$^+$ regulatory T cells in the peripheral blood of human immunodeficiency virus type 1-infected elite suppressors correlates with low CD4$^+$ T-cell activation. *J. Virol.* **82,** 8307–8315.

Chen, X., Howard, O. M., and Oppenheim, J. J. (2007a). Pertussis toxin by inducing IL-6 promotes the generation of IL-17-producing CD4 cells. *J. Immunol.* **178,** 6123–6129.

Chen, X., Zhou, B., Li, M., Deng, Q., Wu, X., Le, X., Wu, C., Larmonier, N., Zhang, W., Zhang, H., Wang, H., and Katsanis, E. (2007b). CD4$^+$CD25$^+$FoxP3$^+$ regulatory T cells suppress *Mycobacterium tuberculosis* immunity in patients with active disease. *Clin. Immunol.* **123,** 50–59.

Chen, G., Liu, J., Wang, Q. H., Wu, Y., Feng, H., Zheng, W., Guo, S. Y., Li, D. M., Wang, J. C., and Cao, Y. M. (2009a). Effects of CD4$^+$CD25$^+$Foxp3$^+$ regulatory T cells on early *Plasmodium yoelii* 17XL infection in BALB/c mice. *Parasitology* **136,** 1107–1120.

Chen, Q., Davidson, T. S., Huter, E. N., and Shevach, E. M. (2009b). Engagement of TLR2 does not reverse the suppressor function of mouse regulatory T cells, but promotes their survival. *J. Immunol.* **183,** 4458–4466.

Chinen, T., Volchkov, P. Y., Chervonsky, A. V., and Rudensky, A. Y. (2010). A critical role for regulatory T cell-mediated control of inflammation in the absence of commensal microbiota. *J. Exp. Med.* **207,** 2323–2330.
Cliffe, L. J., and Grencis, R. K. (2004). The *Trichuris muris* system: A paradigm of resistance and susceptibility to intestinal nematode infection. *Adv. Parasitol.* **57,** 255–307.
Cohen, J. C., and Varmus, H. E. (1979). Endogenous mammary tumour virus DNA varies among wild mice and segregates during inbreeding. *Nature* **278,** 418–423.
Collison, L. W., Chaturvedi, V., Henderson, A. L., Giacomin, P. R., Guy, C., Bankoti, J., Finkelstein, D., Forbes, K., Workman, C. J., Brown, S. A., Rehg, J. E., Jones, M. L., *et al.* (2010). IL-35-mediated induction of a potent regulatory T cell population. *Nat. Immunol.* **11,** 1093–1101.
Coombes, J. L., Siddiqui, K. R., Arancibia-Carcamo, C. V., Hall, J., Sun, C. M., Belkaid, Y., and Powrie, F. (2007). A functionally specialized population of mucosal $CD103^+$ DCs induces $Foxp3^+$ regulatory T cells via a TGF-β and retinoic acid-dependent mechanism. *J. Exp. Med.* **204,** 1757–1764.
Cooper, P. J., Espinel, I., Paredes, W., Guderian, R. H., and Nutman, T. B. (1998). Impaired tetanus-specific cellular and humoral responses following tetanus vaccination in human onchocerciasis: A possible role for interleukin-10. *J. Infect. Dis.* **178,** 1133–1138.
Cooper, P. J., Chico, M. E., Losonsky, G., Sandoval, C., Espinel, I., Sridhara, R., Aguilar, M., Guevara, A., Guderian, R. H., Levine, M. M., Griffin, G. E., and Nutman, T. B. (2000). Albendazole treatment of children with ascariasis enhances the vibriocidal antibody response to the live attenuated oral cholera vaccine CVD 103-HgR. *J. Infect. Dis.* **182,** 1199–1206.
Cooper, P. J., Chico, M., Sandoval, C., Espinel, I., Guevara, A., Levine, M. M., Griffin, G. E., and Nutman, T. B. (2001). Human infection with *Ascaris lumbricoides* is associated with suppression of the interleukin-2 response to recombinant cholera toxin B subunit following vaccination with the live oral cholera vaccine CVD 103-HgR. *Infect. Immun.* **69,** 1574–1580.
Cooper, P. J., Chico, M. E., Rodrigues, L. C., Ordonez, M., Strachan, D., Griffin, G. E., and Nutman, T. B. (2003). Reduced risk of atopy among school-age children infected with geohelminth parasites in a rural area of the tropics. *J. Allergy Clin. Immunol.* **111,** 995–1000.
Correale, J., and Farez, M. (2007). Association between parasite infection and immune responses in multiple sclerosis. *Ann. Neurol.* **61,** 97–108.
Couper, K. N., Blount, D. G., de Souza, J. B., Suffia, I., Belkaid, Y., and Riley, E. M. (2007). Incomplete depletion and rapid regeneration of $Foxp3^+$ regulatory T cells following anti-CD25 treatment in malaria-infected mice. *J. Immunol.* **178,** 4136–4146.
Couper, K. N., Blount, D. G., and Riley, E. M. (2008a). IL-10: The master regulator of immunity to infection. *J. Immunol.* **180,** 5771–5777.
Couper, K. N., Blount, D. G., Wilson, M. S., Hafalla, J. C., Belkaid, Y., Kamanaka, M., Flavell, R. A., de Souza, J. B., and Riley, E. M. (2008b). IL-10 from $CD4^+CD25^-Foxp3^-CD127^-$ adaptive regulatory T cells modulates parasite clearance and pathology during malaria infection. *PLoS Pathog.* **4,** e1000004.
Crellin, N. K., Garcia, R. V., Hadisfar, O., Allan, S. E., Steiner, T. S., and Levings, M. K. (2005). Human $CD4^+$ T cells express TLR5 and its ligand flagellin enhances the suppressive capacity and expression of FOXP3 in $CD4^+CD25^+$ T regulatory cells. *J. Immunol.* **175,** 8051–8059.
Cuéllar, C., Wu, W., and Mendez, S. (2009). The hookworm tissue inhibitor of metalloproteases (Ac-TMP-1) modifies dendritic cell function and induces generation of CD4 and CD8 suppressor T cells. *PLoS Negl. Trop. Dis.* **3,** e439.
Day, C. L., Kaufmann, D. E., Kiepiela, P., Brown, J. A., Moodley, E. S., Reddy, S., Mackey, E. W., Miller, J. D., Leslie, A. J., DePierres, C., Mncube, Z., Duraiswamy, J., *et al.* (2006). PD-1 expression on HIV-specific T cells is associated with T-cell exhaustion and disease progression. *Nature* **443,** 350–354.

de Araujo, F. F., Vitelli-Avelar, D. M., Teixeira-Carvalho, A., Antas, P. R., Assis Silva Gomes, J., Sathler-Avelar, R., Otavio Costa Rocha, M., Eloi-Santos, S. M., Pinho, R. T., Correa-Oliveira, R., and Martins-Filho, O. A. (2011). Regulatory T cells phenotype in different clinical forms of Chagas' disease. *PLoS Negl. Trop. Dis.* **5,** e992.

D'Elia, R., Behnke, J. M., Bradley, J. E., and Else, K. J. (2009). Regulatory T cells: A role in the control of helminth driven intestinal pathology and worm survival. *J. Immunol.* **182,** 2340–2348.

Depaolo, R. W., Tang, F., Kim, I., Han, M., Levin, N., Ciletti, N., Lin, A., Anderson, D., Schneewind, O., and Jabri, B. (2008). Toll-like receptor 6 drives differentiation of tolerogenic dendritic cells and contributes to LcrV-mediated plague pathogenesis. *Cell Host Microbe* **4,** 350–361.

Diana, J., Brezar, V., Beaudoin, L., Dalod, M., Mellor, A., Tafuri, A., von Herrath, M., Boitard, C., Mallone, R., and Lehuen, A. (2011). Viral infection prevents diabetes by inducing regulatory T cells through NKT cell-plasmacytoid dendritic cell interplay. *J. Exp. Med.* **208,** 729–745.

Dietze, K. K., Zelinskyy, G., Gibbert, K., Schimmer, S., Francois, S., Myers, L., Sparwasser, T., Hasenkrug, K. J., and Dittmer, U. (2011). Transient depletion of regulatory T cells in transgenic mice reactivates virus-specific $CD8^+$ T cells and reduces chronic retroviral set points. *Proc. Natl. Acad. Sci. USA* **108,** 2420–2425.

Dittmer, U., He, H., Messer, R. J., Schimmer, S., Olbrich, A. R. M., Ohlen, C., Greenberg, P. D., Stromnes, I. M., Iwashiro, M., Sakaguchi, S., Evans, L. H., Peterson, K. E., *et al.* (2004). Functional impairment of $CD8^+$ T cells by regulatory T cells during persistent retroviral infection. *Immunity* **20,** 293–303.

Dittrich, A. M., Erbacher, A., Specht, S., Diesner, F., Krokowski, M., Avagyan, A., Stock, P., Ahrens, B., Hoffmann, W. H., Hoerauf, A., and Hamelmann, E. (2008). Helminth infection with *Litomosoides sigmodontis* induces regulatory T cells and inhibits allergic sensitization, airway inflammation, and hyperreactivity in a murine asthma model. *J. Immunol.* **180,** 1792–1799.

Doetze, A., Satoguina, J., Burchard, G., Rau, T., Loliger, C., Fleischer, B., and Hoerauf, A. (2000). Antigen-specific cellular hyporesponsiveness in a chronic human helminth infection is mediated by T_h3/T_r1-type cytokines IL-10 and transforming growth factor-β but not by a T_h1 to T_h2 shift. *Int. Immunol.* **12,** 623–630.

Dye, C. (2006). Global epidemiology of tuberculosis. *Lancet* **367,** 938–940.

Ebinuma, H., Nakamoto, N., Li, Y., Price, D. A., Gostick, E., Levine, B. L., Tobias, J., Kwok, W. W., and Chang, K. M. (2008). Identification and in vitro expansion of functional antigen-specific $CD25^+$ $FoxP3^+$ regulatory T cells in hepatitis C virus infection. *J. Virol.* **82,** 5043–5053.

Eggena, M. P., Barugahare, B., Jones, N., Okello, M., Mutalya, S., Kityo, C., Mugyenyi, P., and Cao, H. (2005). Depletion of regulatory T cells in HIV infection is associated with immune activation. *J. Immunol.* **174,** 4407–4414.

Elahi, S., Dinges, W. L., Lejarcegui, N., Laing, K. J., Collier, A. C., Koelle, D. M., McElrath, M. J., and Horton, H. (2011). Protective HIV-specific $CD8^+$ T cells evade Treg cell suppression. *Nat Med* **17,** 989–995.

Elliott, D. E., Setiawan, T., Metwali, A., Blum, A., Urban, J. F., Jr., and Weinstock, J. V. (2004). *Heligmosomoides polygyrus* inhibits established colitis in IL-10-deficient mice. *Eur. J. Immunol.* **34,** 2690–2698.

Enarsson, K., Lundgren, A., Kindlund, B., Hermansson, M., Roncador, G., Banham, A. H., Lundin, B. S., and Quiding-Jarbrink, M. (2006). Function and recruitment of mucosal regulatory T cells in human chronic *Helicobacter pylori* infection and gastric adenocarcinoma. *Clin. Immunol.* **121,** 358–368.

Ertelt, J. M., Rowe, J. H., Mysz, M. A., Singh, C., Roychowdhury, M., Aguilera, M. N., and Way, S. S. (2011). $Foxp3^+$ regulatory T cells impede the priming of protective $CD8^+$ T cells. *J. Immunol.* **187,** 2569–2577.

Everts, B., Perona-Wright, G., Smits, H. H., Hokke, C. H., van der Ham, A. J., Fitzsimmons, C. M., Doenhoff, M. J., van der Bosch, J., Mohrs, K., Haas, H., Mohrs, M., Yazdanbakhsh, M., *et al.* (2009). Omega-1, a glycoprotein secreted by *Schistosoma mansoni* eggs, drives Th2 responses. *J. Exp. Med.* **206,** 1673–1680.

Feleszko, W., Jaworska, J., Rha, R. D., Steinhausen, S., Avagyan, A., Jaudszus, A., Ahrens, B., Groneberg, D. A., Wahn, U., and Hamelmann, E. (2007). Probiotic-induced suppression of allergic sensitization and airway inflammation is associated with an increase of T regulatory-dependent mechanisms in a murine model of asthma. *Clin. Exp. Allergy* **37,** 498–505.

Filippi, C. M., Estes, E. A., Oldham, J. E., and von Herrath, M. G. (2009). Immunoregulatory mechanisms triggered by viral infections protect from type 1 diabetes in mice. *J. Clin. Invest.* **119,** 1515–1523.

Filippi, C. M., Ehrhardt, K., Estes, E. A., Larsson, P., Oldham, J. E., and von Herrath, M. G. (2011). TLR2 signaling improves immunoregulation to prevent type 1 diabetes. *Eur. J. Immunol.* **41,** 1399–1409.

Finney, C. A. M., Taylor, M. D., Wilson, M. S., and Maizels, R. M. (2007). Expansion and activation of $CD4^+CD25^+$ regulatory T cells in *Heligmosomoides polygyrus* infection. *Eur. J. Immunol.* **37,** 1874–1886.

Finney, O. C., Nwakanma, D., Conway, D. J., Walther, M., and Riley, E. M. (2009). Homeostatic regulation of T effector to Treg ratios in an area of seasonal malaria transmission. *Eur. J. Immunol.* **39,** 1288–1300.

Finney, O. C., Riley, E. M., and Walther, M. (2010). Regulatory T cells in malaria—friend or foe? *Trends Immunol.* **31,** 63–70.

Fleming, J. O., and Cook, T. D. (2006). Multiple sclerosis and the hygiene hypothesis. *Neurology* **67,** 2085–2086.

Fleming, J., and Fabry, Z. (2007). The hygiene hypothesis and multiple sclerosis. *Ann. Neurol.* **61,** 85–89.

Fontenot, J. D., Rasmussen, J. P., Williams, L. M., Dooley, J. L., Farr, A. G., and Rudensky, A. Y. (2005). Regulatory T cell lineage specification by the forkhead transcription factor Foxp3. *Immunity* **22,** 329–341.

Fulton, R. B., Meyerholz, D. K., and Varga, S. M. (2010). $Foxp3^+$ CD4 regulatory T cells limit pulmonary immunopathology by modulating the CD8 T cell response during respiratory syncytial virus infection. *J. Immunol.* **185,** 2382–2392.

Furze, R. C., Culley, F. J., and Selkirk, M. E. (2006). Differential roles of the co-stimulatory molecules GITR and CTLA-4 in the immune response to *Trichinella spiralis*. *Microbes Infect.* **8,** 2803–2810.

Gall, A., Horowitz, A., Joof, H., Natividad, A., Tetteh, K., Riley, E., Bailey, R. L., Mabey, D. C., and Holland, M. J. (2011). Systemic effector and regulatory immune responses to chlamydial antigens in trachomatous trichiasis. *Front Microbiol.* **2,** 10.

Ganguly, S., Mukhopadhyay, D., Das, N. K., Chaduvula, M., Sadhu, S., Chatterjee, U., Rahman, M., Goswami, R. P., Guha, S. K., Modak, D., Mallik, S., Gonju, D., *et al.* (2010). Enhanced lesional Foxp3 expression and peripheral anergic lymphocytes indicate a role for regulatory T cells in Indian post-kala-azar dermal leishmaniasis. *J. Invest. Dermatol.* **130,** 1013–1022.

García-Hernández, M. H., Alvarado-Sánchez, B., Calvo-Turrubiartes, M. Z., Salgado-Bustamante, M., Rodríguez-Pinal, C. Y., Gámez-Lopez, L. R., González-Amaro, R., and Portales-Pérez, D. P. (2009). Regulatory T cells in children with intestinal parasite infection. *Parasite Immunol.* **31,** 597–603.

Garg, A., Barnes, P. F., Roy, S., Quiroga, M. F., Wu, S., Garcia, V. E., Krutzik, S. R., Weis, S. E., and Vankayalapati, R. (2008). Mannose-capped lipoarabinomannan- and prostaglandin E2-dependent expansion of regulatory T cells in human *Mycobacterium tuberculosis* infection. *Eur. J. Immunol.* **38,** 459–469.

Gavin, M. A., Rasmussen, J. P., Fontenot, J. D., Vasta, V., Manganiello, V. C., Beavo, J. A., and Rudensky, A. Y. (2007). Foxp3-dependent programme of regulatory T-cell differentiation. *Nature* **445,** 936–940.

Gazzinelli, R. T., Oswald, I. P., Hieny, S., James, S. L., and Sher, A. (1992). The microbicidal activity of interferon-γ-treated macrophages against *Trypanosoma cruzi* involves L-arginine-dependent, nitrogen oxide-mediated mechanism inhibitable by interleukin-10 and transforming growth factor-β. *Eur. J. Immunol.* **22,** 2501–2506.

Ge, Y. Y., Zhang, L., Zhang, G., Wu, J. P., Tan, M. J., Hu, E., Liang, Y. J., and Wang, Y. (2008). In pregnant mice, the infection of *Toxoplasma gondii* causes the decrease of $CD4^+CD25^+$-regulatory T cells. *Parasite Immunol.* **30,** 471–481.

Geuking, M. B., Cahenzli, J., Lawson, M. A., Ng, D. C., Slack, E., Hapfelmeier, S., McCoy, K. D., and Macpherson, A. J. (2011). Intestinal bacterial colonization induces mutualistic regulatory T cell responses. *Immunity* **34,** 794–806.

Gillan, V., and Devaney, E. (2005). Regulatory T cells modulate Th2 responses induced by *Brugia pahangi* third-stage larvae. *Infect. Immun.* **73,** 4034–4042.

Goncalves, R. M., Salmazi, K. C., Santos, B. A., Bastos, M. S., Rocha, S. C., Boscardin, S. B., Silber, A. M., Kallas, E. G., Ferreira, M. U., and Scopel, K. K. (2010). $CD4^+$ $CD25^+$ $Foxp3^+$ regulatory T cells, dendritic cells, and circulating cytokines in uncomplicated malaria: Do different parasite species elicit similar host responses? *Infect. Immun.* **78,** 4763–4772.

Grainger, J. R., Smith, K. A., Hewitson, J. P., McSorley, H. J., Harcus, Y., Filbey, K. J., Finney, C. A. M., Greenwood, E. J. D., Knox, D. P., Wilson, M. S., Belkaid, Y., Rudensky, A. Y., *et al.* (2010). Helminth secretions induce de novo T cell Foxp3 expression and regulatory function through the TGF-β pathway. *J. Exp. Med.* **207,** 2331–2341.

Greene, B. M., Gbakima, A. A., Albiuz, E. J., and Taylor, H. R. (1985). Humoral and cellular immune responses to *Onchocerca volvulus* in humans. *Rev. Infect. Dis.* **7,** 789–795.

Grogan, J. L., Kremsner, P. G., Deelder, A. M., and Yazdanbakhsh, M. (1998). Antigen-specific proliferation and interferon-gamma and interleukin-5 production are down-regulated during *Schistosoma haematobium* infection. *J. Infect. Dis.* **177,** 1433–1437.

Gruden-Movsesijan, A., Ilic, N., Mostarica-Stojkovic, M., Stosic-Grujicic, S., Milic, M., and Sofronic-Milosavljevic, L. (2010). Mechanisms of modulation of experimental autoimmune encephalomyelitis by chronic *Trichinella spiralis* infection in Dark Agouti rats. *Parasite Immunol.* **32,** 450–459.

Guilliams, M., Oldenhove, G., Noel, W., Herin, M., Brys, L., Loi, P., Flamand, V., Moser, M., De Baetselier, P., and Beschin, A. (2007). African trypanosomiasis: Naturally occurring regulatory T cells favor trypanotolerance by limiting pathology associated with sustained type 1 inflammation. *J. Immunol.* **179,** 2748–2757.

Guilliams, M., Bosschaerts, T., Herin, M., Hunig, T., Loi, P., Flamand, V., De Baetselier, P., and Beschin, A. (2008). Experimental expansion of the regulatory T cell population increases resistance to African trypanosomiasis. *J. Infect. Dis.* **198,** 781–791.

Guyot-Revol, V., Innes, J. A., Hackforth, S., Hinks, T., and Lalvani, A. (2006). Regulatory T cells are expanded in blood and disease sites in patients with tuberculosis. *Am. J. Respir. Crit. Care Med.* **173,** 803–810.

Hackl, D., Loschko, J., Sparwasser, T., Reindl, W., and Krug, A. B. (2011). Activation of dendritic cells via TLR7 reduces Foxp3 expression and suppressive function in induced Tregs. *Eur. J. Immunol.* **41,** 1334–1343.

Haeryfar, S. M., DiPaolo, R. J., Tscharke, D. C., Bennink, J. R., and Yewdell, J. W. (2005). Regulatory T cells suppress $CD8^+$ T cell responses induced by direct priming and cross-priming and moderate immunodominance disparities. *J. Immunol.* **174,** 3344–3351.

Hall, C. H. T., Kassel, R., Tacke, R. S., and Hahn, Y. S. (2010). HCV^+ hepatocytes induce human regulatory $CD4^+$ T cells through the production of TGF-β. *PLoS One* **5,** e12154.

Hansen, D. S., and Schofield, L. (2010). Natural regulatory T cells in malaria: Host or parasite allies? *PLoS Pathog.* **6,** e1000771.

Haque, A., Best, S. E., Amante, F. H., Mustafah, S., Desbarrieres, L., de Labastida, F., Sparwasser, T., Hill, G. R., and Engwerda, C. R. (2010). $CD4^+$ natural regulatory T cells prevent experimental cerebral malaria via CTLA-4 when expanded in vivo. *PLoS Pathog* **6,** e1001221.

Hartgers, F. C., Obeng, B. B., Kruize, Y. C., Dijkhuis, A., McCall, M., Sauerwein, R. W., Luty, A. J., Boakye, D. A., and Yazdanbakhsh, M. (2009). Responses to malarial antigens are altered in helminth-infected children. *J. Infect. Dis.* **199,** 1528–1535.

He, H., Messer, R. J., Sakaguchi, S., Yang, G., Robertson, S. J., and Hasenkrug, K. J. (2004). Reduction of retrovirus-induced immunosuppression by in vivo modulation of T cells during acute infection. *J. Virol.* **78,** 11641–11647.

Helmby, H., and Grencis, R. K. (2003). Contrasting roles for IL-10 in protective immunity to different life cycle stages of intestinal nematode parasites. *Eur. J. Immunol.* **33,** 2382–2390.

Hesse, M., Piccirillo, C. A., Belkaid, Y., Prufer, J., Mentink-Kane, M., Leusink, M., Cheever, A. W., Shevach, E. M., and Wynn, T. A. (2004). The pathogenesis of schistosomiasis is controlled by cooperating IL-10-producing innate effector and regulatory T cells. *J. Immunol.* **172,** 3157–3166.

Higgins, S. C., Lavelle, E. C., McCann, C., Keogh, B., McNeela, E., Byrne, P., O'Gorman, B., Jarnicki, A., McGuirk, P., and Mills, K. H. (2003). Toll-like receptor 4-mediated innate IL-10 activates antigen-specific regulatory T cells and confers resistance to *Bordetella pertussis* by inhibiting inflammatory pathology. *J. Immunol.* **171,** 3119–3127.

Himmel, M. E., Hardenberg, G., Piccirillo, C. A., Steiner, T. S., and Levings, M. K. (2008). The role of T-regulatory cells and Toll-like receptors in the pathogenesis of human inflammatory bowel disease. *Immunology* **125,** 145–153.

Hisaeda, H., Maekawa, Y., Iwakawa, D., Okada, H., Himeno, K., Kishihara, K., Tsukumo, S., and Yasutomo, K. (2004). Escape of malaria parasites from host immunity requires $CD4^+CD25^+$ regulatory T cells. *Nat. Med.* **10,** 29–30.

Hisaeda, H., Hamano, S., Mitoma-Obata, C., Tetsutani, K., Imai, T., Waldmann, H., Himeno, K., and Yasutomo, K. (2005). Resistance of regulatory T cells to glucocorticoid-induced [corrected] TNFR family-related protein (GITR) during *Plasmodium yoelii* infection. *Eur. J. Immunol.* **35,** 3516–3524.

Hisaeda, H., Tetsutani, K., Imai, T., Moriya, C., Tu, L., Hamano, S., Duan, X., Chou, B., Ishida, H., Aramaki, A., Shen, J., Ishii, K. J., *et al.* (2008). Malaria parasites require TLR9 signaling for immune evasion by activating regulatory T cells. *J. Immunol.* **180,** 2496–2503.

Hoffmann, K. F., Cheever, A. W., and Wynn, T. A. (2000a). IL-10 and the dangers of immune polarization: Excessive type 1 and type 2 cytokine responses induce distinct forms of lethal immunopathology in murine schistosomiasis. *J. Immunol.* **164,** 6406–6416.

Hoffmann, W. H., Petit, G., Schulz-Key, H., Taylor, D. W., Bain, O., and Le Goff, L. (2000b). *Litomosoides sigmodontis* in mice: Reappraisal of an old model for filarial research. *Parasitol. Today* **16,** 387–389.

Hori, S., Carvalho, T. L., and Demengeot, J. (2002). $CD25^+CD4^+$ regulatory T cells suppress $CD4^+$ T cell-mediated pulmonary hyperinflammation driven by *Pneumocystis carinii* in immunodeficient mice. *Eur. J. Immunol.* **32,** 1282–1291.

Hotez, P. J., Brindley, P. J., Bethony, J. M., King, C. H., Pearce, E. J., and Jacobson, J. (2008). Helminth infections: The great neglected tropical diseases. *J. Clin. Invest.* **118,** 1311–1321.

Hübner, M. P., Stocker, J. T., and Mitre, E. (2009). Inhibition of type 1 diabetes in filaria-infected non-obese diabetic mice is associated with a T helper type 2 shift and induction of $FoxP3^+$ regulatory T cells. *Immunology* **127,** 512–522.

Huehn, J., Siegmund, K., Lehmann, J. C. U., Siewert, C., Haubold, U., Feuerer, M., Debes, G. F., Lauber, J., Frey, O., Przybylski, G. K., Niesner, U., de la Rosa, M., *et al.*

(2004). Developmental stage, phenotype, and migration distinguish naive- and effector/ memory-like $CD4^+$ regulatory T cells. *J. Exp. Med.* **199,** 303–313.

Ince, M. N., Elliott, D. E., Setiawan, T., Metwali, A., Blum, A., Chen, H. L., Urban, J. F., Flavell, R. A., and Weinstock, J. V. (2009). Role of T cell TGF-β signaling in intestinal cytokine responses and helminthic immune modulation. *Eur. J. Immunol.* **39,** 1870–1878.

Iwashiro, M., Messer, R. J., Peterson, K. E., Stromnes, I. M., Sugie, T., and Hasenkrug, K. J. (2001). Immunosuppression by $CD4^+$ regulatory T cells induced by chronic retroviral infection. *Proc. Natl. Acad. Sci. USA* **98,** 9226–9230.

Jangpatarapongsa, K., Chootong, P., Sattabongkot, J., Chotivanich, K., Sirichaisinthop, J., Tungpradabkul, S., Hisaeda, H., Troye-Blomberg, M., Cui, L., and Udomsangpetch, R. (2008). *Plasmodium vivax* parasites alter the balance of myeloid and plasmacytoid dendritic cells and the induction of regulatory T cells. *Eur. J. Immunol.* **38,** 2697–2705.

Jankovic, D., Kullberg, M. C., Feng, C. G., Goldszmid, R. S., Collazo, C. M., Wilson, M., Wynn, T. A., Kamanaka, M., Flavell, R. A., and Sher, A. (2007). Conventional T-bet^+Foxp3^- Th1 cells are the major source of host-protective regulatory IL-10 during intracellular protozoan infection. *J. Exp. Med.* **204,** 273–283.

Jo, E. K., Park, J. K., and Dockrell, H. M. (2003). Dynamics of cytokine generation in patients with active pulmonary tuberculosis. *Curr. Opin. Infect. Dis.* **16,** 205–210.

Johanns, T. M., Ertelt, J. M., Rowe, J. H., and Way, S. S. (2010). Regulatory T cell suppressive potency dictates the balance between bacterial proliferation and clearance during persistent *Salmonella* infection. *PLoS Pathog.* **6,** e1001043.

Kao, J. Y., Zhang, M., Miller, M. J., Mills, J. C., Wang, B., Liu, M., Eaton, K. A., Zou, W., Berndt, B. E., Cole, T. S., Takeuchi, T., Owyang, S. Y., *et al.* (2010). Helicobacter pylori immune escape is mediated by dendritic cell-induced Treg skewing and Th17 suppression in mice. *Gastroenterology* **138,** 1046–1054.

Kaparakis, M., Laurie, K. L., Wijburg, O., Pedersen, J., Pearse, M., van Driel, I. R., Gleeson, P. A., and Strugnell, R. A. (2006). $CD4^+CD25^+$ regulatory T cells modulate the T-cell and antibody responses in *Helicobacter*-infected BALB/c mice. *Infect. Immun.* **74,** 3519–3529.

Karimi, K., Inman, M. D., Bienenstock, J., and Forsythe, P. (2009). *Lactobacillus reuteri*-induced regulatory T cells protect against an allergic airway response in mice. *Am. J. Respir. Crit. Care Med.* **179,** 186–193.

Kaufmann, D. E., and Walker, B. D. (2009). PD-1 and CTLA-4 inhibitory cosignaling pathways in HIV infection and the potential for therapeutic intervention. *J. Immunol.* **182,** 5891–5897.

Kaya, T. I., Eskandari, G., Guvenc, U., Gunes, G., Tursen, U., Burak Cimen, M. Y., and Ikizoglu, G. (2009). $CD4^+CD25^+$ Treg cells in patients with toenail onychomycosis. *Arch. Dermatol. Res.* **301,** 725–729.

Kim, J. M., Rasmussen, J. P., and Rudensky, A. Y. (2007). Regulatory T cells prevent catastrophic autoimmunity throughout the lifespan of mice. *Nat. Immunol.* **8,** 191–197.

King, C. L., Mahanty, S., Kumaraswami, V., Abrams, J. S., Regunathan, J., Jayaraman, K., Ottesen, E. A., and Nutman, T. B. (1993). Cytokine control of parasite-specific anergy in human lymphatic filariasis. Preferential induction of a regulatory T helper type 2 lymphocyte subset. *J. Clin. Invest.* **92,** 1667–1673.

Kinter, A. L., Hennessey, M., Bell, A., Kern, S., Lin, Y., Daucher, M., Planta, M., McGlaughlin, M., Jackson, R., Ziegler, S. F., and Fauci, A. S. (2004). $CD25^+CD4^+$ regulatory T cells from the peripheral blood of asymptomatic HIV-infected individuals regulate $CD4^+$ and $CD8^+$HIV-specific T cell immune responses in vitro and are associated with favorable clinical markers of disease status. *J. Exp. Med.* **200,** 331–343.

Kinter, A., McNally, J., Riggin, L., Jackson, R., Roby, G., and Fauci, A. S. (2007a). Suppression of HIV-specific T cell activity by lymph node $CD25^+$ regulatory T cells from HIV-infected individuals. *Proc. Natl. Acad. Sci. USA* **104,** 3390–3395.

Kinter, A. L., Horak, R., Sion, M., Riggin, L., McNally, J., Lin, Y., Jackson, R., O'Shea, A., Roby, G., Kovacs, C., Connors, M., Migueles, S. A., *et al.* (2007b). $CD25^+$ regulatory T cells isolated from HIV-infected individuals suppress the cytolytic and nonlytic antiviral activity of HIV-specific $CD8^+$ T cells in vitro. *AIDS Res. Hum. Retroviruses* **23,** 438–450.

Kroetz, D. N., and Deepe, G. S., Jr. (2011). An aberrant thymus in $CCR5^{-/-}$ mice is coupled with an enhanced adaptive immune response in fungal infection. *J. Immunol.* **186,** 5949–5955.

Kullberg, M. C., Jankovic, D., Gorelick, P. L., Caspar, P., Letterio, J. J., Cheever, A. W., and Sher, A. (2002). Bacteria-triggered $CD4^+$ T regulatory cells suppress *Helicobacter hepaticus*-induced colitis. *J. Exp. Med.* **196,** 505–515.

Kursar, M., Bonhagen, K., Fensterle, J., Kohler, A., Hurwitz, R., Kamradt, T., Kaufmann, S. H. E., and Mittrücker, H. W. (2002). Regulatory $CD4^+CD25^+$ T cells restrict memory $CD8^+$ T cell responses. *J. Exp. Med.* **196,** 1585–1592.

Kursar, M., Koch, M., Mittrücker, H. W., Nouailles, G., Bonhagen, K., Kamradt, T., and Kaufmann, S. H. E. (2007). Regulatory T cells prevent efficient clearance of *Mycobacterium tuberculosis*. *J. Immunol.* **178,** 2661–2665.

Lahl, K., Loddenkemper, C., Drouin, C., Freyer, J., Arnason, J., Eberl, G., Hamann, A., Wagner, H., Huehn, J., and Sparwasser, T. (2007). Selective depletion of $Foxp3^+$ regulatory T cells induces a scurfy-like disease. *J. Exp. Med.* **204,** 57–63.

Layland, L. E., Rad, R., Wagner, H., and da Costa, C. U. (2007). Immunopathology in schistosomiasis is controlled by antigen-specific regulatory T cells primed in the presence of TLR2. *Eur. J. Immunol.* **37,** 2174–2184.

Layland, L. E., Mages, J., Loddenkemper, C., Hoerauf, A., Wagner, H., Lang, R., and da Costa, C. U. (2010). Pronounced phenotype in activated regulatory T cells during a chronic helminth infection. *J. Immunol.* **184,** 713–724.

Lee, D. C. P., Harker, J. A. E., Tregoning, J. S., Atabani, S. F., Johansson, C., Schwarze, J., and Openshaw, P. J. M. (2010). $CD25^+$ natural regulatory T cells are critical in limiting innate and adaptive immunity and resolving disease following respiratory syncytial virus infection. *J. Virol.* **84,** 8790–8798.

Li, L., Lao, S. H., and Wu, C. Y. (2007a). Increased frequency of $CD4^+CD25^{high}$ Treg cells inhibit BCG-specific induction of IFN-gamma by $CD4^+$ T cells from TB patients. *Tuberculosis (Edinb)* **87,** 526–534.

Li, S., Jones, K. L., Woollard, D. J., Dromey, J., Paukovics, G., Plebanski, M., and Gowans, E. J. (2007b). Defining target antigens for $CD25^+$ $FOXP3^+$ IFN-γ^- regulatory T cells in chronic hepatitis C virus infection. *Immunol. Cell Biol.* **85,** 197–204.

Li, S., Gowans, E. J., Chougnet, C., Plebanski, M., and Dittmer, U. (2008). Natural regulatory T cells and persistent viral infection. *J. Virol.* **82,** 21–30.

Li, S., Floess, S., Hamann, A., Gaudieri, S., Lucas, A., Hellard, M., Roberts, S., Paukovic, G., Plebanski, M., Loveland, B. E., Aitken, C., Barry, S., *et al.* (2009). Analysis of $FOXP3^+$ regulatory T cells that display apparent viral antigen specificity during chronic hepatitis C virus infection. *PLoS Pathog.* **5,** e1000707.

Li, W., Tenner-Racz, K., Racz, P., Janowicz, D. M., Fortney, K. R., Katz, B. P., and Spinola, S. M. (2010). Role played by $CD4^+FOXP3^+$ regulatory T cells in suppression of host responses to *Haemophilus ducreyi* during experimental infection of human volunteers. *J. Infect. Dis.* **201,** 1839–1848.

Lienhardt, C., Azzurri, A., Amedei, A., Fielding, K., Sillah, J., Sow, O. Y., Bah, B., Benagiano, M., Diallo, A., Manetti, R., Manneh, K., Gustafson, P., *et al.* (2002). Active tuberculosis in Africa is associated with reduced Th1 and increased Th2 activity in vivo. *Eur. J. Immunol.* **32,** 1605–1613.

Liu, H., Komai-Koma, M., Xu, D., and Liew, F. Y. (2006). Toll-like receptor 2 signaling modulates the functions of $CD4^+$ $CD25^+$ regulatory T cells. *Proc. Natl. Acad. Sci. USA* **103,** 7048–7053.

Liu, J., Ruckwardt, T. J., Chen, M., Johnson, T. R., and Graham, B. S. (2009). Characterization of respiratory syncytial virus M- and M2-specific CD4 T cells in a murine model. *J. Virol.* **83,** 4934–4941.

Liu, J., Ruckwardt, T. J., Chen, M., Nicewonger, J. D., Johnson, T. R., and Graham, B. S. (2010). Epitope-specific regulatory CD4 T cells reduce virus-induced illness while preserving CD8 T-cell effector function at the site of infection. *J. Virol.* **84,** 10501–10509.

Liu, J. Y., Li, L. Y., Yang, X. Z., Li, J., Zhong, G., Wang, J., Li, L. J., Ji, B., Wu, Z. Q., Liu, H., Yang, X., and Liu, P. M. (2011). Adoptive transfer of DCs isolated from helminth-infected mice enhanced T regulatory cell responses in airway allergic inflammation. *Parasite Immunol.* **33,** 525–534.

Loures, F. V., Pina, A., Felonato, M., Araujo, E. F., Leite, K. R., and Calich, V. L. (2010). Toll-like receptor 4 signaling leads to severe fungal infection associated with enhanced proinflammatory immunity and impaired expansion of regulatory T cells. *Infect. Immun.* **78,** 1078–1088.

Lowenthal, J. W., Corthesy, P., Tougne, C., Lees, R., MacDonald, H. R., and Nabholz, M. (1985). High and low affinity IL 2 receptors: Analysis by IL 2 dissociation rate and reactivity with monoclonal anti-receptor antibody PC61. *J. Immunol.* **135,** 3988–3994.

Lund, J. M., Hsing, L., Pham, T. T., and Rudensky, A. Y. (2008). Coordination of early protective immunity to viral infection by regulatory T cells. *Science* **320,** 1220–1224.

Lundgren, A., Suri-Payer, E., Enarsson, K., Svennerholm, A.-M., and Lundin, B. S. (2003). *Helicobacter pylori*-specific $CD4^{+}$ $CD25^{high}$ regulatory T cells suppress memory T-cell responses to *H. pylori* in infected individuals. *Infect. Immun.* **71,** 1755–1762.

Lundgren, A., Strömberg, E., Sjöling, A., Lindholm, C., Enarsson, K., Edebo, A., Johnsson, E., Suri-Payer, E., Larsson, P., Rudin, A., Svennerholm, A.-M., and Lundin, B. S. (2005). Mucosal FOXP3-expressing $CD4^{+}CD25^{high}$ regulatory T cells in *Helicobacter pylori*-infected patients. *Infect. Immun.* **73,** 523–531.

Mackroth, M. S., Malhotra, I., Mungai, P., Koech, D., Muchiri, E., and King, C. L. (2011). Human cord blood $CD4^{+}$CD25hi regulatory T cells suppress prenatally acquired T cell responses to *Plasmodium falciparum* antigens. *J Immunol* **186,** 2780–2791.

Maizels, R. M. (2005). Infections and allergy—helminths, hygiene and host immune regulation. *Curr. Opin. Immunol.* **17,** 656–661.

Maizels, R. M., and Wiedermann, U. (2009). Immunoregulation by microbes and parasites in the control of allergy and autoimmunity. *In* "The Hygiene Hypothesis and Darwinian Medicine", (G. A. W. Rook, ed.), pp. 45–75. Birkhäuser Verlag, Basel.

Maizels, R. M., Hewitson, J. P., Murray, J., Harcus, Y., Dayer, B., Filbey, K. J., Grainger, J. R., McSorley, H. J., Reynolds, L. A., and Smith, K. A. (2011). Immune modulation and modulators in *Heligmosomoides polygyrus* infection. *Exp. Parasitol.* in press.

Maloy, K. J., Salaun, L., Cahill, R., Dougan, G., Saunders, N. J., and Powrie, F. (2003). $CD4^{+}CD25^{+}$ T_R cells suppress innate immune pathology through cytokine-dependent mechanisms. *J. Exp. Med.* **197,** 111–119.

Mariano, F. S., Gutierrez, F. R., Pavanelli, W. R., Milanezi, C. M., Cavassani, K. A., Moreira, A. P., Ferreira, B. R., Cunha, F. Q., Cardoso, C. R., and Silva, J. S. (2008). The involvement of $CD4^{+}CD25^{+}$ T cells in the acute phase of *Trypanosoma cruzi* infection. *Microbes Infect.* **10,** 825–833.

Martin, S., Agarwal, R., Murugaiyan, G., and Saha, B. (2010). CD40 expression levels modulate regulatory T cells in *Leishmania donovani* infection. *J. Immunol.* **185,** 551–559.

Mazmanian, S. K., Round, J. L., and Kasper, D. L. (2008). A microbial symbiosis factor prevents intestinal inflammatory disease. *Nature* **453,** 620–625.

McGuirk, P., Mahon, B. P., Griffin, F., and Mills, K. H. (1998). Compartmentalization of T cell responses following respiratory infection with *Bordetella pertussis:* Hyporesponsiveness of lung T cells is associated with modulated expression of the co-stimulatory molecule CD28. *Eur. J. Immunol.* **28,** 153–163.

McGuirk, P., McCann, C., and Mills, K. H. G. (2002). Pathogen-specific T regulatory 1 cells induced in the respiratory tract by a bacterial molecule that stimulates interleukin 10 production by dendritic cells: A novel strategy for evasion of protective T helper type 1 responses by *Bordetella pertussis*. *J. Exp. Med.* **195,** 221–231.

McKee, A. S., and Pearce, E. J. (2004). $CD25^+CD4^+$ cells contribute to Th2 polarization during helminth infection by suppressing Th1 response development. *J. Immunol.* **173,** 1224–1231.

McKinley, L., Logar, A. J., McAllister, F., Zheng, M., Steele, C., and Kolls, J. K. (2006). Regulatory T cells dampen pulmonary inflammation and lung injury in an animal model of pneumocystis pneumonia. *J. Immunol.* **177,** 6215–6226.

McSorley, H. J., Harcus, Y. M., Murray, J., Taylor, M. D., and Maizels, R. M. (2008). Expansion of $Foxp3^+$ regulatory T cells in mice infected with the filarial parasite, *Brugia malayi*. *J. Immunol.* **181,** 6456–6466.

Medzhitov, R. (2007). Recognition of microorganisms and activation of the immune response. *Nature* **449,** 819–826.

Mejri, N., Muller, N., Hemphill, A., and Gottstein, B. (2011). Intraperitoneal *Echinococcus multilocularis* infection in mice modulates peritoneal $CD4^+$ and $CD8^+$ regulatory T cell development. *Parasitol. Int.* **60,** 45–53.

Mendez, S., Reckling, S. K., Piccirillo, C. A., Sacks, D., and Belkaid, Y. (2004). Role for $CD4^+$ $CD25^+$ regulatory T cells in reactivation of persistent leishmaniasis and control of concomitant immunity. *J. Exp. Med.* **200,** 201–210.

Mertens, J., Fabri, M., Zingarelli, A., Kubacki, T., Meemboor, S., Groneck, L., Seeger, J., Bessler, M., Hafke, H., Odenthal, M., Bieler, J. G., Kalka, C., *et al.* (2009). *Streptococcus pneumoniae* serotype 1 capsular polysaccharide induces CD8CD28 regulatory T lymphocytes by TCR crosslinking. *PLoS Pathog.* **5,** e1000596.

Metenou, S., Dembele, B., Konate, S., Dolo, H., Coulibaly, S. Y., Coulibaly, Y. I., Diallo, A. A., Soumaoro, L., Coulibaly, M. E., Sanogo, D., Doumbia, S. S., Traore, S. F., *et al.* (2010). At homeostasis filarial infections have expanded adaptive T regulatory but not classical Th2 cells. *J. Immunol.* **184,** 5375–5382.

Metwali, A., Setiawan, T., Blum, A. M., Urban, J., Elliott, D. E., Hang, L., and Weinstock, J. V. (2006). Induction of $CD8^+$ regulatory T cells in the intestine by *Heligmosomoides polygyrus* infection. *Am. J. Physiol. Gastrointest. Liver Physiol.* **291,** G253–G259.

Mikkelsen, S. R., Reckling, S. K., Egan, E. A., and Dean, G. A. (2010). In vivo depletion of $CD4^+$CD25hi regulatory T cells is associated with improved antiviral responses in cats chronically infected with feline immunodeficiency virus. *Virology* **403,** 163–172.

Mikkelsen, S. R., Long, J. M., Zhang, L., Galemore, E. R., VandeWoude, S., and Dean, G. A. (2011). Partial regulatory T cell depletion prior to acute feline immunodeficiency virus infection does not alter disease pathogenesis. *PLoS ONE* **6,** e17183.

Mills, K. H. G. (2004). Regulatory T cells: Friend or foe in immunity to infection? *Nat. Rev. Immunol.* **4,** 841–855.

Minigo, G., Woodberry, T., Piera, K. A., Salwati, E., Tjitra, E., Kenangalem, E., Price, R. N., Engwerda, C. R., Anstey, N. M., and Plebanski, M. (2009). Parasite-dependent expansion of TNF receptor II-positive regulatory T cells with enhanced suppressive activity in adults with severe malaria. *PLoS Pathog.* **5,** e1000402.

Molling, J. W., de Gruijl, T. D., Glim, J., Moreno, M., Rozendaal, L., Meijer, C. J., van den Eertwegh, A. J., Scheper, R. J., von Blomberg, M. E., and Bontkes, H. J. (2007). $CD4^+CD25^{hi}$ regulatory T-cell frequency correlates with persistence of human papillomavirus type 16 and T helper cell responses in patients with cervical intraepithelial neoplasia. *Int. J. Cancer* **121,** 1749–1755.

Monroy, F. G., and Enriquez, F. J. (1992). *Heligmosomoides polygyrus: A* model for chronic gastrointestinal helminthiasis. *Parasitol. Today* **8,** 49–54.

Montagnoli, C., Bacci, A., Bozza, S., Gaziano, R., Mosci, P., Sharpe, A. H., and Romani, L. (2002). B7/CD28-dependent $CD4^+CD25^+$ regulatory T cells are essential components of the memory-protective immunity to *Candida albicans*. *J. Immunol.* **169,** 6298–6308.

Montagnoli, C., Fallarino, F., Gaziano, R., Bozza, S., Bellocchio, S., Zelante, T., Kurup, W. P., Pitzurra, L., Puccetti, P., and Romani, L. (2006). Immunity and tolerance to *Aspergillus* involve functionally distinct regulatory T cells and tryptophan catabolism. *J. Immunol.* **176,** 1712–1723.

Montes, M., Sanchez, C., Verdonck, K., Lake, J. E., Gonzalez, E., Lopez, G., Terashima, A., Nolan, T., Lewis, D. E., Gotuzzo, E., and White, A. C. (2009). Regulatory T cell expansion in HTLV-1 and strongyloidiasis co-infection is associated with reduced IL-5 responses to *Strongyloides stercoralis* antigen. *PLoS Negl. Trop. Dis.* **3,** e456.

Moore, K. W., de Waal Malefyt, R., Coffman, R. L., and O'Garra, A. (2001). Interleukin-10 and the interleukin-10 receptor. *Annu. Rev. Immunol.* **19,** 683–765.

Moreira, A. P., Cavassani, K. A., Massafera Tristao, F. S., Campanelli, A. P., Martinez, R., Rossi, M. A., and Silva, J. S. (2008). CCR5-dependent regulatory T cell migration mediates fungal survival and severe immunosuppression. *J. Immunol.* **180,** 3049–3056.

Moreno-Fernandez, M. E., Rueda, C. M., Rusie, L. K., and Chougnet, C. A. (2011). Regulatory T cells control HIV replication in activated T cells through a cAMP-dependent mechanism. *Blood* **117,** 5372–5380.

Murphy, M. L., Wille, U., Villegas, E. N., Hunter, C. A., and Farrell, J. P. (2001). IL-10 mediates susceptibility to *Leishmania donovani* infection. *Eur. J. Immunol.* **31,** 2848–2856.

Nagase, H., Jones, K. M., Anderson, C. F., and Noben-Trauth, N. (2007). Despite increased $CD4^+Foxp3^+$ cells within the infection site, BALB/c IL-4 receptor-deficient mice reveal $CD4^+$Foxp3-negative T cells as a source of IL-10 in *Leishmania major* susceptibility. *J. Immunol.* **179,** 2435–2444.

Nakagawa, T., Tsuruoka, M., Ogura, H., Okuyama, Y., Arima, Y., Hirano, T., and Murakami, M. (2010). IL-6 positively regulates $Foxp3^+CD8^+$ T cells in vivo. *Int. Immunol.* **22,** 129–139.

Nausch, N., Midzi, N., Mduluza, T., Maizels, R. M., and Mutapi, F. (2011). Regulatory and activated T cells in human *Schistosoma haematobium* infections. *PLoS One* **6,** e16860.

Netea, M. G., Sutmuller, R., Hermann, C., Van der Graaf, C. A. A., Van der Meer, J. W. M., van Krieken, J. H., Hartung, T., Adema, G., and Kullberg, B. J. (2004). Toll-like receptor 2 suppresses immunity against *Candida albicans* through induction of IL-10 and regulatory T cells. *J. Immunol.* **172,** 3712–3718.

Nie, C. Q., Bernard, N. J., Schofield, L., and Hansen, D. S. (2007). $CD4^+$ $CD25^+$ regulatory T cells suppress $CD4^+$ T-cell function and inhibit the development of *Plasmodium berghei*-specific TH1 responses involved in cerebral malaria pathogenesis. *Infect. Immun.* **75,** 2275–2282.

Nilsson, J., Boasso, A., Velilla, P. A., Zhang, R., Vaccari, M., Franchini, G., Shearer, G. M., Andersson, J., and Chougnet, C. (2006). HIV-1-driven regulatory T-cell accumulation in lymphoid tissues is associated with disease progression in HIV/AIDS. *Blood* **108,** 3808–3817.

Noben-Trauth, N., Paul, W. E., and Sacks, D. L. (1999). IL-4- and IL-4 receptor-deficient BALB/c mice reveal differences in susceptibility to *Leishmania major* parasite substrains. *J. Immunol.* **162,** 6132–6140.

Oberg, H. H., Ly, T. T., Ussat, S., Meyer, T., Kabelitz, D., and Wesch, D. (2010). Differential but direct abolishment of human regulatory T cell suppressive capacity by various TLR2 ligands. *J. Immunol.* **184,** 4733–4740.

Ochoa-Reparaz, J., Mielcarz, D. W., Ditrio, L. E., Burroughs, A. R., Begum-Haque, S., Dasgupta, S., Kasper, D. L., and Kasper, L. H. (2010a). Central nervous system demyelinating disease protection by the human commensal *Bacteroides fragilis* depends on polysaccharide A expression. *J. Immunol.* **185,** 4101–4108.

Ochoa-Reparaz, J., Mielcarz, D. W., Wang, Y., Begum-Haque, S., Dasgupta, S., Kasper, D. L., and Kasper, L. H. (2010b). A polysaccharide from the human commensal *Bacteroides fragilis* protects against CNS demyelinating disease. *Mucosal Immunol.* **3,** 487–495.

O'Mahony, C., Scully, P., O'Mahony, D., Murphy, S., O'Brien, F., Lyons, A., Sherlock, G., MacSharry, J., Kiely, B., Shanahan, F., and O'Mahony, L. (2008). Commensal-induced regulatory T cells mediate protection against pathogen-stimulated NF-κB activation. *PLoS Pathog.* **4,** e1000112.

Omer, F. M., de Souza, J. B., and Riley, E. M. (2003). Differential induction of TGF-β regulates proinflammatory cytokine production and determines the outcome of lethal and nonlethal *Plasmodium yoelii* infections. *J. Immunol.* **171,** 5430–5436.

Ordway, D., Henao-Tamayo, M., Harton, M., Palanisamy, G., Troudt, J., Shanley, C., Basaraba, R. J., and Orme, I. M. (2007). The hypervirulent *Mycobacterium tuberculosis* strain HN878 induces a potent TH1 response followed by rapid down-regulation. *J. Immunol.* **179,** 522–531.

Ordway, D. J., Shang, S., Henao-Tamayo, M., Obregon-Henao, A., Nold, L., Caraway, M., Shanley, C. A., Basaraba, R. J., Duncan, C. G., and Orme, I. M. (2011). *Mycobacterium bovis* BCG mediated protection against W-Beijing strains of *Mycobacterium tuberculosis* is diminished concomitant with the emergence of regulatory T cells. *Clin. Vaccine Immunol.* **18,** 1527–1535.

Ostman, S., Rask, C., Wold, A. E., Hultkrantz, S., and Telemo, E. (2006). Impaired regulatory T cell function in germ-free mice. *Eur. J. Immunol.* **36,** 2336–2346.

Othman, A. A., El-Shourbagy, S. H., and Soliman, R. H. (2010). Kinetics of Foxp3-expressing regulatory cells in experimental *Toxocara canis* infection. *Exp. Parasitol.* **127,** 454–459.

Pacholczyk, R., Kern, J., Singh, N., Iwashima, M., Kraj, P., and Ignatowicz, L. (2007). Nonself-antigens are the cognate specificities of Foxp3+ regulatory T cells. *Immunity* **27,** 493–504.

Pandiyan, P., Conti, H. R., Zheng, L., Peterson, A. C., Mathern, D. R., Hernandez-Santos, N., Edgerton, M., Gaffen, S. L., and Lenardo, M. J. (2011). $CD4^+CD25^+Foxp3^+$ regulatory T cells promote Th17 cells in vitro and enhance host resistance in mouse *Candida albicans* Th17 cell infection model. *Immunity* **34,** 422–434.

Park, S. K., Cho, M. K., Park, H. K., Lee, K. H., Lee, S. J., Choi, S. H., Ock, M. S., Jeong, H. J., Lee, M. H., and Yu, H. S. (2009). Macrophage migration inhibitory factor homologs of anisakis simplex suppress Th2 response in allergic airway inflammation model via $CD4^+CD25^+Foxp3^+$ T cell recruitment. *J Immunol* **182,** 6907–6914.

Pasquali, P., Thornton, A. M., Vendetti, S., Pistoia, C., Petrucci, P., Tarantino, M., Pesciaroli, M., Ruggeri, F., Battistoni, A., and Shevach, E. M. (2010). $CD4^+CD25^+$ T regulatory cells limit effector T cells and favor the progression of brucellosis in BALB/c mice. *Microbes Infect.* **12,** 3–10.

Pearce, E. J., and MacDonald, A. S. (2002). The immunobiology of schistosomiasis. *Nat. Rev. Immunol.* **2,** 499–511.

Peng, G., Guo, Z., Kiniwa, Y., Voo, K. S., Peng, W., Fu, T., Wang, D. Y., Li, Y., Wang, H. Y., and Wang, R. F. (2005). Toll-like receptor 8-mediated reversal of $CD4^+$ regulatory T cell function. *Science* **309,** 1380–1384.

Piessens, W. F., McGreevy, P. B., Piessens, P. W., McGreevy, M., Koiman, I., Saroso, J. S., and Dennis, D. T. (1980). Immune responses in human infections with *Brugia malayi*. Specific cellular unresponsiveness to filarial antigens.. *J. Clin. Invest.* **65,** 172–179.

Piessens, W. F., Partono, F., Hoffman, S. L., Ratiwayanto, S., Piessens, P. W., Palmieri, J. R., Koiman, I., Dennis, D. T., and Carney, W. P. (1982). Antigen-specific suppressor T lymphocytes in human lymphatic filariasis. *N. Engl. J. Med.* **307,** 144–148.

Polk, B. F., and Kasper, D. L. (1977). *Bacteroides fragilis* subspecies in clinical isolates. *Ann. Intern. Med.* **86,** 569–571.

Popescu, I., Macedo, C., Abu-Elmagd, K., Shapiro, R., Hua, Y., Thomson, A. W., Morelli, A. E., Storkus, W. J., and Metes, D. (2007). EBV-specific $CD8^+$ T cell reactivation

in transplant patients results in expansion of $CD8^+$ type-1 regulatory T cells. *Am. J. Transplant.* **7,** 1215–1223.
Punkosdy, G. A., Blain, M., Glass, D. D., Lozano, M. M., O'Mara, L., Dudley, J. P., Ahmed, R., and Shevach, E. M. (2011). Regulatory T-cell expansion during chronic viral infection is dependent on endogenous retroviral superantigens. *Proc. Natl. Acad. Sci. USA* **108,** 3677–3682.
Qin, X. J., Shi, H. Z., Liang, Q. L., Huang, L. Y., and Yang, H. B. (2008). $CD4^+CD25^+$ regulatory T lymphocytes in tuberculous pleural effusion. *Chin. Med. J. (Engl)* **121,** 581–586.
Quinn, K. M., McHugh, R. S., Rich, F. J., Goldsack, L. M., de Lisle, G. W., Buddle, B. M., Delahunt, B., and Kirman, J. R. (2006). Inactivation of $CD4^+$ $CD25^+$ regulatory T cells during early mycobacterial infection increases cytokine production but does not affect pathogen load. *Immunol. Cell Biol.* **84,** 467–474.
Rad, R., Brenner, L., Bauer, S., Schwendy, S., Layland, L., da Costa, C. P., Reindl, W., Dossumbekova, A., Friedrich, M., Saur, D., Wagner, H., Schmid, R. M., *et al.* (2006). $CD25^+/Foxp3^+$ T cells regulate gastric inflammation and *Helicobacter pylori* colonization in vivo. *Gastroenterology* **131,** 525–537.
Raghavan, S., Fredriksson, M., Svennerholm, A. M., Holmgren, J., and Suri-Payer, E. (2003). Absence of $CD4^+CD25^+$ regulatory T cells is associated with a loss of regulation leading to increased pathology in *Helicobacter pylori*-infected mice. *Clin. Exp. Immunol.* **132,** 393–400.
Randall, L. M., Amante, F. H., McSweeney, K. A., Zhou, Y., Stanley, A. C., Haque, A., Jones, M. K., Hill, G. R., Boyle, G. M., and Engwerda, C. R. (2008). Common strategies to prevent and modulate experimental cerebral malaria in mouse strains with different susceptibilities. *Infect. Immun.* **76,** 3312–3320.
Rausch, S., Huehn, J., Kirchhoff, D., Rzepecka, J., Schnoeller, C., Pillai, S., Loddenkemper, C., Scheffold, A., Hamann, A., Lucius, R., and Hartmann, S. (2008). Functional analysis of effector and regulatory T cells in a parasitic nematode infection. *Infect. Immun.* **76,** 1908–1919.
Rausch, S., Huehn, J., Loddenkemper, C., Hepworth, M. R., Klotz, C., Sparwasser, T., Hamann, A., Lucius, R., and Hartmann, S. (2009). Establishment of nematode infection despite increased Th2 responses and immunopathology after selective depletion of $Foxp3^+$ cells. *Eur. J. Immunol.* **39,** 3066–3077.
Rehermann, B. (2009). Hepatitis C virus versus innate and adaptive immune responses: A tale of coevolution and coexistence. *J. Clin. Invest.* **119,** 1745–1754.
Reiner, S. L., and Locksley, R. M. (1995). The regulation of immunity to *Leishmania major*. *Annu. Rev. Immunol.* **13,** 151–177.
Repa, A., Grangette, C., Daniel, C., Hochreiter, R., Hoffmann-Sommergruber, K., Thalhamer, J., Kraft, D., Breiteneder, H., Mercenier, A., and Wiedermann, U. (2003). Mucosal co-application of lactic acid bacteria and allergen induces counter-regulatory immune responses in a murine model of birch pollen allergy. *Vaccine* **22,** 87–95.
Ribeiro-Rodrigues, R., Resende Co, T., Rojas, R., Toossi, Z., Dietze, R., Boom, W. H., Maciel, E., and Hirsch, C. S. (2006). A role for $CD4^+CD25^+$ T cells in regulation of the immune response during human tuberculosis. *Clin. Exp. Immunol.* **144,** 25–34.
Richer, M. J., Straka, N., Fang, D., Shanina, I., and Horwitz, M. S. (2008). Regulatory T-cells protect from type 1 diabetes after induction by coxsackievirus infection in the context of transforming growth factor-β. *Diabetes* **57,** 1302–1311.
Robertson, S. J., Messer, R. J., Carmody, A. B., and Hasenkrug, K. J. (2006). In vitro suppression of $CD8^+$ T cell function by Friend virus-induced regulatory T cells. *J. Immunol.* **176,** 3342–3349.
Rodrigues, V., Jr., Santana da Silva, J., and Campos-Neto, A. (1998). Transforming growth factor-β and immunosuppression in experimental visceral leishmaniasis. *Infect. Immun.* **66,** 1233–1236.

Rodrigues, O. R., Marques, C., Soares-Clemente, M., Ferronha, M. H., and Santos-Gomes, G. M. (2009). Identification of regulatory T cells during experimental *Leishmania infantum* infection. *Immunobiology* **214,** 101–111.

Roelants, G. E., Pearson, T. W., Mayor-Withey, K. S., and Lundin, L. B. (1979). Immune depression in *Trypanosoma congolense*-infected mice. *Adv. Exp. Med. Biol.* **114,** 661–666.

Round, J. L., and Mazmanian, S. K. (2010). Inducible Foxp3^{+} regulatory T-cell development by a commensal bacterium of the intestinal microbiota. *Proc. Natl. Acad. Sci. USA* **107,** 12204–12209.

Round, J. L., Lee, S. M., Li, J., Tran, G., Jabri, B., Chatila, T. A., and Mazmanian, S. K. (2011). The Toll-like receptor 2 pathway establishes colonization by a commensal of the human microbiota. *Science* **332,** 974–977.

Rouse, B. T., and Horohov, D. W. (1986). Immunosuppression in viral infections. *Rev. Infect. Dis.* **8,** 850–873.

Rouse, B. T., Sarangi, P. P., and Suvas, S. (2006). Regulatory T cells in virus infections. *Immunol. Rev.* **212,** 272–286.

Rowe, J. H., Ertelt, J. M., Aguilera, M. N., Farrar, M. A., and Way, S. S. (2011). Foxp3^{+} regulatory T cell expansion required for sustaining pregnancy compromises host defense against prenatal bacterial pathogens. *Cell Host Microbe* **10,** 54–64.

Rubtsov, Y. P., Niec, R. E., Josefowicz, S., Li, L., Darce, J., Mathis, D., Benoist, C., and Rudensky, A. Y. (2010). Stability of the regulatory T cell lineage in vivo. *Science* **329,** 1667–1671.

Rudensky, A. Y. (2011). Regulatory T cells and Foxp3. *Immunol. Rev.* **241,** 260–268.

Rushbrook, S. M., Ward, S. M., Unitt, E., Vowler, S. L., Lucas, M., Klenerman, P., and Alexander, G. J. (2005). Regulatory T cells suppress in vitro proliferation of virus-specific CD8^{+} T cells during persistent hepatitis C virus infection. *J. Virol.* **79,** 7852–7859.

Sakaguchi, S., Yamaguchi, T., Nomura, T., and Ono, M. (2008). Regulatory T cells and immune tolerance. *Cell* **133,** 775–787.

Salhi, A., Rodrigues, V., Jr., Santoro, F., Dessein, H., Romano, A., Castellano, L. R., Sertorio, M., Rafati, S., Chevillard, C., Prata, A., Alcais, A., Argiro, L., *et al.* (2008). Immunological and genetic evidence for a crucial role of IL-10 in cutaneous lesions in humans infected with *Leishmania braziliensis*. *J. Immunol.* **180,** 6139–6148.

Sarangi, P. P., Sehrawat, S., Suvas, S., and Rouse, B. T. (2008). IL-10 and natural regulatory T cells: Two independent anti-inflammatory mechanisms in herpes simplex virus-induced ocular immunopathology. *J. Immunol.* **180,** 6297–6306.

Sartono, E., Kruize, Y. C. M., Kurniawan, A., van der Meide, P. H., Partono, F., Maizels, R. M., and Yazdanbakhsh, M. (1995). Elevated cellular responses and interferon-γ release after long-term diethylcarbamazine treatment of patients with human lymphatic filariasis. *J. Infect. Dis.* **171,** 1683–1687.

Sather, B. D., Treuting, P., Perdue, N., Miazgowicz, M., Fontenot, J. D., Rudensky, A. Y., and Campbell, D. J. (2007). Altering the distribution of Foxp3$^{(+)}$ regulatory T cells results in tissue-specific inflammatory disease. *J. Exp. Med.* **204,** 1335–1347.

Satoguina, J., Mempel, M., Larbi, J., Badusche, M., Loliger, C., Adjei, O., Gachelin, G., Fleischer, B., and Hoerauf, A. (2002). Antigen-specific T regulatory-1 cells are associated with immunosuppression in a chronic helminth infection (onchocerciasis). *Microbes Infect.* **4,** 1291–1300.

Satoguina, J. S., Weyand, E., Larbi, J., and Hoerauf, A. (2005). T regulatory-1 cells induce IgG4 production by B cells: Role of IL-10. *J. Immunol.* **174,** 4718–4726.

Satoguina, J. S., Adjobimey, T., Arndts, K., Hoch, J., Oldenburg, J., Layland, L. E., and Hoerauf, A. (2008). Tr1 and naturally occurring regulatory T cells induce IgG4 in B cells through GITR/GITR-L interaction, IL-10 and TGF-β. *Eur. J. Immunol.* **38,** 3101–3113.

Sayi, A., Kohler, E., Toller, I. M., Flavell, R. A., Muller, W., Roers, A., and Muller, A. (2011). TLR-2-activated B cells suppress *Helicobacter*-induced preneoplastic gastric immunopathology by inducing T regulatory-1 cells. *J. Immunol.* **186,** 878–890.

Scholzen, A., Mittag, D., Rogerson, S. J., Cooke, B. M., and Plebanski, M. (2009). *Plasmodium falciparum*-mediated induction of human $CD25^{hi}Foxp3^{hi}$ CD4 T cells is independent of direct TCR stimulation and requires IL-2, IL-10 and TGFβ. *PLoS Pathog.* **5,** e1000543.

Scholzen, A., Minigo, G., and Plebanski, M. (2010). Heroes or villains? T regulatory cells in malaria infection. *Trends Parasitol.* **26,** 16–25.

Schramm, G., Mohrs, K., Wodrich, M., Doenhoff, M. J., Pearce, E. J., Haas, H., and Mohrs, M. (2007). IPSE/alpha-1, a glycoprotein from *Schistosoma mansoni* eggs, induces IgE-dependent, antigen-independent IL-4 production by murine basophils in vivo. *J. Immunol.* **178,** 6023–6027.

Scott-Browne, J. P., Shafiani, S., Tucker-Heard, G., Ishida-Tsubota, K., Fontenot, J. D., Rudensky, A. Y., Bevan, M. J., and Urdahl, K. B. (2007). Expansion and function of Foxp3-expressing T regulatory cells during tuberculosis. *J. Exp. Med.* **204,** 2159–2169.

Segura, M., Su, Z., Piccirillo, C., and Stevenson, M. M. (2007). Impairment of dendritic cell function by excretory-secretory products: A potential mechanism for nematode-induced immunosuppression. *Eur. J. Immunol.* **37,** 1887–1904.

Sehrawat, S., Suvas, S., Sarangi, P. P., Suryawanshi, A., and Rouse, B. T. (2008). In vitro-generated antigen-specific $CD4^+CD25^+$ $Foxp3^+$ regulatory T cells control the severity of herpes simplex virus-induced ocular immunoinflammatory lesions. *J. Virol.* **82,** 6838–6851.

Setiawan, T., Metwali, A., Blum, A. M., Ince, M. N., Urban, J. F., Jr., Elliott, D. E., and Weinstock, J. V. (2007). *Heligmosomoides polygyrus* promotes regulatory T-cell cytokine production in the murine normal distal intestine. *Infect. Immun.* **75,** 4655–4663.

Seyerl, M., Kirchberger, S., Majdic, O., Seipelt, J., Jindra, C., Schrauf, C., and Stockl, J. (2010). Human rhinoviruses induce IL-35-producing Treg via induction of B7-H1 (CD274) and sialoadhesin (CD169) on DC. *Eur. J. Immunol.* **40,** 321–329.

Shafiani, S., Tucker-Heard, G., Kariyone, A., Takatsu, K., and Urdahl, K. B. (2010). Pathogen-specific regulatory T cells delay the arrival of effector T cells in the lung during early tuberculosis. *J. Exp. Med.* **207,** 1409–1420.

Shi, B., Wang, Z., Jin, H., Chen, Y. W., Wang, Q., and Qian, Y. (2009). Immunoregulatory *Cordyceps sinensis* increases regulatory T cells to Th17 cell ratio and delays diabetes in NOD mice. *Int. Immunopharmacol.* **9,** 582–586.

Shimizu, J., Yamazaki, S., Takahashi, T., Ishida, Y., and Sakaguchi, S. (2002). Stimulation of $CD25^+CD4^+$ regulatory T cells through GITR breaks immunological self-tolerance. *Nat. Immunol.* **3,** 135–142.

Sing, A., Rost, D., Tvardovskaia, N., Roggenkamp, A., Wiedemann, A., Kirschning, C. J., Aepfelbacher, M., and Heesemann, J. (2002). *Yersinia* V-antigen exploits toll-like receptor 2 and CD14 for interleukin 10-mediated immunosuppression. *J. Exp. Med.* **196,** 1017–1024.

Singh, K. P., Gerard, H. C., Hudson, A. P., Reddy, T. R., and Boros, D. L. (2005). Retroviral Foxp3 gene transfer ameliorates liver granuloma pathology in *Schistosoma mansoni* infected mice. *Immunology* **114,** 410–417.

Smith, K. A., Hochweller, K., Hämmerling, G. J., Boon, L., Macdonald, A. S., and Maizels, R. M. (2011). Chronic helminth infection mediates tolerance in vivo through dominance of $CD11c^{lo}$ $CD103^-$ DC population. *J. Immunol.* **186,** 7098–7109.

Smits, H. H., and Yazdanbakhsh, M. (2007). Chronic helminth infections modulate allergen-specific immune responses: Protection against development of allergic disorders? *Ann. Med.* **39,** 428–439.

Smyk-Pearson, S., Golden-Mason, L., Klarquist, J., Burton, J. R., Jr., Tester, I. A., Wang, C. C., Culbertson, N., Vandenbark, A. A., and Rosen, H. R. (2008). Functional suppression by $FoxP3^+CD4^+CD25^{high}$ regulatory T cells during acute hepatitis C virus infection. *J. Infect. Dis.* **197,** 46–57.

Steeg, C., Adler, G., Sparwasser, T., Fleischer, B., and Jacobs, T. (2009). Limited role of $CD4^{+}Foxp3^{+}$ regulatory T cells in the control of experimental cerebral malaria. *J. Immunol.* **183,** 7014–7022.

Steel, C., and Nutman, T. B. (2003). CTLA-4 in filarial infections: Implications for a role in diminished T cell reactivity. *J. Immunol.* **170,** 1930–1938.

Steinfelder, S., Andersen, J. F., Cannons, J. L., Feng, C. G., Joshi, M., Dwyer, D., Caspar, P., Schwartzberg, P. L., Sher, A., and Jankovic, D. (2009). The major component in schistosome eggs responsible for conditioning dendritic cells for Th2 polarization is a T2 ribonuclease (omega-1). *J. Exp. Med.* **206,** 1681–1690.

Su, Z., Segura, M., Morgan, K., Loredo-Osti, J. C., and Stevenson, M. M. (2005). Impairment of protective immunity to blood-stage malaria by concurrent nematode infection. *Infect. Immun.* **73,** 3531–3539.

Suffia, I., Reckling, S. K., Salay, G., and Belkaid, Y. (2005). A role for CD103 in the retention of $CD4^{+}CD25^{+}$ Treg and control of *Leishmania major* infection. *J. Immunol.* **174,** 5444–5455.

Suffia, I. J., Reckling, S. K., Piccirillo, C. A., Goldszmid, R. S., and Belkaid, Y. (2006). Infected site-restricted $Foxp3^{+}$ natural regulatory T cells are specific for microbial antigens. *J. Exp. Med.* **203,** 777–788.

Suffner, J., Hochweller, K., Kuhnle, M. C., Li, X., Kroczek, R. A., Garbi, N., and Hammerling, G. J. (2010). Dendritic cells support homeostatic expansion of $Foxp3^{+}$ regulatory T cells in Foxp3.LuciDTR mice. *J. Immunol.* **184,** 1810–1820.

Sugimoto, K., Ikeda, F., Stadanlick, J., Nunes, F. A., Alter, H. J., and Chang, K. M. (2003). Suppression of HCV-specific T cells without differential hierarchy demonstrated ex vivo in persistent HCV infection. *Hepatology* **38,** 1437–1448.

Summers, R. W., Elliott, D. E., and Weinstock, J. V. (2005). Why *Trichuris suis* should prove safe for use in inflammatory bowel diseases. *Inflamm. Bowel Dis.* **11,** 783–784.

Summers, R. W., Elliott, D. E., and Weinstock, J. V. (2006). Therapeutic colonization with Trichuris suis. *Arch. Pathol. Lab. Med.* **130,** 1753author reply 1753–1754.

Sun, C. M., Hall, J. A., Blank, R. B., Bouladoux, N., Oukka, M., Mora, J. R., and Belkaid, Y. (2007). Small intestine lamina propria dendritic cells promote de novo generation of Foxp3 T reg cells via retinoic acid. *J. Exp. Med.* **204,** 1775–1785.

Sutmuller, R. P., den Brok, M. H., Kramer, M., Bennink, E. J., Toonen, L. W., Kullberg, B. J., Joosten, L. A., Akira, S., Netea, M. G., and Adema, G. J. (2006). Toll-like receptor 2 controls expansion and function of regulatory T cells. *J. Clin. Invest.* **116,** 485–494.

Suvas, S., Kumaraguru, U., Pack, C. D., Lee, S., and Rouse, B. T. (2003). $CD4^{+}CD25^{+}$ T cells regulate virus-specific primary and memory $CD8^{+}$ T cell responses. *J. Exp. Med.* **198,** 889–901.

Suvas, S., Azkur, A. K., Kim, B. S., Kumaraguru, U., and Rouse, B. T. (2004). $CD4^{+}CD25^{+}$ regulatory T cells control the severity of viral immunoinflammatory lesions. *J. Immunol.* **172,** 4123–4132.

Tang, C. L., Lei, J. H., Wang, T., Lu, S. J., Guan, F., Liu, W. Q., and Li, Y. L. (2011). Effect of $CD4^{+}$ $CD25^{+}$ regulatory T cells on the immune evasion of *Schistosoma japonicum*. *Parasitol. Res.* **108,** 477–480.

Taylor, M., Le Goff, L., Harris, A., Malone, E., Allen, J. E., and Maizels, R. M. (2005). Removal of regulatory T cell activity reverses hyporesponsiveness and leads to filarial parasite clearance in vivo. *J. Immunol.* **174,** 4924–4933.

Taylor, J. J., Mohrs, M., and Pearce, E. J. (2006). Regulatory T cell responses develop in parallel to Th responses, and control the magnitude and phenotype of the Th effector population. *J. Immunol.* **176,** 5839–5847.

Taylor, M. D., Harris, A., Babayan, S., Bain, O., Culshaw, A., Allen, J. E., and Maizels, R. M. (2007). CTLA-4^{+} and $CD4^{+}CD25^{+}$ regulatory T cells inhibit protective immunity to filarial parasites in vivo. *J. Immunol.* **179,** 4626–4634.

Taylor, M. D., van der Werf, N., Harris, A., Graham, A. L., Bain, O., Allen, J. E., and Maizels, R. M. (2009). Early recruitment of natural $CD4^+Foxp3^+$ regulatory T cells by infective larvae determines the outcome of filarial infection. *Eur. J. Immunol.* **39,** 192–206.

Teixeira-Carvalho, A., Martins-Filho, O. A., Peruhype-Magalhaes, V., Silveira-Lemos, D., Malaquias, L. C., Oliveira, L. F., Silveira, A. M., Gazzinelli, A., Gazzinelli, G., and Correa-Oliveira, R. (2008). Cytokines, chemokine receptors, $CD4^+CD25^{HIGH+}$ T-cells and clinical forms of human schistosomiasis. *Acta Trop.* **108,** 139–149.

Tetsutani, K., Ishiwata, K., Ishida, H., Tu, L., Torii, M., Hamano, S., Himeno, K., and Hisaeda, H. (2009). Concurrent infection with *Heligmosomoides polygyrus* suppresses anti-*Plasmodium yoelii* protection partially by induction of $CD4^+CD25^+Foxp3^+$ Treg in mice. *Eur. J. Immunol.* **39,** 2822–2830.

Thornton, A. M., Korty, P. E., Tran, D. Q., Wohlfert, E. A., Murray, P. E., Belkaid, Y., and Shevach, E. M. (2010). Expression of Helios, an Ikaros transcription factor family member, differentiates thymic-derived from peripherally induced $Foxp3^+$ T regulatory cells. *J. Immunol.* **184,** 3433–3441.

Toka, F. N., Suvas, S., and Rouse, B. T. (2004). $CD4^+$ $CD25^+$ T cells regulate vaccine-generated primary and memory $CD8^+$ T-cell responses against herpes simplex virus type 1. *J. Virol.* **78,** 13082–13089.

Tortorella, D., Gewurz, B. E., Furman, M. H., Schust, D. J., and Ploegh, H. L. (2000). Viral subversion of the immune system. *Annu. Rev. Immunol.* **18,** 861–926.

Trandem, K., Anghelina, D., Zhao, J., and Perlman, S. (2010). Regulatory T cells inhibit T cell proliferation and decrease demyelination in mice chronically infected with a coronavirus. *J. Immunol.* **184,** 4391–4400.

Trautmann, L., Janbazian, L., Chomont, N., Said, E. A., Gimmig, S., Bessette, B., Boulassel, M. R., Delwart, E., Sepulveda, H., Balderas, R. S., Routy, J. P., Haddad, E. K., *et al.* (2006). Upregulation of PD-1 expression on HIV-specific $CD8^+$ T cells leads to reversible immune dysfunction. *Nat. Med.* **12,** 1198–1202.

Tsunemi, S., Iwasaki, T., Imado, T., Higasa, S., Kakishita, E., Shirasaka, T., and Sano, H. (2005). Relationship of CD4+CD25+ regulatory T cells to immune status in HIV-infected patients. *AIDS* **19,** 879–886.

Turner, J. D., Jackson, J. A., Faulkner, H., Behnke, J., Else, K., Kamgno, J., Boussinesq, M., and Bradley, J. E. (2008). Intensity of intestinal infection with multiple worm species is related to regulatory cytokine output and immune hyporesponsiveness. *J. Infect. Dis.* **197,** 1204–1212.

Vahlenkamp, T. W., Tompkins, M. B., and Tompkins, W. A. (2004). Feline immunodeficiency virus infection phenotypically and functionally activates immunosuppressive $CD4^+CD25^+$ T regulatory cells. *J. Immunol.* **172,** 4752–4761.

van der Kleij, D., Latz, E., Brouwers, J. F. H. M., Kruize, Y. C. M., Schmitz, M., Kurt-Jones, E. A., Espevik, T., de Jong, E. C., Kapsenberg, M. L., Golenbock, D. T., Tielens, A. G. M., and Yazdanbakhsh, M. (2002). A novel host-parasite lipid cross talk: Schistosomal lysophosphatidylserine activates Toll-like receptor 2 and affects immune polarization. *J. Biol. Chem.* **277,** 48122–48129.

van Maren, W. W. C., Jacobs, J. F., de Vries, I. J., Nierkens, S., and Adema, G. J. (2008). Toll-like receptor signalling on Tregs: To suppress or not to suppress? *Immunology* **124,** 445–452.

van Maren, W. W. C., Nierkens, S., Toonen, L. W., Bolscher, J. M., Sutmuller, R. P., and Adema, G. J. (2011). Multifaceted effects of synthetic TLR2 ligand and *Legionella pneumophilia* on Treg-mediated suppression of T cell activation. *BMC Immunol.* **12,** 23.

Van, L. P., Bardel, E., Gregoire, S., Vanoirbeek, J., Schneider, E., Dy, M., and Thieblemont, N. (2011). Treatment with the TLR7 agonist R848 induces regulatory T-cell-mediated suppression of established asthma symptoms. *Eur. J. Immunol.* **41,** 1992–1999.

Vigario, A. M., Gorgette, O., Dujardin, H. C., Cruz, T., Cazenave, P. A., Six, A., Bandeira, A., and Pied, S. (2007). Regulatory CD4+ CD25+ Foxp3+ T cells expand during experimental Plasmodium infection but do not prevent cerebral malaria. *Int. J. Parasitol.* **37,** 963–973.

Walsh, C. M., Smith, P., and Fallon, P. G. (2007). Role for CTLA-4 but not $CD25^+$ T cells during *Schistosoma mansoni* infection of mice. *Parasite Immunol.* **29,** 293–308.

Walther, M., Tongren, J. E., Andrews, L., Korbel, D., King, E., Fletcher, H., Andersen, R. F., Bejon, P., Thompson, F., Dunachie, S. J., Edele, F., de Souza, J. B., *et al.* (2005). Upregulation of TGF-β, FOXP3, and $CD4^+CD25^+$ regulatory T cells correlates with more rapid parasite growth in human malaria infection. *Immunity* **23,** 287–296.

Walther, M., Jeffries, D., Finney, O. C., Njie, M., Ebonyi, A., Deininger, S., Lawrence, E., Ngwa-Amambua, A., Jayasooriya, S., Cheeseman, I. H., Gomez-Escobar, N., Okebe, J., *et al.* (2009). Distinct roles for $FOXP3^+$ and $FOXP3^-$ $CD4^+$ T cells in regulating cellular immunity to uncomplicated and severe *Plasmodium falciparum* malaria. *PLoS Pathog.* **5,** e1000364.

Wammes, L. J., Hamid, F., Wiria, A. E., de Gier, B., Sartono, E., Maizels, R. M., Luty, A. J., Fillie, Y., Brice, G. T., Supali, T., Smits, H. H., and Yazdanbakhsh, M. (2010). Regulatory T cell in human geohelminth infection suppress immune responses to BCG and *Plasmodium falciparum*. *Eur. J. Immunol.* **40,** 437–442.

Wang, X., Zhou, S., Chi, Y., Wen, X., Hoellwarth, J., He, L., Liu, F., Wu, C., Dhesi, S., Zhao, J., Hu, W., and Su, C. (2009). $CD4^+CD25^+$ Treg induction by an HSP60-derived peptide SJMHE1 from *Schistosoma japonicum* is TLR2 dependent. *Eur. J. Immunol.* **39,** 3052–3065.

Watanabe, K., Mwinzi, P. N., Black, C. L., Muok, E. M., Karanja, D. M., Secor, W. E., and Colley, D. G. (2007). T regulatory cell levels decrease in people infected with *Schistosoma mansoni* on effective treatment. *Am. J. Trop. Med. Hyg.* **77,** 676–682.

Weber, M. S., Benkhoucha, M., Lehmann-Horn, K., Hertzenberg, D., Sellner, J., Santiago-Raber, M. L., Chofflon, M., Hemmer, B., Zamvil, S. S., and Lalive, P. H. (2010). Repetitive pertussis toxin promotes development of regulatory T cells and prevents central nervous system autoimmune disease. *PLoS One* **5,** e16009.

Wei, G., and Tabel, H. (2008). Regulatory T cells prevent control of experimental African trypanosomiasis. *J. Immunol.* **180,** 2514–2521.

Weiss, L., Donkova-Petrini, V., Caccavelli, L., Balbo, M., Carbonneil, C., and Levy, Y. (2004). Human immunodeficiency virus-driven expansion of $CD4^+CD25^+$ regulatory T cells, which suppress HIV-specific CD4 T-cell responses in HIV-infected patients. *Blood* **104,** 3249–3256.

Wherry, E. J., Blattman, J. N., Murali-Krishna, K., van der Most, R., and Ahmed, R. (2003). Viral persistence alters CD8 T-cell immunodominance and tissue distribution and results in distinct stages of functional impairment. *J. Virol.* **77,** 4911–4927.

Wilson, M. S., Taylor, M., Balic, A., Finney, C. A. M., Lamb, J. R., and Maizels, R. M. (2005). Suppression of allergic airway inflammation by helminth-induced regulatory T cells. *J. Exp. Med.* **202,** 1199–1212.

Wu, Y., Wang, Q. H., Zheng, L., Feng, H., Liu, J., Ma, S. H., and Cao, Y. M. (2007). *Plasmodium yoelii*: Distinct $CD4^+CD25^+$ regulatory T cell responses during the early stages of infection in susceptible and resistant mice. *Exp. Parasitol.* **115,** 301–304.

Wu, J. J., Chen, G., Liu, J., Wang, T., Zheng, W., and Cao, Y. M. (2010). Natural regulatory T cells mediate the development of cerebral malaria by modifying the pro-inflammatory response. *Parasitol. Int.* **59,** 232–241.

Xu, D., Liu, H., Komai-Koma, M., Campbell, C., McSharry, C., Alexander, J., and Liew, F. Y. (2003). $CD4^+CD25^+$ regulatory T cells suppress differentiation and functions of Th1 and Th2 cells, *Leishmania major* infection, and colitis in mice. *J. Immunol.* **170,** 394–399.

Yang, G., Liu, A., Xie, Q., Guo, T. B., Wan, B., Zhou, B., and Zhang, J. Z. (2007a). Association of $CD4^+CD25^+Foxp3^+$ regulatory T cells with chronic activity and viral clearance in patients with hepatitis B. *Int. Immunol.* **19,** 133–140.

Yang, J., Zhao, J., Yang, Y., Zhang, L., Yang, X., Zhu, X., Ji, M., Sun, N., and Su, C. (2007b). *Schistosoma japonicum* egg antigens modulate airway inflammation by producing $CD4^{+}CD25^{+}$ T cells in a murine model of asthma. *Immunology* **120,** 8–18.

Yazdanbakhsh, M., Paxton, W. A., Kruize, Y. C. M., Sartono, E., Kurniawan, A., van het Wout, A., Selkirk, M. E., Partono, F., and Maizels, R. M. (1993). T cell responsiveness correlates differentially with antibody isotype levels in clinical and asymptomatic filariasis. *J. Infect. Dis.* **167,** 925–931.

Zaccone, P., Burton, O. T., Gibbs, S. E., Miller, N., Jones, F. M., Schramm, G., Haas, H., Doenhoff, M. J., Dunne, D. W., and Cooke, A. (2011). The *S. mansoni* glycoprotein ω-1 induces Foxp3 expression in NOD mouse $CD4^{+}$ T cells. *Eur. J. Immunol.* **41,** 2709–2718.

Zanin-Zhorov, A., Cahalon, L., Tal, G., Margalit, R., Lider, O., and Cohen, I. R. (2006). Heat shock protein 60 enhances $CD4^{+}$ $CD25^{+}$ regulatory T cell function via innate TLR2 signaling. *J. Clin. Invest.* **116,** 2022–2032.

Zelinskyy, G., Robertson, S. J., Schimmer, S., Messer, R. J., Hasenkrug, K. J., and Dittmer, U. (2005). $CD8^{+}$ T-cell dysfunction due to cytolytic granule deficiency in persistent Friend retrovirus infection. *J. Virol.* **79,** 10619–10626.

Zelinskyy, G., Kraft, A. R., Schimmer, S., Arndt, T., and Dittmer, U. (2006). Kinetics of $CD8^{+}$ effector T cell responses and induced $CD4^{+}$ regulatory T cell responses during Friend retrovirus infection. *Eur. J. Immunol.* **36,** 2658–2670.

Zelinskyy, G., Dietze, K., Sparwasser, T., and Dittmer, U. (2009a). Regulatory T cells suppress antiviral immune responses and increase viral loads during acute infection with a lymphotropic retrovirus. *PLoS Pathog.* **5,** e1000406.

Zelinskyy, G., Dietze, K. K., Husecken, Y. P., Schimmer, S., Nair, S., Werner, T., Gibbert, K., Kershaw, O., Gruber, A. D., Sparwasser, T., and Dittmer, U. (2009b). The regulatory T-cell response during acute retroviral infection is locally defined and controls the magnitude and duration of the virus-specific cytotoxic T-cell response. *Blood* **114,** 3199–3207.

Zhang, M., Liu, M., Luther, J., and Kao, J. Y. (2010). *Helicobacter pylori* directs tolerogenic programming of dendritic cells. *Gut Microbes* **1,** 325–329.

Zhao, J., Fett, C., Trandem, K., Fleming, E., and Perlman, S. (2011). IFN-γ- and IL-10-expressing virus epitope-specific $Foxp3^{+}$ T reg cells in the central nervous system during encephalomyelitis. *J. Exp. Med.* **208,** 1571–1577.

Zuany-Amorim, C., Sawicka, E., Manlius, C., Le Moine, A., Brunet, L. R., Kemeny, D. M., Bowen, G., Rook, G., and Walker, C. (2002). Suppression of airway eosinophilia by killed *Mycobacterium vaccae*-induced allergen-specific regulatory T-cells. *Nat. Med.* **8,** 625–629.

CHAPTER 4

Biological Functions of Regulatory T Cells

Ethan M. Shevach

Contents

Laboratory of Immunology, National Institutes of Allergy and Infectious Diseases, National Institutes of Health, Bethesda, Maryland, USA

Advances in Immunology, Volume 112
ISSN 0065-2776, DOI: 10.1016/B978-0-12-387827-4.00004-8

Abstract The subpopulation of $CD4^+$ T lymphocytes that co-express the transcription factor Foxp3 plays a unique role as regulatory T lymphocytes (Tregs) that modulate many aspects of the immune response. Multiple mechanisms have been proposed for the suppressor function of $CD4^+Foxp3^+$ T cells based on *in vitro* studies, but much less is known about how Tregs suppress immune responses *in vivo*. Both polyclonal Tregs and antigen-specific Tregs are capable of exerting potent suppressive effects *in vivo*, and it is likely that they mediate their biologic functions using different mechanisms. Antigen-specific Tregs primarily target dendritic cells and inhibit dendritic cell functions including the expression of costimulatory molecules and the presentation of antigen early during the generation of the immune response. The end result is a complete inhibition of both the expansion and the differentiation of T effector cells. Polyclonal Tregs also act on dendritic cells, but at a later phase, and do not inhibit expansion of T effector cells, but appear to modulate differentiation and cell trafficking. The cell surface molecules involved in the interaction of Tregs with dendritic cells, as well as the biochemical pathways modified by this interaction remain to be fully elucidated. A complete understand of the biological functions of Tregs *in vivo* should facilitate the development of pharmacologic and biologic agents that can be used to modulate Treg function in a therapeutic setting.

1. INTRODUCTION

Studies over the past 15 years (Hori *et al.*, 2003; Sakaguchi *et al.*, 2008) have demonstrated that potentially autoreactive T cells can be regulated by a distinct lineage of $CD4^+$ T suppressor cells or as they are more commonly known T regulatory cells (Tregs). Tregs are generated during the normal course of T cell development in the thymus and are termed natural (n) Tregs, but may also develop extrathymically in certain situations and are termed induced (i) iTregs (Curotto de Lafaille and Lafaille, 2009). Although $CD4^+$ Tregs were originally identified (Sakaguchi *et al.*, 1995) by their expression of high levels of the activation antigen (CD25), it is now clear that the transcription factor, Foxp3, is most optimal marker for this population (Josefowicz and Rudensky, 2009). We have recently reviewed the multiple mechanisms used by Tregs to suppress immune responses *in vitro* (Shevach, 2009), and this review will focus exclusively on the biological functions and mechanisms potentially used by both polyclonal Tregs and antigen-specific Tregs *in vivo*. Although multiple mechanisms for Treg suppressor function have been shown *in vitro*, it is unclear whether the same or different mechanisms are used by Tregs *in vivo*. In fact, one must question the significance of studies performed in the fixed milieu of the tissue culture dish. *In vivo* Tregs must home to various parts of the body and interact directly or indirectly with other components of the immune system. T effector (Teff) cells can easily evade the suppressive activity of Tregs *in vivo*.

There are a number of other major differences between the properties of Tregs *in vivo* and the *in vitro* situation. Although Tregs are anergic or nonresponsive *in vitro* due to their failure to produce IL-2, Tregs expand after engagement of their T cell receptor (TCR) by its cognate antigen in a manner indistinguishable from conventional T cells (Klein *et al.*, 2003; Walker *et al.*, 2003). While most *in vitro* studies of Treg suppressor function do not implicate suppressor cytokines such as IL-10 or TGFβ as mediators of suppression, studies in several *in vivo* models have shown that suppressor cytokines can be major contributors to the suppressive effects of Tregs (Belkaid and Rouse, 2005; McGeachy *et al.*, 2005). While the major effect of Tregs *in vitro* is to suppress the production of IL-2 by Teff, the role of IL-2 in the antigen-driven expansion of $CD4^+$ T cells has not been elucidated and antigen-specific cells from IL-2 deficient $^{(-/-)}$ expand normally *in vivo* when triggered via their TCR (Shevach *et al.*, 2006). Taken together, it is very difficult to translate from the *in vitro* studies of Treg function to the *in vivo* situation. As a major goal of research in this area is the manipulation of Treg function *in vivo* in disease, a clear understanding of Treg biology *in vivo* is required.

2. BIOLOGIC FUNCTIONS OF POLYCLONAL TREGS *IN VIVO*

The role of Tregs in the control of organ-specific autoimmune diseases was first defined by studies in mice that had been thymectomized on day 3 of life (d3Tx). D3Tx mice develop organ-specific autoimmunity that can be prevented by reconstitution of the mice with $CD4^+Foxp3^+$ Tregs between days 10 and 12 of life. A similar spectrum of autoimmune disease develops when immunodeficient (*nu/nu* or $RAG^{-/-}$ mice) are reconstituted with $CD4^+Foxp3^-$ T cells and cotransfer of $Foxp3^+$ cells prevents the development of disease. Polyclonal Tregs have also been shown to inhibit the immune response to alloantigens (Kingsley *et al.*, 2002), prevent graft versus host disease (Taylor *et al.*, 2002), and prevent the immune response to tumors (Shimizu *et al.*, 1999). Although the ability of nTregs to inhibit the development of these disease states is well documented, the mechanisms by which polyclonal Tregs inhibit immune responses *in vivo* are poorly understood. A number of possible mechanisms for the suppressive function of Tregs *in vivo* have been proposed including whether they inhibit the priming and expansion of Teff cells, block the migration of Teff cells from the site of immunization (e.g., the draining lymph node [LN] to the target organ), or inhibit the differentiation of naïve precursors into pathogenic Teff (Th1, Th2, or Th17) cells. While Tregs have been used in many *in vivo* models, I will focus this discussion on model systems in which it is possible to analyze the effects of Treg suppression on Teff cells on a single cell basis.

2.1. Polyclonal Tregs inhibits the induction of disease in immunodeficient mice

Our initial studies in this area exclusively used autoimmune gastritis (AIG) as the animal model for probing the mechanisms of action of polyclonal Tregs *in vivo* (DiPaolo *et al.*, 2005). We selected AIG as a model because it is one of the more common diseases that develops in mice after d3Tx or transfer of $CD4^+Foxp3^-$ T cells to *nu/nu* recipients on a BALB/c background. Further, it is one of few models of spontaneous autoimmune disease in which the target antigen, the proton pump of the gastric parietal cell (H/K ATPase), has been clearly defined. In addition, it is an ideal animal model for the human disease, chronic gastritis, which results in the development of pernicious anemia secondary to the destruction of intrinsic factor producing gastric parietal cells by human $CD4^+$ Teff that also recognize the H/K ATPase (D'elios *et al.*, 2001). Analysis of the function of Tregs in AIG has been facilitated by the development of a number of strains of TCR transgenic (Tg) mice expressing TCRs derived from T cell clones specific for the H/K ATPase derived from mice with AIG (Shevach *et al.*, 2006). One of these Tg lines termed TxA23 expresses a TCR specific for the peptide H/K ATPase α-$\text{chain}_{630-641}$. It is difficult to study Treg function in the TxA23 mouse itself as it develops very severe AIG early in life (day 10). However, low numbers of naïve Tg expressing $CD4^+Foxp3^-$ thymocytes cells efficiently transfer AIG to *nu/nu* recipients, and we have used this population as a source of naïve Teff in many of our studies. This model has several advantages as the target antigen is expressed and presented at physiological levels as opposed to being introduced artificially as a Tg and immunization with exogenous antigen is not required.

Initially, we cotransferred Teff and polyclonal Tregs from normal BALB/c donors to adult *nu/nu* mice, and this allowed us to examine the suppressive effects of the Tregs throughout the course of the disease process. One complication of studies using *nu/nu* mice as recipients in this model is lymphopenia-induced proliferation. We attempted to modulate this process by also coinjecting a large number of Treg-depleted splenocytes. Cotransfer of polyclonal Tregs with a small number of naïve autoantigen-specific T cells was quite effective in inhibiting the development of AIG as measured by suppression of parietal cell destruction and inhibition of autoantibody production. Tregs had no effect on the migration of the autoreactive T cells into the gastric LN or into the target organ and did not inhibit the antigen-driven expansion of the H/K ATPase-specific cells in the gastric LN. The proliferation of the Teff cells was identical at both 6 and 12 days after transfer either in the presence or in the absence of Tregs as measured by dilution of CFSE-labeled Teff. It should also be noted that although Tregs suppressed the destruction of gastric parietal cells, extensive mononuclear cell infiltration was observed in both untreated and Treg-treated mice. These results resembled a benign inflammatory response in which the Tregs keep the Teff cells in a nonpathogenic state. The polyclonal Tregs rapidly migrated to the gastric LN in the target organ in the presence of the antigen-specific Teff, but the Tregs could also be detected in other LNs and spleen. The ratio of Tregs to Teff cells in the gastric LN increased over time suggesting that the proliferation and/or homing of the Tregs was influenced in part by the presence of autoantigen-stimulated Teff in the gastric LN. Suppression of disease by Tregs was not transient, as treated mice remained disease-free for as long as 100 days after cell transfer.

Although the expansion and contraction of TxA23 T cells within the gastric LN were not altered in Treg-treated animals (Fig. 4.1), a marked decrease in the total cellularity of the gastric LN was observed in Treg-treated mice beginning about 10 days after cell transfer. This decrease in cellularity was not confined to a specific cell type suggesting that the presence of the polyclonal Tregs led to inhibition of the recruitment and/or expansion of other cell types. The primary effect Treg treatment appeared to be inhibition of the differentiation of the autoreactive cells to Th1 Teff cells, as reflected by a decrease in interferon-γ (IFNγ) production in supernatants from peptide-stimulated gastric LN cells between 7–14 days after transfer and a decrease in antigen-stimulated IFNγ and T-bet mRNA expression as measured by real-time PCR in cells infiltrating the stomach 28 and 42 days after cell transfer. It is also likely that the decrease in cellularity of the gastric LN in recipients of the Tregs was secondary to decreased cytokine/chemokine production by the Teff.

2.2. Polyclonal Tregs suppress disease induced by fully activated Teff in immunodeficient mice

The studies above utilized naïve autoantigen-specific T cells as a source of Teff, but a more clinically relevant approach should examine whether Tregs can inhibit an ongoing autoimmune disease process. The aims of the next series of studies were first to assess the capacity of Th1, Th2, and Th17 Teff cells derived from the TxA23 Tg mouse to induce AIG after transfer to *nu/nu* recipients and then to determine the susceptibility of each of the fully differentiated Teff cell types to suppression by polyclonal Tregs *in vivo* (Stummvoll *et al.*, 2008). We also examined whether Tregs could modulate cytokine production by fully differentiated Teff cells

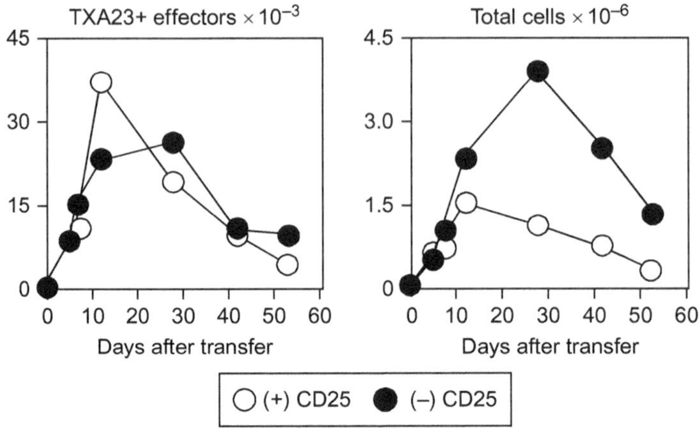

FIGURE 4.1 Polyclonal Tregs do not suppress the expansion of Teff-mediating AIG. Teff cell recognizing the H/K ATPase in gastric parietal cell were transferred with (open circles) or without (closed circles) polyclonal nTregs into *nu/nu* recipients. The number of transferred Teff in the gastric LN was enumerated at the indicated time point (left panel). The total number of cells present in the gastric LN was counted at the indicated time point (right panel).

in vivo. Naïve $CD4^+Foxp3^-$ T cells were isolated from the thymi of TxA23 mice and stimulated *in vitro* under standard priming conditions to differentiate into Th1, Th2, or Th17 Teff cells. The cytokine producing potential of each group was evaluated by intracellular staining following stimulation with PMA/ionomycin. Th1 cells contained 67% IFNγ producers and low percentages of IL-4 or IL-17 producers; Th2 cells contained 41% IL-4 producers, but no IFNγ or IL-17 producers, while Th17 cells were highly enriched for IL-17 producers and under the conditions used in our studies very few IFNγ or IL-4 producing cells.

After transfer, all three Teff populations preferentially accumulated in the gastric LN. All three populations also induced AIG 4 weeks after transfer with significant destruction of gastric parietal cells that was highest in recipients of Th17 cells. After 6 weeks, almost all of the recipients in all three groups demonstrated severe AIG. Most animals in each group also produced antiparietal cell antibodies. Th1-mediated inflammatory infiltrates consisted mainly of $CD3^+$ T cells, whereas Th2-mediated infiltrates also contained large numbers of granulocytes, primarily eosinophils, and were associated with elevated serum IgE levels. Surprisingly, the highest number of infiltrating polymorphonuclear cells, including a very high percentage of granulocytes and eosinophils, was observed in recipients of Th17 cells. The enhanced aggressive disease seen with Th17 cells could have been secondary to their increased proliferative responses *in vivo* or perhaps to enhanced homing to the target organ. Most other studies of the cellular composition of Th17-associated infiltrates demonstrated that they were composed of neutrophils. It is unclear whether the predominantly eosinophilic infiltration observed in our studies is a unique characteristic of Th17-induced AIG or whether studies of other Th17-associated diseases overlooked the presence of eosinophils because they are hard to distinguish from neutrophils without special staining.

The differentiated Th2, Th2, and Th17 lines used to induce AIG-exhibited variable patterns of stability after transfer and exposure to autoantigen *in vivo*. All the cell lines exhibited the appropriate cytokine profile when tested before transfer. Th1 cells maintained their phenotype for as long as 6 weeks after transfer, whereas infiltrates from Th2 recipients contained a significant number of IFNγ producing cells after 6 weeks. It is possible that some of these IFNγ producers arose from uncommitted precursors in the starting population that responded to environmental signals in this AIG model that seem to favor Th1 differentiation. Alternatively, some plasticity might exist in the committed Th2 population that facilitates conversion to Th1 cells under these environmental conditions. A significant number of the cells recovered from recipients of Th17 cells continued to produce IL-17, but an equal number of IFNγ producing cells was also recovered. It is likely that these IFNγ-producing cells represent Th17 cells that have further differentiated in the presence of IL-23 to produce IFNγ (Hirota *et al.*, 2011). Importantly, the lack of stability in cytokine profile seen during the 6 weeks of the study had little effect on the striking differences in the inflammatory infiltrates.

We used the same Teff : Treg ratios that we used in the previous cotransfer studies with naïve TxA23 cells. Polyclonal Tregs markedly suppressed the disease induced by Th1 cells, moderately suppressed Th2-induced disease, but exerted very little suppression of Th17-induced disease (Fig. 4.2). In contrast to the results seen with naïve Teff, significant inhibition of Teff cell expansion was seen when polyclonal Treg were cotransferred with Th1 and Th2 Teff populations. The inhibition of expansion correlated with protection from destructive AIG. If one assumes that Treg-mediated suppression is independent of the type of Teff present, it is possible that Th17 Teff might overcome suppression by Tregs because of their enhanced proliferative capacity potentially out numbering Tregs for access to dendritic cell (DC)-presented autoantigen. However, there are several other important differences between the studies with the differentiated Teff and the naïve Teff. Naïve Teff could be detected in the gastric LN as early as 3 days after transfer, and significant expansion could be observed by day 5. In contrast, the fully differentiated Teff could barely be detected in the gastric LN after 1 week.

The differentiated Teff expressed low levels of CD62L, the molecule that is required for entry into LN. It is also possible that the Teff was initially stimulated in the stomach itself and then trafficked to the gastric LN.

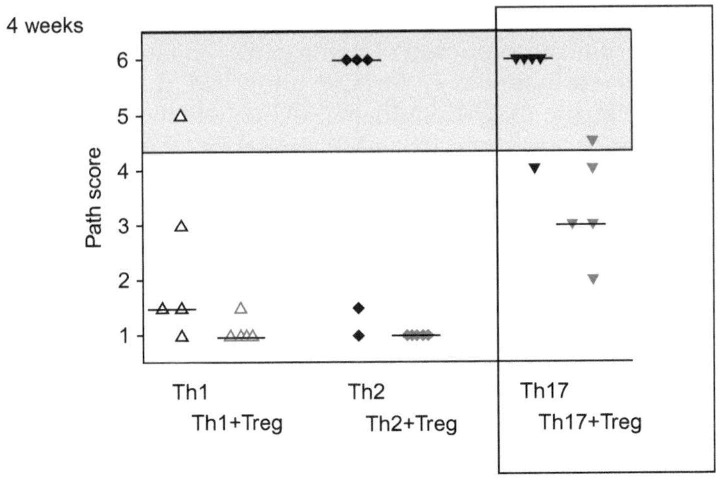

FIGURE 4.2 Polyclonal Tregs inhibits the induction of AIG by fully differentiated Teff. Antigen-specific Th1, Th2, and Th17 Teff recognizing the H/K ATPase in gastric parietal cells were cotransferred into *nu/nu* recipients with or without polyclonal Tregs. Disease score was determined 4 weeks after transfer.

In contrast, the cotransferred Tregs expressed high levels of CD62L and could be detected in all LN at early time points after transfer. It thus remains possible that a low percentage of the Tregs, that expressed TCRs specific for the H/K ATPase, was retained in the gastric LN and expanded. These expanded antigen-specific cells could now exert their effects on DC resulting in an inhibition of Teff cell expansion in a manner similar to antigen-specific Tregs (see Section 3.4). This hypothesis is consistent with the studies of Godebu *et al.* (2008) that have demonstrated that polyclonal T cells induced *in vitro* with TGFβ can protect against the development of diabetes for prolonged periods of time. One year after treatment, the iTregs that persisted in the treated animals had a restricted T cell repertoire that used one single TCR β-chain that is consistent with selection by antigen. Although Tregs induced significant suppression of expansion of Th1 and Th2 Teff cells, significant number of Teff could be detected in protected mice. When these cells were recovered and stimulated *ex vivo*, no significant differences were observed in cytokine production when compared to Teff isolated from non-Treg-treated mice. Thus, exposure of the Teff cells to polyclonal T cells *in vivo* did not render them anergic, and they were fully competent to proliferate and produce cytokines *ex vivo*. The primary effect of the Tregs is thus likely to be on T cell expansion. This result is consistent with studies (Samy *et al.*, 2005) in the d3Tx model of oophoritis in which the continuous presence of Tregs was required for protection from disease.

2.3. Effects of polyclonal Tregs on T cell activation in immunocompetent mice

One of the major problems with the study of the mechanisms of action of polyclonal Tregs in the AIG model is that both naïve and fully differentiated Teff populations fail to induce disease when transferred to normal immunocompetent mice. Presumably, this failure to induce disease is secondary to the presence of polyclonal Tregs in the normal recipients. Ideally, it would be of interest to transfer Teff to normal mice that have been depleted of Tregs, but we have not been successful in establishing such a model either using antibody-mediated depletion of $CD4^{+}CD25^{+}$ T cells or complete elimination of Foxp3-expressing cells by using diptheria toxin to treat mice expressing the human diptheria toxin receptor (DTR) under control of the Foxp3 promoter. We therefore have developed a model system that allows us to examine different steps of the T cell activation process *in vivo*. It is now widely accepted that the initial event in T cell priming is the capture of antigen by DCs that migrate to the draining LN where they present the antigen to naïve T cells. The naïve antigen-specific cells become activated, proliferate, and during the course of expansion differentiate into primed Teff cells capable of producing different cytokines depending on the cytokine milieu in the draining LN. Once the antigen-specific cells have been primed, they can leave the LN and traffic through the circulation to enter tissues. Each of the steps in the process represents a potential checkpoint for the suppressive effects of Tregs.

Our initial approach to this question involved an analysis of the potential mechanisms of action used by polyclonal Tregs to suppress autoimmune disease, specifically, experimental autoimmune encephalomyelitis (EAE), the animal model of multiple sclerosis in man. Several years ago, Kohm *et al.* (2002) demonstrated that the administration of $1–2 \times 10^{6}$ polyclonal Tregs from normal donors to normal recipients at the time they were immunized with a peptide derived from myelin oligodendrocyte glycoprotein (MOG) markedly reduced the incidence and severity of EAE. We repeated these studies, but instead of freshly explanted Tregs, we first expanded the Tregs *in vitro* by TCR stimulation in the presence of IL-2 (Davidson and Shevach, 2011). One day following transfer, the normal recipient mice were immunized for the induction of EAE. Treg-treated mice displayed significantly reduced clinical severity when compared to non-Treg-treated controls. Conversely, we could also demonstrate that endogenous Tregs played an important role *in vivo* as partial depletion of Tregs with an anti-CD25 antibody prior to immunization consistently results in an exacerbated clinical course.

Overall, these studies demonstrated that a modest augmentation or a reduction in the number of Tregs in a normal mouse results in a dramatic alteration in the course of an autoimmune disease. To determine potential mechanisms of Treg action in this model, we evaluated the phenotype of Teff that had trafficked to the brain. We isolated Teff from the spinal cord of mice with EAE that had or had not received polyclonal Tregs at the time of immunization, restimulated them in culture with PMA/ionomycin, and evaluated their capacity to produce cytokines by intracellular staining. Mice that had received polyclonal Tregs had a 50% reduction in the percentage of $CD4^+$ T cells that infiltrated the central nervous system. Surprisingly, on a per cell basis, the percentages of $CD4^+$ T cells that produced either IFNγ or IL-17 were the same in infiltrates from the Treg-treated and the nontreated mice. No differences in the production of other cytokines or in the expression of memory/activation markers could be detected in the Teff populations isolated from the treated and untreated mice. Thus, the reduced clinical disease strongly correlated *only* with the number of Teff cells that had migrated to the brain and infiltrated the CNS.

We next developed a model that allowed us to monitor the fate and differentiation of the Teff on a single cell basis (Davidson and Shevach, 2011). Congenically marked CFSE-labeled TCR Tg naïve T cells were adoptively transferred to normal recipients with or without congenically distinct preactivated polyclonal Tregs. The following day, the mice were immunized with peptide in CFA and the numbers and activation status of the transferred Tg T cells were analyzed at various time points. We hypothesized that polyclonal Tregs would limit the expansion of the Teff in the draining LN as this would explain why we observed fewer Teff cells in the CNS in the animals with EAE that had been immunized in the presence of an excess of Tregs. Surprisingly, cotransfer of Tregs did not inhibit Teff proliferation as measured by CFSE dilution and paradoxically resulted in a twofold increase in the absolute number of Teff cells present in the draining LN (Fig. 4.3). There was no difference in the expression of activation markers on the Teff immunized in the presence or absence of Tregs, and most importantly, there was no difference in the percentages of Teff cells producing IFNγ and IL-17. Similar results were observed when the mice were immunized under different conditions including antigen in incomplete adjuvant or with peptide-pulsed DC. The twofold enhancement of Teff in the draining LN was also observed in spleen when the mice were immunized with peptide antigen intravenously under tolerogenic conditions. When we selectively examined the fate of the Tregs under these conditions, there was significant proliferation of the Tregs, but they did not divide to the same extent as the Teff. Teff cell expansion greatly outpaced Tregs expansion, becoming 97% of the total transferred cells on day 5 after immunization. No reduction in the percentage of cells expressing Foxp3 or in the level of expression of Foxp3 was observed even when the mice were immunized with CFA.

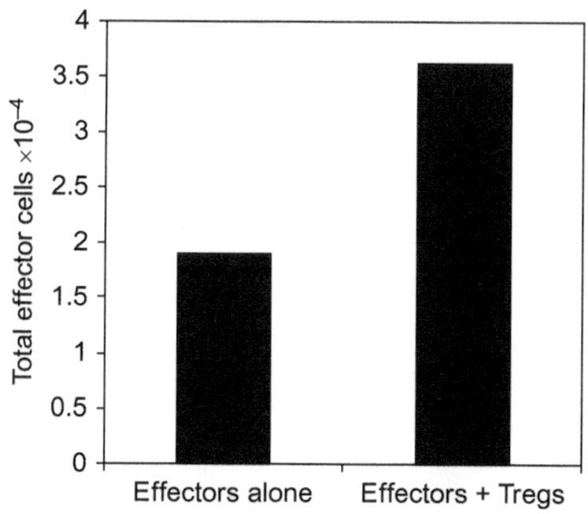

FIGURE 4.3 Polyclonal Tregs enhance the number of Teff in the draining LN. Naïve antigen-specific TCR Tg cells were cotransferred with or without polyclonal Tregs into normal recipients. Five days after immunization with the cognate antigen in IFA, the absolute number of Teff cells in the draining LN was determined.

The Treg-induced increase in the number of antigen-specific T cells in the draining LN appears to superficially contradict the results of our studies in the EAE model that demonstrated a decrease in the number of Teff in the target organ in the presence of an excess of Tregs. However, the total number of T cells in the LN is determined not only by *in situ* proliferation and expansion but also by the contribution of entry and exit from the LN. We therefore determined the relative proportion of transferred T cells in the LN and peripheral blood. In mice that received Teff cells only, the ratio of Teff in the LN to Teff cells in the blood was approximately 2:1. In contrast, in the mice that received Teff and polyclonal Tregs, the ratio was 10:1. In multiple experiments, we consistently found a greater number of cells in the LN and fewer cells in the blood of mice that had received Tregs at all time points studied. As this result favored a model in which Tregs altered the trafficking of Teff cells, we used a modified, delayed hypersensitivity model in which we could precisely control the location of a tissue dwelling antigen (Fig. 4.4). Antigen-specific TCR Tg T cells were cotransferred with or without Tregs, and the recipients were immunized 1 day later with the target peptide in CFA. One week later, the recipients were challenged in the ear with the target peptide in saline. The absolute number of Teff cells that homed to the ear after 24 h was determined following dissociation of the ear tissue. We observed a marked decrease number of donor cells in the ears of mice that had received the Tregs. This result is most consistent with the hypothesis that in the presence of polyclonal Tregs, fewer cells leave the ear to enter the circulation and few cells are therefore available to respond to antigen at a distant site.

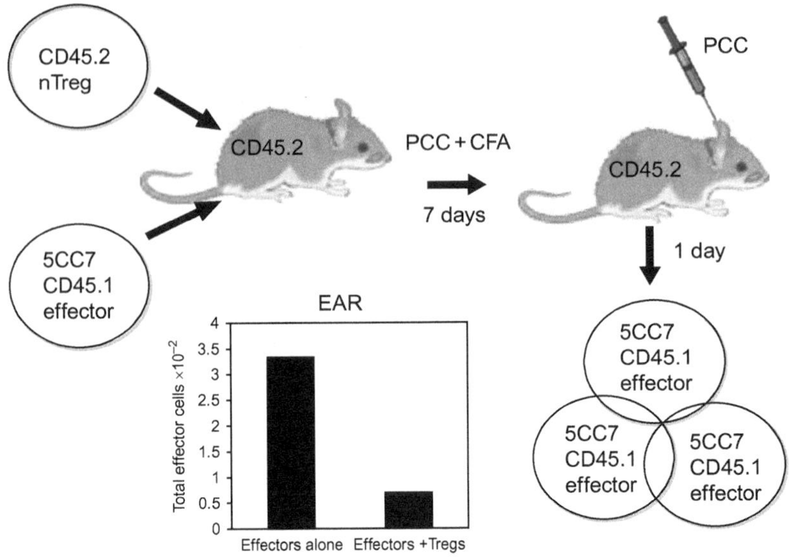

FIGURE 4.4 Polyclonal Tregs limit the number of Teff in the target organ following antigen challenge. Antigen-specific Teff was cotransferred with or without polyclonal Tregs into normal recipients. Seven days after immunization, the mice were challenged with the pigeon cytochrome *C* (PCC) in the ear. Twenty hours later, the absolute number of Teff in the ear was determined. (For color version of this figure, the reader is referred to the Web version of this chapter.)

To further probe for potential mechanisms by which Tregs would modulate Teff cell trafficking, we isolated Teff after immunization from mice that did or did not receive the Treg cotransfer. On microarray mRNA analysis, there were very few differences in gene expression of cell surface markers, cytokines, chemokines, or chemokine receptor genes. This result suggested that the presence of an excess number of Tregs did not induce global or dramatic changes in the immune response but likely targeted very specific pathways. One gene that was decreased fourfold in the Treg cotransfer group was the chemokine receptor, CXCR4, and this result was confirmed at the protein level. We also observed a decrease in expression of Syndecan-4 in the Treg cotransfer group, and this molecule has also been shown to be involved in cell motility (Longley *et al.*, 1999).

An additional molecule that plays a critical role in the exit of primed T cells from the LN is the sphingosine 1-phosphate receptor 1 (S1P1). S1P1 levels are downregulated on T cells following entry into the LN, but as T cells are primed and differentiated, they upregulate S1P1 (Grigorova *et al.*, 2009; Thangada *et al.*, 2010). The elevated levels of S1P1 allow the primed T cell to respond to high levels of S1P in the circulation and exit the LN in response to this concentration gradient. We observed a dramatic decrease in S1P1 levels on Teff cells that had been primed in the presence of an excess of Tregs compared to the levels seen on T cells primed in the absence of Tregs. Although these data are quite consistent with the data we have observed, it needs to be further validated. It also remains unclear whether the Tregs act directly to regulate S1P1 expression on the responder Teff cells or indirectly by acting on another cell type such as the antigen-presenting DC.

Taken together, all of the studies we have performed in this model strongly suggest that polyclonal Tregs primarily function by altering the trafficking potential of Teff cells rather than altering priming, expansion, or differentiation. This conclusion is also consistent with the results of the EAE studies that demonstrated that fewer cell could be found in the brain of mice that received an excess of Tregs, but those cells that were present were phenotypically indistinguishable from Teff cells that were primed in the absence of Tregs. By trapping Teff cells in the LN, Tregs would limit the number of potentially pathogenic autoreactive T cells that would be able to migrate into tissues and subsequently induce tissue damage. It should be noted that all of these studies have been performed in a prevention model, and we have not yet extended these studies to determine the effects of polyclonal Tregs on immune responses at defined intervals after priming. However, most organ-specific autoimmune diseases are characterized by epitope spreading where tissue damage induces the release of autoantigens that were not involved in the initial priming of Teff cells. Thus, one might propose that the delayed administration of Tregs would trap T cells in the LN after they are activated during the process of epitope spreading and prevent their ability to access the autoantigen in the target organ.

2.4. Effect of polyclonal Tregs on antibody production

As Tregs can inhibit immune responses to pathogens during natural infection, it has been suggested that depleting Tregs would improve vaccine efficacy. For example, depletion of Tregs before vaccination against herpes virus (Suvas *et al.*, 2003) or infection with malaria parasites (Hisaeda *et al.*, 2004) enhanced vaccine efficacy. Depletion of Tregs can also augment immune response to a tumor vaccine (Steitz *et al.*, 2001). In general, the efficacy of vaccines can be greatly improved by adjuvants that enhance and modify the magnitude and duration of the induced protective immunity. Cholera toxin (CT) is a potent mucosal and systemic adjuvant particularly when it is administered at mucosal sites (Vendetti *et al.*, 2009). CT mainly augments a Th2 response resulting in an increase in antigen-specific systemic and mucosal antibodies. One might predict that removal of Tregs would greatly augment the immune response observed when CT was used as an adjuvant. Indeed, serum antibody levels are higher in mice immunized intranasally with CT and tetanus toxoid (TT) than in mice immunized only with TT. However, when Tregs were partially depleted from mice using an anti-CD25 antibody at the time of intranasal immunization with the combination of CT and TT, depletion of Tregs did not enhance the anti-TT IgG response as predicted but resulted in a significant decrease in the titer of specific antibody (Fig. 4.5). Even after five immunizations, when anti-TT IgA can be readily measured in mucosal compartments, the titers of mucosal anti-TT-specific IgA were much lower in mice that had been depleted of Tregs (Vendetti *et al.*, 2010).

Similar results were observed when $CD4^+Foxp3^-$ cells were transferred alone or cotransferred with $CD4^+Foxp3^+$ T cells into *nu/nu* recipients. Adoptive transfer of $CD4^+Foxp3^+$ T cells enhanced the antibody response induced by CT delivered by the mucosal route. Similar results were observed when transferred small numbers of $Foxp3^+$ T cells to fully immunocompetent mice using the same protocol described above for suppression of EAE. In these studies, we observed a strong correlation between the augmentation of the antibody titer and the number of Tregs that had been transferred. A deleterious effect of Tregs depletion has also been reported to impair the early protective immune responses to viral infection by modulating the capacity of Teff to home to sites of the infection (Lund *et al.*, 2008). In contrast, most of the augmenting

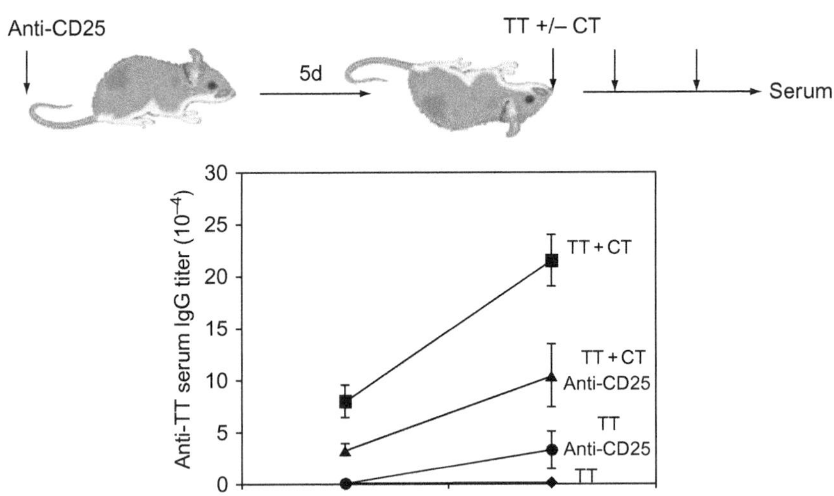

FIGURE 4.5 Depletion of polyclonal Tregs decreased the antibody response. Tregs were depleted from normal mice by injection of a depleting anti-CD25. Five days later, the mice were immunized intranasally with tetanus toxoid (TT) with or without CT. The mice were boosted two additional times by intranasal immunization. One week after the last boost, the titer of anti-TT antibody was determined in the serum. (For color version of this figure, the reader is referred to the Web version of this chapter.)

effects of Tregs depletion have been observed in studies of the cellular immune response against intracellular bacteria, parasites, and some viruses. Even though Tregs depletion was reported to accelerate the cellular immune response to BCG vaccination, protection against tuberculosis was not enhanced (Quinn *et al.*, 2008). Taken together, these data indicate that Tregs are required for the adjuvant activity of CT to become fully manifest and are consistent with a model in which Foxp3^{+} Tregs cannot only function as suppressor cells, but under certain conditions of immunization may exert T helper effects depending on the class of the immune response being assayed and the microenvironment in which it is generated.

A number of possible mechanisms may mediate the enhancement of immune responses in the presence of Tregs. It is possible that the Tregs influenced the quality of the immune response by modulating cytokine production (Th1 vs. Th2). However, we detected increase in both IgG_1 and IgG_{2a} antibody subclasses and observed a comparable cytokine profile in the presence or absence of Tregs indicating that Tregs modulated the magnitude, but not the quality of the immune response. Importantly, the enhanced antibody responses were closely correlated with an increased accumulation of activated antigen-specific T cells in the draining LN. This result is very similar to the cotransfer studies described above (Section 2.3) which resulted in a decrease emigration of activated T cells from the site of priming and a reduced number of Teff reaching the target organ. Thus, the enhanced accumulation of antigen-specific T cells is not unique to any specific immunization protocol and is a general function of polyclonal Tregs. How could this accumulation of Teff result in augmentation of humoral immune responses? It is possible that Treg-mediated enhancement of Teff cells results in enhanced germinal center formation and enhanced systemic and mucosal antibody responses. It had also been recently reported that adoptively transferred Tregs into T cell-deficient mice result in their differentiation into follicular T helper cells that induce germinal center formation (Tsuji *et al.*, 2009). Future studies will be needed to address these possible mechanisms in our model system of mucosal immunization with CT as the adjuvant.

If Tregs can function in a manner similar to helper T cells for the enhancement of antibody responses, one must raise the possibility of potential deleterious effects in the therapeutic use of Tregs for the treatment of autoimmune diseases that are accompanied by high levels of pathogenic autoantibody production such as systemic lupus erythematosis. It has been proposed that one of the earliest events during the course of a normal immune response is activation of polyclonal Tregs (O'Gorman *et al.*, 2009). Although this process of early activation may be a mechanism to minimize the risk of autoimmunity during the immune response to a foreign antigen, our findings suggest that this activation may permit the establishment of a microenvironment that would promote the enhancement of autoantibody responses.

3. BIOLOGIC EFFECTS OF ANTIGEN-SPECIFIC TREGS *IN VIVO*

3.1. Isolation of antigen-specific Foxp3$^+$ T cells

Although polyclonal Tregs can exert potent suppressor functions, one might predict that Tregs specific for autoantigens or foreign antigens would be more potent suppressor cells and potentially would utilize different methods for mediating their suppressive effects. In general, it has been very difficult, but not impossible, to isolate Foxp3$^+$ T cells from the T cell repertoire of normal mice. Ideally, antigen specificity must be defined at the single cell level by measuring the binding of peptide–MHC class II tetramers. The majority of studies that have analyzed the function of antigen-specific Foxp3$^+$ T cells have used TCR Tg mice on a conventional background in which a small proportion of the TCR Tg cells expresses Foxp3 and have been shown to exert antigen-specific Tregs function. As TCR Tg Foxp3$^+$ T cells cannot be readily isolated from TCR Tg mice on a RAG$^{-/-}$ background, it is widely assumed that Foxp3$^+$ T cells are selected into the Tregs lineage because they express an endogenous TCR α-chain with high affinity for a self-antigen that mediates Tregs selection. Although several studies have used Tregs of this type, interpretation of these studies is potentially complicated because the TCR Tg Foxp3$^+$ T cells can be activated both by antigens recognized by the Tg TCR (Tg TCR α-chain and Tg TCR β-chain) and potentially by the unknown antigens recognized by a second TCR (endogenous TCR α-chain and Tg TCR β-chain). One other difficulty with the use of these cell populations is that they are frequently present at very low frequencies in TCR Tg mice, are difficult to obtain with a high degree of purity by bead selection technologies, and can only be isolated in low numbers by FACS sorting. A second approach for obtaining antigen-specific Tregs is to isolate these cells from doubly Tg mice that express a TCR Tg and its cognate antigen as a Tg. In many model systems, this results in the generation of large number of Foxp3$^+$ TCR Tg$^+$ cells secondary to selection of these cells in the thymus. In certain TCR-antigen double Tg models, Foxp3$^+$ T cells will also arise when the mice are on a RAG$^{-/-}$ background (Klein *et al.*, 2003). Although it is likely that Foxp3$^+$ T cells from these doubly Tg mice are *bona fide* Tregs, it should be pointed out that expression of the Tg encoding the antigen Tg is under control of a widely expressed promoter and some of the Foxp3$^+$ T cells might therefore be generated at peripheral sites. These peripherally generated Foxp3$^+$ T cells may be distinct from thymic-generated Foxp3$^+$ T cells. Lastly, one always isolates these cells from an antigen-rich environment in which they may be partially activated. Studies with antigen-specific Tregs isolated from doubly Tg mice will be described in Section 3.6.

3.2. Generation of antigen-specific Foxp3$^+$ T cells *in vitro*

A number of studies have reported that peripheral CD4$^+$Foxp3$^-$ T cells can be induced *in vitro* to express Foxp3 when they are activated via their TCR in the presence of TGF-β (Chen *et al.*, 2003). We initially examined the cytokine and costimulatory molecule requirements required for the generation of what have been termed iTregs (Davidson *et al.*, 2007). In general, we start with a homogenous population of CD4$^+$Foxp3$^-$ T cells isolated from TCR Tg mice on a RAG$^{-/-}$ background that lack an endogenous population of Foxp3$^+$ T cells. Following stimulation with plate-bound anti-CD3 and anti-CD28 in the presence of TGFβ, a significant number of the cells express Foxp3 after 24 h prior to the initiation of cell proliferation and the percentage of Foxp3$^+$ T cells progressively increases and reaches a maximum 80–90% at 96 h of stimulation (Fig. 4.6). For maximal induction of Foxp3, exogenous IL-2 can eliminate the need for anti-CD28 costimulation; in the presence of costimulation, exogenous IL-2 was not required. However, in the absence of exogenous IL-2, neutralization of endogenous IL-2 abrogated the capacity of TGFβ to induce Foxp3 expression.

We also attempted to induce Foxp3 in the presence of DC and the specific peptide recognized by the TCR on the Foxp3$^-$ T cells. In some studies, the induction of Foxp3 expression was as strong as with plate-bound anti-CD3 and anti-CD28, but in other studies, fewer Foxp3$^+$ T cells

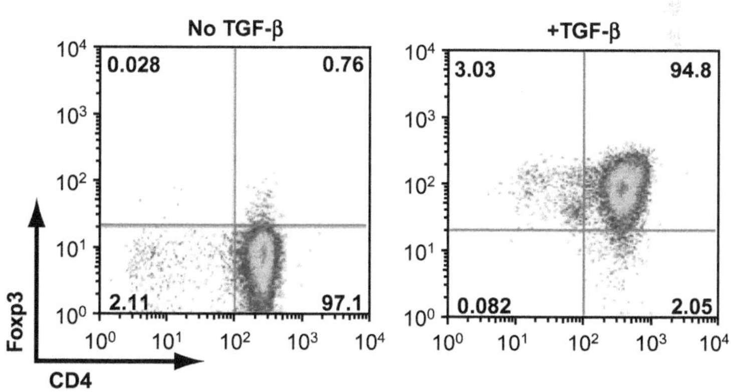

FIGURE 4.6 Efficient induction of iTregs *in vitro*. Naïve Foxp3$^-$ T cells were stimulated *in vitro* with plate-bound anti-CD3, anti-CD28, and IL-2 with or without TGFβ. The expanded T cells were stained for the intracellular expression of Foxp3 after 96 h of stimulation. (For color version of this figure, the reader is referred to the Web version of this chapter.)

were induced. It is likely that soluble factors produced by DC such as IL-6 that can inhibit the induction of Foxp3 played some role in the variability of the results. When high concentrations of anti-CD3 were used to coat the plate, anti-CD3 alone effectively induced Foxp3 expression, but at concentrations of 0.4 μg/ml or less, Foxp3 induction required anti-CD28-mediated costimulation. The requirement for IL-2 for induction of Foxp3 expression was unique, as no other cytokine utilizing the γc chain of the IL-2 receptor complex could induce Foxp3. We also failed to observe Foxp3 induction in the presence of IL-6, IL-10, IL-12, or IL-18. We initially evaluated the suppressive function of the TGFβ iTregs *in vitro* by coculturing them at various ratios with $CD4^+Foxp3^-$ T cells, anti-CD3, and antigen-presenting cells. Proliferation was markedly suppressed even at low ratios of Tregs to responder cells. The suppressive activity of the iTregs was not abrogated by the addition of neutralizing antibodies to either IL-10 or TGFβ. In contrast, T cells stimulated under identical conditions in the absence of TGFβ actually augmented proliferative responses. When restimulated with anti-CD3/CD28, iTregs produced very little IL-2, IL-4, IFNγ, and TNFα. Surprisingly, the production of IL-10 that can be produced by nTregs was also suppressed in the iTregs. Thus, functionally, iTregs have almost all the properties of nTregs.

3.3. Suppression of immune responses *in vivo* by iTregs

As one approach to analyze the suppressive capacity of iTregs *in vivo*, we assayed the capacity of polyclonal iTreg to prevent the fatal lymphoproliferative/multiorgan inflammation syndrome that develops in scurfy mice that have a deletion in the forkhead domain of Foxp3 and fail to develop nTregs. As adoptive transfer of nTregs into neonatal scurfy mice prevents the development of disease (Fontenot *et al.*, 2003), we evaluated whether polyclonal iTregs derived from a healthy adult mouse can protect scurfy mice from the development of disease. iTregs were generated by the protocol described above with plate-bound anti-CD3 and anti-CD28 in the presence of TGFβ. iTregs were injected into male scurfy mice on day 1 or 2 of life (Huter *et al.*, 2008a).

Untreated scurfy mice or scurfy mice that received activated $Foxp3^-$ T cells showed overt signs of disease by day 12–15 of life, and most of these mice became moribund between days 20 and 30 of life. iTreg-treated scurfy mice did not show any external signs of disease on day 21 after transfer and had no signs of immunopathology in the skin, liver, or lung at this time point. Treated mice also had a marked reduction in cell numbers in LNs and spleen (Fig. 4.7). Importantly, the percentage of iTregs detected in the LN of scurfy mice was higher in the iTreg-treated wild-type (WT) controls. There was clearly more expansion of the iTregs in the scurfy mouse than in the noninflammatory environment.

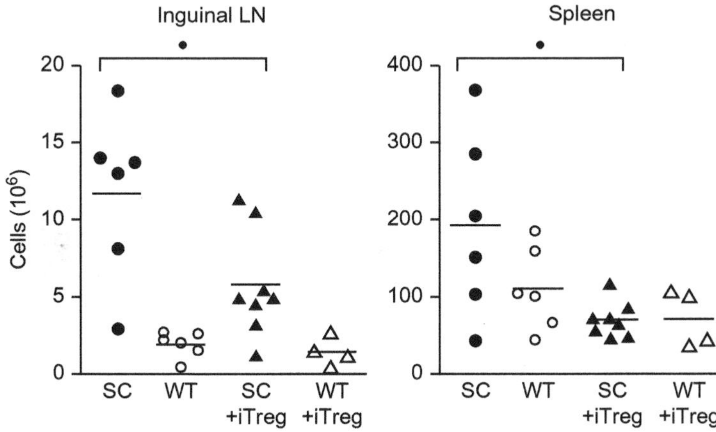

FIGURE 4.7 Polyclonal iTregs prevent the lymphoproliferation that develops in scurfy mice. WT or scurfy mice were left untreated or injected with polyclonal iTregs on day 1 of life. The absolute number of cells in spleen or inguinal LN was determined on day 21 of life.

The iTregs in the inguinal LN of the scurfy mice and in the skin maintained high level of expression of Foxp3 (~90%), while much lower percentages of Foxp3 expressing transferred iTregs could be detected in WT animals. Polyclonal iTregs could also prevent the induction of autoimmune disease observed after transfer of cells from 7-day-old scurfy mice to $RAG^{-/-}$ recipients. These results demonstrate that polyclonal iTregs can suppress all the manifestations of the severe autoimmune disease that develops in the scurfy mouse. While it appeared in these studies that the TCR repertoire of the iTregs generated from normal peripheral T cells from adult mice was sufficient to prevent disease in the scurfy mouse, these studies and the previous studies with nTregs (Fontenot *et al.*, 2003) were carried out for a short period of time and terminated on day 30 of life. Haribhai *et al.* (2011) have used a similar model of reconstitution of newborn scurfy mice and demonstrated that neither nTregs nor iTregs were sufficient to establish a long-term state of tolerance in scurfy mice. It appeared that iTregs were necessary in addition to nTregs to expand the TCR repertoires of the regulatory T cell populations.

3.4. Treatment of organ-specific autoimmune disease with antigen-specific iTregs

Polyclonal nTregs were quite effective at preventing the induction of AIG when cotransferred with either naïve TxA23 T cells or fully differentiated TxA23 cells (DiPaolo *et al.*, 2005; Stummvoll *et al.*, 2008). It is widely believed that nTregs are selected based on recognition of self-peptides, but it is not clear whether nTregs recognize organ-specific autoantigens, or whether they are activated by ubiquitous self-peptide–MHC class II complexes and suppress organ-specific disease by a bystander mechanism. Some studies using Tregs from a TCR Tg mouse have indicated that organ-specific nTregs are much more efficient at preventing autoimmunity than a polyclonal population of nTregs of unknown antigen specificities (Tang *et al.*, 2004). To explore the potential of antigen-specific iTregs in the therapy of AIG (DiPaolo *et al.*, 2007), we stimulated $CD4^+Foxp3^-$ thymocytes from TxA23 mice and activated them in the induction protocol described above (Section 3.2) in the presence of anti-CD3/CD28, IL-2, and TGFβ. Under these culture conditions, >90% of the TxA23 $Foxp3^-$ cells were induced to express Foxp3 and almost all of the induced cells expressed the Tg TCR. We then evaluated the ability of the antigen-specific iTregs to inhibit the transfer of AIG into *nu/nu* recipients. The iTregs were potent suppressors of the induction of pathology in the gastric mucosa and also markedly inhibited the production of antiparietal cell antibodies.

To more fully understand the mechanisms used by iTregs to mediate suppression of disease, we harvested the gastric LN at various time points after transfer and measured total cellularity as well as the specific number of Teff. A week following transfer of Teff alone we noted a 10-fold increase in the total cellularity of the gastric LN and a marked expansion of Teff (150 cells on day 2 after transfer to 50,000 on day 7). We observed a 20- to 40-fold decrease in the expansion of the Teff in the presence of the iTreg (Fig. 4.8) and very little increase in the total cellularity of the LN. The iTregs did not inhibit the migration of the Teff into the node as similar numbers of Teff were observed on day 2 after transfer. iTregs also did not kill DC by a granzyme B-dependent mechanism, as iTregs generated in the presence of TGFβ do not express granzyme B.

The most likely explanation for the marked inhibition of Teff expansion at the site of priming is that the antigen-specific iTregs inhibited the priming and proliferation of these cells. When we examined the initial proliferation of CFSE-labeled Teff, there was a 50-fold increase in the fraction of Teff that had not divided in animals that had been treated with the iTregs. The low level of cell division seen in the gastric LN of iTreg-treated mice was similar to that seen in peripheral LN that does not contain H/K ATPase presenting DC. A significant decrease in the expression of CD69 was also observed on the Teff from iTreg-treated animals. Taken together, these data are most consistent with a model in which iTregs are inhibiting the initial priming and subsequent expansion of autoantigen-specific Teff very early in the course of the immune response.

One unique aspect of this model of AIG is that a small number of DCs in the gastric LN express the H/K ATPase (Scheinecker *et al.*, 2002).

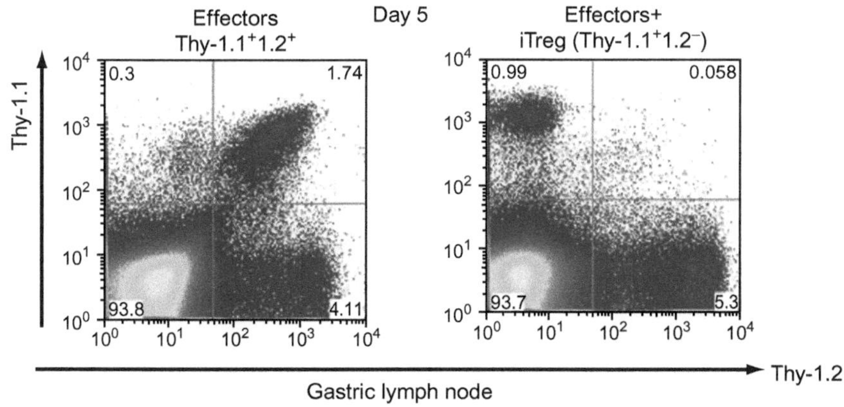

FIGURE 4.8 iTregs inhibit the expansion of Teff in response to endogenous autoantigen. Naïve T cells recognizing the H/K ATPase were injected into *nu/nu* mice with or without iTregs that had been generated from the same naïve T cell pool. On day 5 after transfer, the presence of Teff (Thy-1.1^{+}1.2^{+}) and iTregs (Thy-1.1^{+}1.2^{-}) was analyzed in the gastric LN. (For color version of this figure, the reader is referred to the Web version of this chapter.)

To determine whether the iTregs suppress antigen presentation by DC in the gastric LN, we isolated DC from the gastric LN of mice that 48 h previously had received Teff alone, Teff cells and iTregs, or DC from noninjected mice. As a sensitive test for the presenting function of these DC populations, we measured their capacity to stimulate a cell line derived from TxA23 mice in the absence of added peptide or H/K ATPase. DC from unmanipulated *nu/nu* mice stimulated proliferation and cytokine production by the T cell line and DC isolated from mice that received Teff alone stimulated even more proliferation and cytokine production. This enhanced stimulatory capacity is likely due to the mutual activation (e.g., CD40/CD40L) between Teff and DC during their interaction *in vivo*. In contrast, DCs isolated from the gastric LN of mice that received Teff and iTregs stimulated less proliferation and cytokine production than either the group that received DC from unmanipulated mice or DC from mice that received Teff alone (Fig. 4.9). The same pattern was observed when DCs were fixed immediately after isolation indicating that the reduction in stimulation was not due to secreted inhibitory factors.

It therefore appears that the major mechanism by which antigen-specific iTregs exert their suppressive function *in vivo* is by reducing the ability of DC to prime autoreactive T cells, thereby preventing the induction of autoimmune disease. The iTregs exert their negative effects by decreasing the stimulatory capacity of DC rather than by competing with Teff for antigen as has been proposed in some models (Tadoko *et al.*, 2006; Tang *et al.*, 2006) or by acting directly on Teff cells to block their

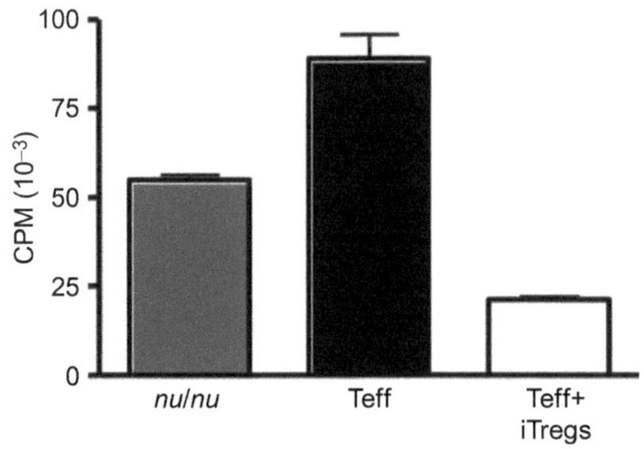

FIGURE 4.9 DCs isolated from animals that had been treated with iTregs are disabled in their capacity to present autoantigen. *Nu/nu* mice were left untreated, injected with antigen-specific Teff alone, or injected with a mixture of antigen-specific Teff and antigen-specific iTregs. DCs were isolated from the gastric LN 48 h later and mixed with a T cell line specific for the H/K ATPase. Exogenous antigen was not added. Proliferation of the cell line was measured 72 h later.

interaction with DC. A direct approach to the analysis of the potential effects of iTregs on DCs would have been to analyze both membrane molecule expression and gene expression on DCs isolated from animals that had been treated with iTregs alone. However, only a small subset of H/K ATPase-bearing DC is present in the gastric LN, and it is impossible to isolate and purify significant numbers of these cells for analysis. We have examined the effects of antigen-specific iTregs on DC *in vitro* (see Section 3.7).

Antigen-specific iTregs have been used in several other disease models. Weber *et al.* (2006) induced diabetes by transferring islet antigen-specific TCR Tg Th1 cells into NOD–Scid mice. Cotransfer of T cells of the same specificity that had been induced to become iTregs by TCR stimulation in the presence of TGFβ protected mice from disease for prolonged periods of time. It was postulated that Tregs inhibited Th1 cells via Fas-L-dependent cytotoxicity, but no direct evidence that this pathway was operative *in vivo* was obtained. Selvaraj and Geiger (2008) demonstrated that iTregs specific for MOG_{35-55} were as effective as nTregs in suppressing disease development following immunization of mice with MOG_{35-55} in CFA. The iTregs were also able to act therapeutically when given up to 10 days after immunization. The absolute number of $CD4^+$ T cells infiltrating the CNS after iTregs treatment was lower than after no treatment. iTregs from IL-$10^{-/-}$ mice could not protect. iTregs treatment suppressed the proliferative responses of recipient cells to MOG, but this nonresponsiveness *in vitro* could be overcome by addition of IL-2 suggesting that the iTreg-induced anergy in the MOG-specific

T cells in the host. Zhang *et al.* (2010) converted TCR Tg T cells specific for proteolipid protein ($PLP_{130-151}$) into iTregs. These iTregs inhibited EAE when transferred into $PLP_{139-151}$-primed recipients but did not inhibit disease in animals primed with $PLP_{178-191}$ or a mixture of the two peptides suggesting that the effects of the iTregs are exquisitively antigen-specific and that they do not exert bystander suppression. Similar results were observed in a cell transfer model. iTregs specific for $PLP_{139-151}$ failed to inhibit the proliferation of cotransferred MOG_{35-55}-specific T cells in mice immunized with a mixture of MOG_{35-55} and $PLP_{139-151}$ in CFA. The results of these studies if they can be generalized to other models have important implications for the therapeutic use of antigen-specific Tregs (Section 4).

3.5. Antigen-specific iTregs inhibit Th17-induced AIG

We have not yet examined the therapeutic use of iTregs to reverse or stop the progression of ongoing autoimmune disease. However, the capacity of iTregs to inhibit the presentation of autoantigens by DC raises the possibility that the process of epitope spreading might be susceptible to iTreg-mediated downregulation. As described above (Section 2.2), cotransfer of polyclonal $Foxp3^+$ Tregs markedly suppressed disease induced by fully differentiated Th1 and Th2 cells lines, while the disease induced by Th17 lines was resistant to suppression even at very high Treg to effector ratio (20:1). The resistance of Th17-induced AIG to Treg-mediated suppression raised the possibility that Th17 cells are intrinsically not responsive to suppression by Tregs or that the inflammatory milieu generated by Th17 cells induced the production of cytokines that rendered the Teff cells resistant to suppression or neutralized the suppressive effects of Tregs. As we have shown that autoantigen-specific iTregs are more potent suppressors of AIG, we evaluated the capacity of antigen-specific iTregs to suppress Th17-induced AIG (Huter *et al.*, 2008b).

Transfer of fully differentiated autoantigen-specific Th17 cells induced destructive AIG. While both polyclonal nTregs and iTregs had no effect on the development of disease, transfer of antigen-specific iTregs completely prevented destructive tissue pathology and significantly reduced the enhanced cellularity in the gastric mucosa seen in the mice that received Th17 cells alone. Further, iTregs reduced both the percentage and the absolute number of Th17 cells in the gastric LN. The suppressor function of the iTregs compared to the polyclonal nTregs was not secondary to the activation of the former *in vitro* prior to transfer, as preactivated nTregs also failed to prevent AIG. One possible explanation for the enhanced potency of the iTregs is that they selectively expand at the site of autoantigen presentation. Indeed, we found much greater proliferation of the antigen-specific iTregs compared to the polyclonal iTregs or nTregs in the gastric LN, while all three populations proliferated at the same rate in the inguinal LN. We also tested whether antigen-specific iTregs were capable of suppressing disease in a therapy model. When we coinjected iTregs 6 days after transfer of the Th17 Teff, marked inhibition of destructive gastritis as well as expansion of the Th17 Teff was observed when assayed 6 weeks after transfer (Fig. 4.10).

Although some studies have suggested that Th17-induced inflammatory responses would be resistant to Treg-mediated suppression secondary to the production of high levels of IL-6 and TNFα that could potentially neutralize or block Tregs function, in our hands, antigen-specific iTregs could easily suppress the induction of AIG and were also of considerable therapeutic benefit when administered after initiation of disease. Similar results were obtained when ovalbumin (OVA)-specific T cells were injected into $RAG^{-/-}$ mice that expressed a Tg-encoding OVA as a secreted protein (Lohr *et al.*, 2006). The recipient animals developed weight loss and a severe skin disease that appeared to be mediated by IL-17 producing cells. When they cotransferred OVA-specific nTregs from a TCR Tg mouse which expressed membrane OVA as a Tg, the nTregs completely prevented the weight loss and skin inflammation. The most striking effect of the nTregs was a profound inhibition of T cell accumulation in the skin that was detectable as early as 5 days after transfer. While the total number of IL-17 producers was decreased and the production of IFNγ was completely inhibited, an enhanced frequency of IL-17 producers was seen among the remaining Teff. No increase in IL-17 producer frequencies was seen with lower numbers of cotransferred nTregs.

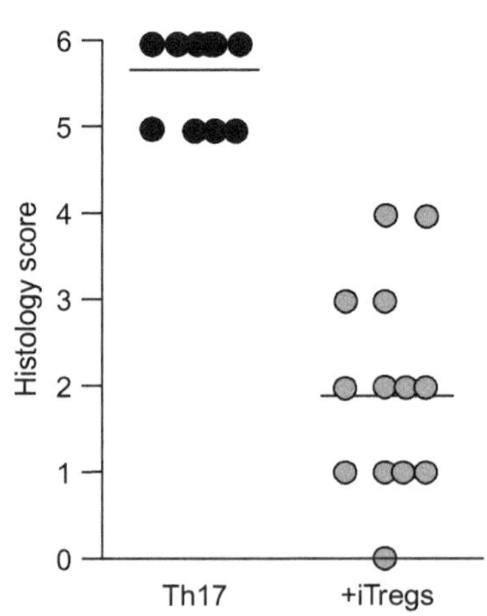

FIGURE 4.10 Fully differentiated Th17 Teff cells specific for the H/K ATPase were injected into *nu/nu* recipients. Six days later, antigen-specific iTregs were injected into some of the recipients. The disease score was determined 6 weeks later.

This result is consistent with the possibility that under certain conditions, Tregs can actually induce or enhance IL-17 production perhaps secondary to the secretion of TGFβ or expression of latent TGFβ on their cell surface (Xu *et al.*, 2007). In contrast to the AIG model, transfer of nTregs in this model 6 days after transfer of the Teff did not result in prevention of disease.

3.6. Mechanistic analysis of the biologic effects of antigen-specific Tregs *in vivo*

Although antigen-specific iTregs have proven prophylactic and therapeutic value in a number of disease models (see Section 3.3), it is important to analyze their function and potential mechanisms of action in normal mice in the absence of lymphopenia or ongoing inflammation. Even naïve T cells can exert regulatory function in lymphopenic recipients (Barthlott *et al.*, 2003). Klein *et al.* (2003) were the first to initiate a detailed *in vivo* analysis of the properties of antigen-specific Tregs. They obtained Tregs from TCR/antigen doubly Tg mice and transferred them alone or in combination with naïve T cells of the same antigen specificity to normal immunocompetent mice. They observed that both Tregs and naïve antigen-specific T cells expanded in similar fashion 8 days after immunization. Thus, in contrast to the failure of Tregs to proliferate *in vitro* following TCR stimulation (Thornton and Shevach, 1998), Tregs readily expand *in vivo* when stimulated via their TCR. In fact, the failure of Tregs to expand *in vitro* is probably an artifact of the *in vitro* culture conditions, as any cell population that is critically important for immune homeostasis must have the capacity to expand under the appropriate circumstances *in vivo*. It should also be noted that the factors (cytokines) driving the expansion of the Tregs *in vivo* under these conditions are still completely uncharacterized.

In cotransfer studies, Tregs did not inhibit the proliferation of naïve T cells *in vivo* in the draining LN when analyzed 90 h after immunization (Klein *et al.*, 2003). After this time point, in the absence of the Tregs, the Teff continued to expand, while in the presence of the Tregs, their numbers remained stable. The percent of cytokine-producing cells 8 days after priming was identical in the groups that did or did not receive Tregs. The conclusions drawn from this study were that Tregs had a late effect *in vivo* on the expansion of Teff but had no effect on their differentiation into cytokine-producing Teff cells. One difficulty with the interpretation of these studies is that at late time points, Teff may begin to emigrate from the site of immunization in the LN. It thus remains possible that the cytokine-producing cells detected in the LN were a selected population of Teff cells that had somehow escaped suppression. Billiard *et al.* (2006) used an identical cotransfer model but observed marked suppression of the expansion of the antigen-specific Teff cells as well as marked inhibition of cytokine production (IL-2 and IFNγ) 96 h after cell transfer. A correlation between Treg-mediated suppression and Treg division was observed in these studies, and this is consistent with the view that Treg activation is a key requirement for effective immunosuppression *in vivo*. The differences between this study and the study of Klein *et al.* (2003) are not clear but may reflect the type of immunization and the ratio of Treg to naïve T cells in the transfer inoculum.

A number of other studies have suggested that Tregs inhibit the activation of Teff at early time points after priming. Sumoza-Toledo *et al.* (2006) studied whether Tregs could impair the development of the immune synapse at the interface between a T cell and a DC that occurs after the engagement of relatively few TCR molecules. *In vitro*, antigen-specific Tregs obtained from a TCR/antigen doubly Tg mouse inhibited the induction of CD69 expression on Teff at 20 h and greatly inhibited IL-2 gene expression. Notably, when immune synapse formation was analyzed 30 min after the interaction of T cells with the DC, the Tregs modified the type of immune synapse that is established between the naïve T cells and the antigen-loaded DC by inhibiting the recruitment of protein kinase Cθ (PKCθ) to the contact between the T cells and the peptide-loaded DC. PKCθ is critical to several different pathways of T cell activation and ultimately controls the production of IL-2 and the upregulation of CD69. No evidence for bystander suppression was seen in these studies, as Tregs specific for one antigen did not impair the recruitment of PKCθ to the synapse formed by the interaction of a different antigen-specific T cells specific for its cognate peptide presented by the same DC.

The rapid modification of immune synapse formation *in vitro* by antigen-specific Tregs raised the issue of whether a similar rapid process of suppression occurred *in vivo* upon interaction of Tregs, Teff with their target antigen on the surface of the DC. Hanig and Lutz (2008) cotransferred peptide-pulsed bone marrow DC, that had been matured with lipopolysaccharide (LPS), with antigen-specific Teff with or without antigen-specific Tregs to normal recipients. Clustering of DC with both Teff and Tregs was observed in the spleen at 24 h after cell transfer. Similarly, when all populations were injected simultaneously, no effects of Tregs on total spleen cell size and Teff numbers were observed for the first 48 h. Maximum expansion of spleen size and numbers of Teff were detected at 72 h after injection, and modest (25%) suppression was seen at this time point when Tregs were cotransferred. Lower numbers of Teff in the memory phase of the response were seen when Tregs was coinjected. In contrast, when Tregs were transferred 10–24 h before the Teff, cluster formation was completely inhibited and marked suppression of Teff cell expansion was seen. Thus, in this model using antigen-pulsed bone marrow DC that had been prematured *in vitro* by treatment with LPS, Treg-mediated suppression required that the Tregs be preactivated *in vivo*. When Tregs and DCs were injected 24 h before the Teff cells, marked down regulation of the expression of CD80/CD86, PDL-1, and PD-L2 was observed 48 h later on DC matured with TNF or LPS. No changes of ICOS-L, CD70, CD40, or MHC class II expression were observed. They conclude that Treg activity *in vivo* that leads to a block in Teff proliferation may be mediated by downregulation of specific costimulatory molecules from the DC surface that, in turn, indirectly controls the Teff expansion. However, Tregs could not inhibit Teff cell expansion and modulation of DC cell surface antigen expression, when the DCs were matured with the combination of LPS and anti-CD40. Thus, certain modes of DC activation may result in a stimulatory DC that is completely resistant to Treg-mediated suppression.

To analyze the function of antigen-specific iTregs *in vivo* (Davidson and Shevach, 2011), we used a cell transfer system very similar to the one developed for the analysis of polyclonal Treg function *in vivo* (Section 2.3). We generated iTregs from $CD4^{+}Foxp3^{-}$ T cells derived from TCR Tg mice on a $RAG^{-/-}$ background by stimulation with plate-bound anti-CD3 and anti-CD28 in the presence of TGFβ and IL-2. We first cotransferred the iTregs together with congenically marked, CFSE-labeled naïve T cells from the same TCR Tg donor to a normal recipient and immunized the recipient with the target peptide in CFA or IFA. Four to five days later, we determined the state of activation, differentiation, and expansion of the Teff in the draining LN. In marked contrast to the results obtained with polyclonal nTregs or iTregs, the expansion of the naïve Teff was inhibited by $>90\%$ either when measured by the total percentage of donor Teff in the draining LN, by CFSE dilution or when the absolute number of Teff cells was counted (Fig. 4.11). The small number of Teff cell recovered from the draining node failed to upregulate CD44 expression indicating that their activation was also markedly impaired. Further, when the recovered Teff cells were restimulated with PMA/ionomycin to measure cytokine production, expanded Teff from mice that had not received iTregs produced both IFNγ and IL-17, while the few recovered T cells from mice iTreg-treated mice failed to produce either of these cytokines. Thus, iTregs profoundly inhibited all aspects of the T cell activation cascade *in vivo*.

We could also perform these studies using congenically marked iTreg. In contrast, to the complete lack of cell expansion of the Teff in iTreg-treated animals, iTregs from these mice readily proliferated as measured by CFSE dilution as first shown in the studies of Klein *et al.* (2003). The profound inhibition of Teff activation suggested that the iTregs were inhibiting antigen presentation by acting on DC. One possibility is that the iTregs killed DC by utilizing the perforin–granzyme B pathway, as

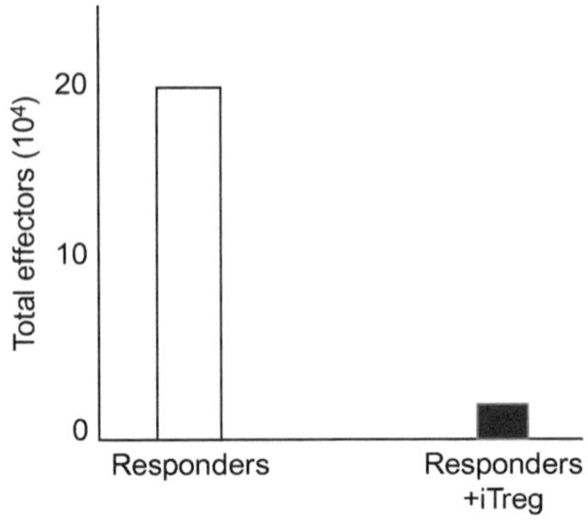

FIGURE 4.11 Naïve antigen-specific T cells were transferred into normal mice with or without iTregs specific for the same antigen. The absolute number of Teff in the draining LN was determined 4 days later.

nTregs have been shown to inhibit antigen presentation by activated B cells by this pathway *in vitro* (Zhao *et al.*, 2006). However, iTregs did not express intracellular granzyme B. iTregs have been shown to produce IL-10 after restimulation *in vitro* in some studies, and IL-10 can function as a potent inhibitor of costimulatory molecule expression on DC (Ding *et al.*, 1993). Expanded iTregs failed to produce IL-10 when they were restimulated *ex vivo*. To more accurately address the possibility that DCs were the targets for iTreg-mediated suppression in this model, we used the three-cell transfer model described by Hanig and Lutz (2008) but used antigen-pulsed splenic DCs that were not subjected to further maturation *in vitro*. We then performed the triple transfer of antigen-specific iTregs, CFSE-labeled, naïve Teff, and antigen-pulsed spleen cells and gated on the transferred Teff in the spleen 4–5 days later. Marked expansion of Teff cells was observed in recipients that received antigen-pulsed splenic DC without Tregs, and marked inhibition of Teff cell expansion was seen in recipients of the iTregs.

The major advantage of the three-cell transfer system is that it allows for careful control of all aspects of T cell activation *in vivo* as one can vary the ratio of Teff:Treg, and vary the strength of the TCR signal by varying the numbers of transferred DC, or the amount of antigen used to pulse the DC. This protocol also allowed us to recover the antigen-presenting DC at different time points after cell transfer. Since Teff cell proliferation was markedly inhibited, we reasoned that iTreg-mediated suppression of DC function should be observed at an early time point after cell transfer. We reisolated the DCs from the spleen 18 h after cell transfer and tested their capacity to activate fresh naïve TCR Tg T cells specific for the target peptide. DCs from animals that did not receive either Teff or iTregs were quite efficient at activating the proliferation of naïve T responder cells, and similar results were observed with DCs recovered from recipients that only received Teff. However, DCs recovered from animals that received both Teff and iTregs were markedly deficient in their capacity to activate naïve responder cells. This result strongly suggests that the iTregs modified DC function resulting in disabling the capacity of the DCs to present antigen. These results are identical to our previous studies with antigen-specific iTreg-mediated suppression of AIG (Section 3.3). This conclusion is still preliminary as we have not yet performed studies with DCs isolated from animals that only received iTregs in the absence of Teff. It should be noted that the number of DCs recovered animals that received iTregs and Teff was always less than the number of DCs recovered from animals that received only DCs or DCs and Teff. One possibility is that the iTregs inhibit chemokine production by the DCs resulting in less recruitment of the transferred DCs to the site of immunization or that the iTregs either directly or indirectly mediate DC cell death (Morlacchi *et al.*, 2011). In contrast to the results of Hanig and Lutz (2008), we did not observe any reduction in the expression of the costimulatory molecules CD80/CD86, CD40, or MHC class II expression when we analyzed DCs from mice that had been cotransferred with Tregs and Teff compared to DCs from mice that received Teff alone.

From the data we have obtained thus far in this model, it is still difficult to rule out direct effects of the iTregs on the responder T cells *in vivo*. It is likely that both cell types could interact with the same antigen-presenting DC and complex interactions between all three cells that would occur in this milieu. We have recently demonstrated that Teff production of IL-2 is required for the maintenance of Foxp3 expression in the transferred iTregs (Chen *et al.*, 2011). As described above, when antigen-specific Teff and iTregs were cotransferred into normal mice and the recipients immunized the recipients with antigen in IFA, marked suppression of Teff proliferation was observed and the iTregs maintained high levels of expression of Foxp3. However, when we treated the recipient animals with anti-IL-2, the iTregs lost Foxp3 expression and a significant reversal of Treg-mediated suppression was observed. Thus, IL-2 presumably produced by the Teff is required for the maintenance of Foxp3 expression and initiation of suppressor function. This study strongly supports the concept that at least soluble factors produced by Teff can have profound effects on Tregs in the milieu of the antigen-presenting DC. We have not yet evaluated if IL-2 production by Teff would be required to maintain Foxp3 expression and suppressor function of antigen-specific nTregs.

While most studies of antigen-specific Treg function *in vivo* have focused on inhibition of T cells activation and Teff cell expansion, Morlacchi *et al.* (2011) have used an adoptive transfer model in which naïve antigen-specific T cells are transferred into normal BALB/c mice in the presence of absence of antigen-specific Tregs of the same antigen specificity. At low TCR signal strength, they observed that blockade of both the proliferation and the production of the chemokines, CCL3 and CCL4, by DC in the LN was significantly reduced, while proliferation of the transferred naïve T cells was not blocked. At higher signal strength, suppression of T cell proliferation was abrogated, but the magnitude of the immune response was still reduced because chemokine production by DCs still remains suppressed. As noted above, a controversial point regarding Treg suppression is whether it is completely antigen-specific or whether Tregs can mediate bystander suppression. One interpretation of their results is that the bystander effect of Tregs on immune responses directed toward unrelated antigens observed *in vivo* may be secondary to a decrease in recruitment of such T cells and in the magnitude of the response rather than a direct effect on their proliferation or priming.

3.7. Biologic effects of Tregs on DC functions

One of the major questions in Treg biology that remains unanswered is the target cell for Treg-mediated suppression *in vivo*. The results of many of the studies described above, particularly those that have used antigen-specific Tregs, are most compatible with a model in which Tregs downmodulate DC costimulatory molecule expression or disable the antigen-presenting function of the DCs. It also remains unclear whether any component of Treg-mediated suppression *in vivo* is directed against the Teff, or whether all the effects of Treg-mediated suppression on the Teff cell are indirect. Lastly, a related issue is whether Tregs can only exert their suppressive functions on immature DCs by preventing their activation and expression of costimulatory molecules on their cell surface, or whether Tregs can also suppress fully mature DC. As can be see in the studies of Hanig and Lutz (2008) described above, the nature of the stimuli utilized in the process of DC maturation may also be critical.

It is appropriate to begin this discussion with a review of the large number of studies that have analyzed the effects of Treg-mediated suppression *in vitro*. Most of these studies have reached similar conclusions, but one must question whether the results have any bearing on Treg-mediated suppression *in vivo*. Cederbom *et al.* (2000) were the first to claim that the primary effects of Tregs on DC were to prevent the upregulation of costimulatory molecules during the process of DC maturation. Misra *et al.* (2004) obtained similar results in studies of human Tregs and human monocyte-derived DCs. DCs recovered from Treg containing cultures were poor stimulators of an allogeneic mixed lymphocyte response. In their studies, the Tregs continued to mediate their suppressive functions even when the DCs were prestimulated via CD40L/CD40 interaction. Treg-mediated suppression in these studies required that the Tregs be activated for 24 h in the presence of anti-CD3 and IL-2. One problem with the interpretation of these studies is that they compared the effects of activated Teff on DC with activated Tregs. As activated Teff expresses one of the major inducers of DC activation, the CD40L, while activated Tregs express much lower levels or none at all, it remains possible that the Teff induced maturation, but that the Tregs did not suppress maturation, but were just poor inducers.

Velldhoen *et al.* (2006) used a different approach to the analysis of Treg-mediated suppressive effects on BMDC *in vitro*. They compared the effects of naïve T cells and Tregs on the production of cytokines by DCs. In the presence of a DC activator, LPS, naïve T cells induced IL-6 production but suppressed IL-10 production, while Tregs had the opposite effect. In cultures containing mixtures of naïve T cells and Tregs, the effects of the Tregs were dominant. Tregs also reduced the expression of DC maturation markers (CD80, CD86, CD40, and MHC class II) under noninflammatory conditions but had no effect on maturation marker expression under inflammatory conditions. Similar results were seen in the studies of Serra *et al.* (2003) who demonstrated that DCs activated with LPS, CpG DNA, or anti-CD40 were refractory to the regulatory activity of Tregs. One possible explanation for the ability of Tregs to exert suppression even under proinflammatory conditions was that Tregs continued to enhance the production of IL-10 that may keep nonstimulated DCs immature even in an inflammatory environment (Velldhoen *et al.*, 2006).

DiPaolo *et al.* (2007) examined the effects of antigen-specific iTregs on splenic DC in the presence of continuous antigen *in vitro*. In contrast to the studies described above, iTregs induced downregulation of CD80/CD86 expression on the DC but did not modulate the expression of CD40 or MHC class II (Fig. 4.12). Surprisingly, the magnitude of reduction in CD80/CD86 expression was significantly greater when both Teff and iTregs were present in the cultures. This result raised the possibility that Treg-mediated suppression of DC function is optimal only after a signal from the Teff cells, that is, the Teff cells license the DC for susceptibility to suppression. It is also possible that the Teff cells directly act on the Tregs to enhance their suppressive capacity, or that the Tregs directly suppress some aspect of Teff function. Onishi *et al.* (2008) studied the effects of antigen-activated nTregs in cell cultures with DC and noted that the nTregs outcompeted naïve T cells in forming aggregates around DCs. As was seen in the studies of DiPaolo *et al.* (2007), the iTregs specifically downregulated the expression of CD80/CD86 and did not modulate the expression of other DC maturation markers. The downregulation of CD80 and CD86 was observed even in the presence of strong DC-activating stimuli including TNFα, type I IFN, IFNγ, LPS, and zymosan. Blockade of CTLA-4 with a Fab anti-CTLA-4 abrogated Treg-mediated suppression.

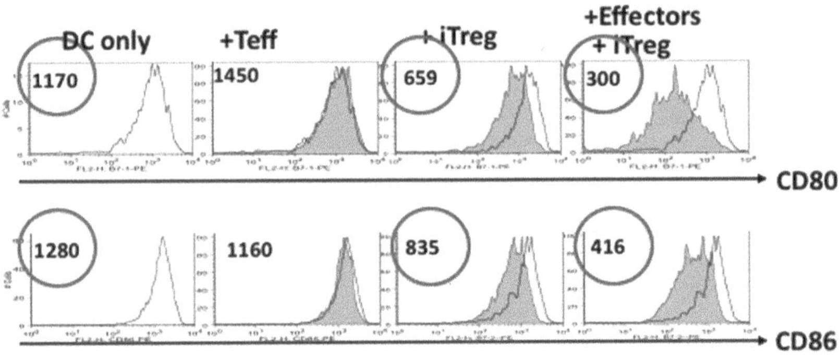

FIGURE 4.12 Antigen-specific iTregs modulate the expression of costimulatory molecules on DCs *in vitro*. Splenic DCs pulsed with peptide, washed, and cultured for 18 h alone, or in the presence of antigen-specific Teff, iTregs, or a mixture of Teff and iTregs. The expression of CD80 and CD86 was quantitated by FACS analysis. The numbers in the circles indicate the mean fluorescence intensity. (For color version of this figure, the reader is referred to the Web version of this chapter.)

Tregs from both CTLA-4$^{-/-}$ and LFA-1$^{-/-}$ mice did not induce downregulation of CD80/CD86 expression. The conclusions drawn from this study were that Tregs mediate suppression of DC function in two steps. In step 1, the higher levels of LFA-1 on Tregs allowed them to preferentially form aggregates with DC, and thereby hinder interactions of the Teff with the DC. In step 2, CTLA-4 mediates active downmodulation of CD80/CD86.

The results of these *in vitro* studies were most consistent with a model in which Treg-mediated inhibition of upregulation of CD80/CD86 on DC or the downmodulation of CD80/CD86 expression were absolutely essential for the suppressive effects of Tregs and that the expression of CTLA-4 on the membrane of Tregs mediated these functions. However, CTLA-4 has primarily been characterized as an inhibitor of T cell activation that functioned in a cell intrinsic fashion, and it remained unclear how it could function in a cell extrinsic manner by acting on DCs. Qureshi *et al.* (2011) have recently characterized a unique pathway for the cell extrinsic function of CLTA-4 on Tregs termed trans-endocytosis, in which CTLA-4 captures its ligands (CD80/CD86) from opposing cells. After removal, CD80/CD86 are degraded inside CTLA-4 expressing cells resulting in impaired costimulation via CD28. Trans-endocytosis differs from the more general process of intercellular exchange of molecules termed trogocytosis in which transferred proteins are detected at the cell surface. However, CTLA-4-mediated trans-endocytosis is not a unique property of Tregs, as both Foxp3^{+} and activated Foxp3^{-} T cells are capable of CD86 acquisition. While these investigators have clearly demonstrated that trans-endocytosis occurs *in vivo* and occurs in precisely the same settings in which CTLA-4-dependent regulation is operative, it remains unresolved whether this process is the major pathway utilized by Tregs to mediate suppression in physiological settings.

Only a small number of studies have attempted to analyze the role of Tregs on DC function *in vivo*. Oldenhove *et al.* (2003) injected mature antigen-pulsed splenic DCs into unmanipulated mice or mice that had been depleted by treatment with anti-CD25. The splenic DCs were matured by culture in the absence of exogenous stimuli. Immature DC failed to prime. Significant enhancement of IFNγ production was seen in Treg-depleted mice. IFNγ production by $CD8^+$ T cells was also strongly enhanced by Treg depletion. In contrast, Th2 priming was inhibited in the absence of Tregs. Enhancement of IFNγ production in Treg-depleted mice was also observed when mice were immunized with DC that had been activated via TLR ligands. Although this study suggests that polyclonal Tregs exert a basal level of suppression on the function of DC, these studies offered no mechanistic insight as to the molecular basis for suppression. As one of the major properties of Treg suppression is inhibition of IL-2 production, it is surprising that depletion of Tregs had no effect on Th2 priming, as priming of Th2 responses is even more dependent on IL-2 than priming of Th1 responses.

None of the above studies have directly analyzed Treg function *in vivo* under physiological conditions and most have not taken into account that an intact lymphoid organ is needed for the development of an immune response. Two studies (Tadoko *et al.*, 2006; Tang *et al.*, 2006) have used two-photon laser-scanning microscopy to directly analyze the role of Tregs in priming immune responses *in vivo*. After entering the LN through high endothelial venules, T cells travel through a region near the paracortical T cell zone and B cell follicles where tissue emigrant DCs are found. In the absence of antigen, T cells move freely within the T cell zones of the LN. In contrast, in the presence of antigen, T cells interact with the antigen-loaded DC first by a process termed swarming that ultimately results in arrest of their locomotion for a period of hours before they again become motile. Taken together, these studies demonstrated that both Tregs and Teff formed similar clusters with DCs. In the absence of Tregs, cluster formation by Teff was more sustained. Cluster formation by Tregs preceded the inhibition of Teff activation. Tregs appeared to block Teff cell priming by decreasing the contact time between DCs and the naïve $CD4^+$ T cells. Surprisingly, no direct interactions between Tregs and Teff cells were observed in these studies indicating that the major suppressive effects of the Tregs were directed to the DCs. While most of these studies were performed with antigen-specific Tregs, polyclonal Tregs in both studies decreased the arrest stage, but in one study (Tang *et al.*, 2006) did not block swarming. In the study of Tang *et al.* (2006), simultaneous transfer of Tregs and Teff did not result in inhibition of Teff cell proliferation but blocked their differentiation to Th1 cells. The results of these studies are consistent with models that propose that Tregs cells can preferentially interact with DCs and either prevent the subsequent interaction of DCs with Teff or ultimately lead to inactivation of the DCs. It remains unclear why Tregs that express the same TCR as the Teff would preferentially interact with DCs. One explanation is the expression of higher levels of molecules mediating cell adhesion such as LFA-1 (Onishi *et al.*, 2008) or neuropilin-1 (Sarris *et al.*, 2008).

4. CONCLUSIONS AND SPECULATIONS

The ideal model for the study of the biologic functions of Foxp3$^+$ Tregs would be the creation of a mouse that completely lacked Foxp3$^+$ T cells but still expressed a normal, diverse TCR repertoire and was physiologically normal in all other respects. Unfortunately, deletion of Foxp3$^+$ Treg in adult mice by injection of diptheria toxin into mice that express the DTR selectively in Foxp3$^+$ T cells results in the rapid induction of fatal autoimmune disease (Kim *et al.*, 2007) or a transient depletion followed by rapid reconstitution of the Foxp3$^+$ population (Lahl *et al.*, 2007). While many groups have attempted to deplete Foxp3$^+$ Tregs by treatment of mice with a depleting anti-CD25 antibody, such treatment fails to deplete Foxp3$^+$CD25$^-$ T cells and, in general, only results in a partial depletion of Foxp3$^+$CD25$^+$ T cells. Interpretation of studies using this approach should be interpreted with caution.

We have attempted to distinguish between the biologic effects of polyclonal Foxp3$^+$ T cells and antigen-specific Foxp3$^+$ T cells. It should be noted that most of the studies with antigen-specific Foxp3$^+$ T cells have been performed with Foxp3$^+$ T cells isolated from TCR Tg mice or with iTregs induced from conventional T cells derived from TCR Tg mice. It has been very difficult to identify Foxp3$^+$ T cells specific for foreign antigens in the normal T cell repertoire. As the TCR repertoire of Foxp3$^+$ T cell is likely biased toward self-antigens, the relative frequency of foreign antigen-specific T cells is probably significantly lower in the Foxp3$^+$ compared to the Foxp3$^-$ T cell pools. Future studies and improvements in technologies to identify CD4$^+$ antigen-binding T cells will be needed to resolve this issue.

From the multiple studies presented above, it is clear that polyclonal Foxp3$^+$ T cells are not only needed to maintain normal T cell homeostasis in general but also can preferentially inhibit an immune response induced by immunization or by the transfer of primed or naïve T cells specific for an autoantigen. In many of these studies, the primary effect of polyclonal T cells was to inhibit differentiation from the naïve state to fully differentiated cytokine-producing Teff (DiPaolo *et al.*, 2005; Oldenhove *et al.*, 2003).

In other studies, polyclonal Tregs appeared to modulate Teff cell trafficking (Davidson and Shevach, 2011) or to modulate chemokine production by DC (Morlacchi *et al.*, 2011). In contrast, antigen-specific Tregs appeared to inhibit all aspects of T cell expansion and differentiation. It remains unclear from all of the studies discussed above whether the suppressive effects of either polyclonal or antigen-specific Tregs are directed to a single cell type (e.g., the DC) or to multiple cell types (e.g., DCs and Teff).

The prevalent view largely based on *in vitro* studies is that Treg are capable of using multiple mechanisms to suppress immune responses (Shevach, 2009). Could one target cell actually be responsible for Treg-mediated suppression *in vivo*? A unifying hypothesis to explain many of the studies described in this review would be that the primary target of the Treg *in vivo* is the DC. A large fraction of polyclonal Tregs exist in an activated state and are continuously dividing *in vivo* (Fisson *et al.*, 2003; Stephens *et al.*, 2007) secondary to activation by self-peptide–MHC complexes on DCs. Presumably, the basal level of suppression mediated by this subpopulation is critical to maintaining immune homeostasis. Polyclonal Tregs exhibit a broad pattern of trafficking following transfer and do not selectively home to the site of priming early during the immune response. However, they could be recruited to a site of inflammation following activation of the antigen-specific Teff. Their suppressor function might be enhanced by further stimulation by elevated levels of their self-ligands at the site. If their suppressive activity is primarily directed to the DC, they would not inhibit Teff cell expansion but could suppress the production of cytokines and/or chemokines responsible for Teff cell differentiation (e.g., IL-12) and amplification of the response. In contrast, antigen-specific Tregs would preferentially home to the site of their target antigen, interact with the DC before the Teff (Tang *et al.*, 2006; Tadoko *et al.*, 2006), receive a stronger signal during the recognition of high levels of their cognate antigen, and profoundly downregulate all aspects of DC function including costimulatory molecule expression and antigen presentation. These two pathways are not mutually exclusive, as some studies have shown that polyclonal Tregs can also block the interaction of Teff with antigen-presenting DCs (Tang *et al.*, 2006; Tadoko *et al.*, 2006). The strength of the antigenic stimulus might also play a role in determining which DC functions are downregulated (Morlacchi *et al.*, 2011).

What remains to be determined are the cellular and molecular mechanisms used by the Tregs to modulate DC function. At present, a number of studies strongly support the view that the major pathway used by Tregs to suppress DCs involves the interaction of CTLA-4 on the DC with its target ligands CD80/CD86 resulting in decreased costimulation (Onishi *et al.*, 2008; Qureshi *et al.*, 2011). This mechanism would surely be applicable to the function of antigen-specific Tregs but is less likely to be one used later in the response by polyclonal Tregs. It would also explain the ability of antigen-specific Tregs to mediate bystander suppression by acting on DCs expressing both the cognate antigen recognized by Tregs and other tissue-derived antigens expressed on the same DC. However, the critical role of CTLA-4, as the major molecular mediator of Treg suppression, is still not firmly established. Further, the ability of both $Foxp3^+$ and activated $Foxp3^-$ T cells to mediate trans-endocytosis of CD80/CD86 (Qureshi *et al.*, 2011) raises questions about the relevance of the uniqueness of this suppressor mechanism for the function of $Foxp3^+$ Tregs. Even if the DCs are not the sole target of Tregs, they certainly are likely to be one of the major targets. Further analysis of the molecular mechanisms utilized by Tregs to modulate the function of DCs will be crucial to the development of both pharmacologic and biologic agents that can be used to enhance or suppress Treg function in autoimmunity and tumor immunity, respectively.

ACKNOWLEDGMENTS

The studies in this chapter describing experiments from my laboratory were supported by funds from the Intramural Program of the National Institutes of Allergy and Infectious Diseases. I would like to thank all of the postdoctoral fellows involved in these studies particularly R. DiPaolo, T. Davidson, E. Huter, G. Stummvoll, and Q. Chen for their contributions.

REFERENCES

Barthlott, T., Kassiotis, G., and Stockinger, B. (2003). T cell regulation as a side effect of homeostasis and competition. *J. Exp. Med.* **197**, 451–460.

Belkaid, Y., and Rouse, B. T. (2005). Natural regulatory T cells in infectious disease. *Nat. Immunol.* **6**, 353–354.

Billiard, F., Litvinova, E., Saadoun, D., Djelti, F., Klatzmann, D., Cohen, J. L., Marodon, G., and Salomon, B. L. (2006). Regulatory and effector T cell activation levels are prime determinants of in vivo immune regulation. *J. Immunol.* **177**, 2167–2174.

Cederbom, L., Hall, H., and Ivars, F. (2000). $CD4^+CD25^+$ regulatory T cell down-regulate co-stimulatory molecules on antigen-presenting cells. *Eur. J. Immunol.* **30**, 1538–1543.

Chen, W., Jin, W., Hardegen, N., Lei, K. J., Li, L., Marinos, G., McGrady, G., and Wahl, S. M. (2003). Conversion of peripheral $CD4^+CD25^-$ naïve T cells to $CD4^+CD25^+$ regulatory T cells by TGF-beta induction of transcription factor Foxp3. *J. Exp. Med.* **198**, 1875–1886.

Chen, Q., Kim, Y. C., Laurence, A., Punkosdy, G. A., and Shevach, E. M. (2011). IL-2 controls the stability of Foxp3 expression in TGF-β-induced $Foxp3^+$ T cells in vivo. *J. Immunol.* **186**, 6329–6337.

Curotto de Lafaille, M. A., and Lafaille, J. J. (2009). Natural and adaptive Foxp3+ regulatory T cells: More of the same or a division of labor? *Immunity* **30**, 626–635.

Davidson, T. S., and Shevach, E. M. (2011). Polyclonal Tregs modulate T effector cell trafficking. *Eur. J. Immunol.* **41**, 2862–2870.

Davidson, T. S., DiPaolo, R. J., Andersson, J., and Shevach, E. M. (2007). Cutting Edge: IL-2 is essential for TGF-β-induction of Foxp3+ T regulatory cells. *J. Immunol.* **178**, 4022–4026.

D'elios, M. M., Bergman, M. P., Azzurri, A., Amedei, A., Benagiano, M., De Pont, J. J., Cianchi, F., Vandenbroucke-Grauls, C. M., Romagnani, S., Appelmelk, B. J., and Del Prete, G. (2001). H+, K+-ATPase (proton pump) is the target autoantigen of Th1-type cytotoxic T cells in autoimmune gastritis. *Gastroenterology* **120,** 377–386.

Ding, L., Linsley, P. S., Huang, L. Y., Germain, R. N., and Shevach, E. M. (1993). IL-10 inhibits macrophage costimulatory activity by selectively inhibiting the up-regulation of B7 expression. *J. Immunol.* **151,** 1224–1234.

DiPaolo, R. J., Glass, D. D., Bijwaard, K. E., and Shevach, E. M. (2005). $CD4^+CD25^+$ T cells prevent the development of organ-specific autoimmune disease by inhibiting the differentiation of autoreactive effector T cells. *J. Immunol.* **175,** 7135–7142.

DiPaolo, R. J., Brinster, C., Davidson, T. S., Andersson, J., Glass, D., and Shevach, E. M. (2007). Autoantigen-specific TGFβ-induced $Foxp3^+$ regulatory T cells prevent autoimmunity by inhibiting dendritic cells from activating T cells. *J. Immunol.* **179,** 4685–4693.

Fisson, S., Darrasse-Jeze, G., Litvinova, E., Septier, F., Klatzmann, D., Liblau, R., and Salomon, B. L. (2003). Continuous activation of autoreactive $CD4^+CD25^+$ regulatory T cells in the steady state. *J. Exp. Med.* **198,** 737–746.

Fontenot, J. D., Gavin, M. A., and Rudensky, A. Y. (2003). Foxp3 programs the development and function of $CD4^+CD25^+$ regulatory T cells. *Nat. Immunol.* **4,** 330–336.

Godebu, E., Summers-Torres, D., Lin, M. M., Baaten, B. J. G., and Bradley, L. M. (2008). Polyclonal adaptive regulatory CD4 cells that can reverse type I diabetes become oligoclonal long-term protective memory cells. *J. Immunol.* **181,** 1798–1805.

Grigorova, I. L., Scheab, S. R., Phan, T. H. M., Okada, T., and Cyster, J. G. (2009). Cortical sinus probing, S1P1-dependent entry and flow-based capture of egressing T cells. *Nat. Immunol.* **10,** 58–65.

Hanig, J., and Lutz, M. B. (2008). Suppression of mature dendritic cell function by regulatory T cells in vivo is abrogated by CD40 licensing. *J. Immunol.* **180,** 1405–1413.

Haribhai, D., Williams, J. B., Jia, S., Nickerson, D., Schmitt, E. G., Edwards, B., Ziegelbauer, J., Yassai, M., Li, S. H., Relland, L. M., Wise, P. M., Chen, A., *et al.* (2011). A requisite role for induced regulatory T cells in tolerance based on expanding antigen receptor diversity. *Immunity* **35,** 109–122.

Hirota, K., Duarte, J. H., Veldhoen, M., Hornsby, E., Li, Y., Cua, D. J., Ahifors, H., Wilhelm, C., Tolaini, M., Menzel, U., Garefalaki, A., Potocnik, A. J., *et al.* (2011). Fate mapping of IL-17-producing cells in inflammatory responses. *Nat. Immunol.* **12,** 255–263.

Hisaeda, H., Maekawa, Y., Iwakawa, D., Okada, H., Himeno, K., Kishihara, K., and Yasutomo, K. (2004). Escape of malaria parasites from host immunity requires $CD4^+CD25^+$ regulatory T cells. *Nat. Med.* **10,** 29–30.

Hori, S., Takahashi, T., and Sakaguchi, S. (2003). Control of autoimmunity by naturally arising regulatory $CD4^+$ T cells. *Adv. Immunol.* **81,** 331–371.

Huter, E. N., Punkosdy, G. A., Glass, D. D., Cheng, L. I., Ward, J. M., and Shevach, E. M. (2008a). TGF-β-induced $Foxp3^+$ regulatory T cells rescue scurfy mice. *Eur. J. Immunol.* **38,** 1814–1821.

Huter, E. N., Stumvoll, G. H., DiPaolo, R. J., Glass, D. D., and Shevach, E. M. (2008b). Cutting edge: Antigen-specific TGFβ-induced regulatory T cells suppress Th17-mediated autoimmune disease. *J. Immunol.* **181,** 8209–8213.

Josefowicz, S. Z., and Rudensky, A. (2009). Control of regulatory T cell lineage commitment and maintenance. *Immunity* **30,** 616–625.

Kim, J. M., Rasmussen, J. P., and Rudensky, A. Y. (2007). Regulatory T cells prevent catastrophic autoimmunity throughout the lifespan of the mouse. *Nat. Immunol.* **8,** 191–197.

Kingsley, C. I., Karim, M., Bushell, A. R., and Wood, K. J. (2002). $CD25^+CD4^+$ regulatory T cells prevent graft rejection: CTLA-4- and IL-10-dependent immunoregulation of alloresponses. *J. Immunol.* **168,** 1080–1086.

Klein, L., Khazaie, K., and von Boehmer, H. (2003). *In vivo* dynamics of antigen-specific regulatory T cell not predicted from behavior *in vitro*. *Proc. Natl. Acad. Sci. USA* **100,** 8886–8891.
Kohm, A., Carpentier, P. A., Anger, H. A., and Miller, S. D. (2002). Cutting edge: $CD4^+CD25^+$ regulatory T cells suppress antigen-specific autoreactive immune responses and central nervous system inflammation during active experimental autoimmune encephalomyelitis. *J. Immunol.* **169,** 4712–4716.
Lahl, K., Loddenkemper, C., Drouit, C., Freyer, J., Arnason, J., Eberl, G., Hamman, A., Wagner, H., Huehn, J., and Sparwasser, T. (2007). Selective depletion of $Foxp3^+$ regulatory T cells induces a scurfy-like disease. *J. Exp. Med.* **204,** 57–63.
Lohr, J., Knoechel, B., Wang, J. J., Villarino, A. V., and Abbas, A. K. (2006). Role of IL-17 and regulatory T lymphocytes in a systemic autoimmune disease. *J. Exp. Med.* **203,** 2785–2791.
Longley, R. L., Woods, A., Fleetwood, A., Cowling, G. J., Gallagher, J. T., and Couchman, J. T. (1999). Control of morphology, cytoskeleton and migration by syndecan-4. *Cell Sci.* **112,** 3421–3431.
Lund, J. M., Hsing, L., Pham, T. T., and Rudensky, A. Y. (2008). Coordination of early protective to viral infection by regulatory T cells. *Science* **320,** 1220–1224.
McGeachy, M. J., Stephens, L. A., and Anderton, S. M. (2005). Natural recovery and protection from autoimmune encephalomyelitis: Contribution of $CD4^+CD25^+$ regulatory cells within the central nervous system. *J. Immunol.* **175,** 3025–3032.
Misra, N., Bayry, J., Lacroix-Desmazes, S., Kazatchkine, M. D., and Kaveri, S. V. (2004). Cutting Edge: Human $CD4^+CD25^+$ T cells restrain the maturation and antigen-presenting function of dendritic cells. *J. Immunol.* **172,** 4676–4680.
Morlacchi, S., Dal Secco, V., Soldani, C., Glaichenhaus, N., Viola, A., and Sarukhan, A. (2011). Regulatory T cells target chemokine secretion by dendritic cells independently of their capacity to regulate T cell proliferation. *J. Immunol.* **186,** 6807–6814.
O'Gorman, W. E., Dooms, H., Thorne, S. H., Kuswanto, W. F., Simonds, E. F., Krutzik, P. O., Nolan, G., and Abbas, A. K. (2009). The initial phase of an immune response functions to activate regulatory T cells. *J. Immunol.* **183,** 332–339.
Oldenhove, G., Heusch, M. D., Urbain-Vansanten, G., Urbain, J., Maliszewski, C., Leo, U., and Moser, M. (2003). $CD4^+CD25^+$ regulatory T cells control T helper cell type 1 responses to foreign antigens induced by mature dendritic cells in vivo. *J. Exp. Med.* **198,** 259–266.
Onishi, Y., Fehervari, Z., Yamaguchi, T., and Sakaguchi, S. (2008). $Foxp3^+$ natural regulatory cells preferentially form aggregates on dendritic cells *in vitro* and actively inhibit their maturation. *Proc. Natl. Acad. Sci. USA* **105,** 10113–10118.
Quinn, K. M., Rich, F. J., Goldsak, L. M., de Lisle, G. W., Buddle, B. M., Delahunt, B., and Kirman, J. R. (2008). Accelerating the immune response by inactivating CD4+CD25+ IT regulatory T cells prior to BCG vaccination does not enhance protection against TB. *Eur. J. Immunol.* **38,** 695–705.
Qureshi, O. S., Zheng, Y., Nakamura, K., Attridge, K., Manzotti, C., Schmidt, E. M., Baker, J., Jeffery, L. E., Kaur, S., Briggs, Z., Hou, T. Z., Futter, C. E., *et al.* (2011). Trans-endocytosis of CD80 and CD86: A molecular basis for the cell-extrinsic function of CTLA-4. *Science* **332,** 600–603.
Sakaguchi, S., Sakaguchi, N., Asano, M., Itoh, M., and Toda, M. (1995). Immunologic self-tolerance maintained by activated T cells expressing IL-2 receptor α-chains (CD25): Breakdown of a single mechanism of self-toleracne causes various autoimmune diseases. *J. Immunol.* **155,** 1151–1164.
Sakaguchi, S., Yamaguchi, T., Namura, T., and Ono, M. (2008). Regulatory T cells and immune tolerance. *Cell* **135,** 775–787.
Samy, E. T., Parker, L. A., Sharp, C. P., and Tung, K. S. (2005). Continuous control of autoimmune disease by antigen-dependent polyclonal $CD4^+CD25^+$ regulatory T cells in the regional lymph node. *J. Exp. Med.* **202,** 771–781.

Sarris, M., Anderson, K. G., Randow, F., Mayr, L., and Betz, A. G. (2008). Neuropilin-1 expression on regulatory T cells enhances their interactions with dendritic cells during antigen recognition. *Immunity* **28,** 402–413.

Scheinecker, C., McHugh, R., Shevach, E. M., and Germain, R. E. (2002). Constitutive presentation of a natural tissue autoantigen exclusively by dendritic cells in the draining lymph node. *J. Exp. Med.* **196,** 1079–1090.

Selvaraj, R. K., and Geiger, T. L. (2008). Mitigation of experimental allergic encephalomyelitis by TGF-β induced Foxp3$^+$ regulatory T lymphocytes through the induction of anergy and infectious tolerance. *J. Immunol.* **180,** 2830–2838.

Serra, P., Amrani, A., Yamanouchi, J., Han, B., Thiessen, S., Utsugi, T., Verdaguer, J., and Santamaria, P. (2003). CD40 ligation releases immature dendritic cells from the control of regulatory CD4$^+$CD25$^+$ T cells. *Immunity* **19,** 877–889.

Shevach, E. M. (2009). Mechanisms of Foxp3$^+$ T regulatory cell-mediated suppression. *Immunity* **30,** 636–645.

Shevach, E. M., DiPaolo, R. A., Andersson, J., Zhao, D. M., Stephens, G. L., and Thornton, A. M. (2006). The lifestyle of naturally occurring CD4$^+$CD25$^+$ regulatory T cells. *Immunol. Rev.* **212,** 60–73.

Shimizu, J., Yamazaki, S., and Sakaguchi, S. (1999). Induction of tumor immunity by removing CD25+CD4+ T cells: A common basis between tumor immunity and autoimmunity. *J. Immunol.* **163,** 5211–5218.

Steitz, J., Bruck, J., Lenz, J., Knop, J., and Tuting, T. (2001). Depletion of CD25$^+$ CD4$^+$ T cells and treatment with tyrosinase-related protein 2-transduced dendritic cells enhance the interferon alpha-induced, CD8$^+$ T-cell-dependent immune defense of B16 melanoma. *Cancer Res.* **61,** 8643–8646.

Stephens, G. L., Andersson, J., and Shevach, E. M. (2007). Distinct subsets of FoxP3+ regulatory T cells participate in the control of immune responses. *J. Immunol.* **178,** 6901–6911.

Stummvoll, G. H., DiPaolo, R. J., Huter, E., Davidson, T. S., Glass, D., Ward, J. M., and Shevach, E. M. (2008). Th1, Th2, and Th17 effector T cell-induced autoimmune gastritis differs in pathological pattern and in susceptibility to suppression by regulatory T cells. *J. Immunol.* **181,** 1908–1916.

Sumoza-Toledo, A., Eaton, A. D., and Sarukhan, A. (2006). Regulatory T cells inhibit protein kinase Cθ recruitment to the immune synapse of naïve T cells with the same antigen specificity. *J. Immunol.* **176,** 5779–5787.

Suvas, S., Kumaraguru, U., Pack, C. D., Lee, S., and Rouse, B. T. (2003). CD4+ CD25+T cells regulate virus-specific primary and memory CD8+ T cell responses. *J. Exp. Med.* **198,** 889–901.

Tadoko, C. E., Shakar, G., Shen, S., Ding, Y., Lino, A. C., Maraver, A., Lafaille, J. J., and Dustin, M. L. (2006). Regulatory T cells inhibit stable contacts between CD4+ T cells and dendritic cells in vivo. *J. Exp. Med.* **203,** 505–511.

Tang, Q., Henriksen, K. J., Bi, M., Finger, E. B., Szot, G., Ye, J., Masteller, E. J., McDevitt, H., Bonayhadi, M., and Bluestone, J. E. (2004). In vitro-expanded antigen-specific regulatory T cells suppress autoimmune diabetes. *J. Exp. Med.* **199,** 1467–1477.

Tang, Q., Adams, J. Y., Tooley, A. J., Bi, M., Fife, B. T., Serra, P., Santamaria, P., Locksley, R. M., Krummerl, M. F., and Bluestone, J. A. (2006). Visualizing regulatory T cell control of autoimmune responses in nonobese diabetic mice. *Nat. Immunol.* **7,** 83–92.

Taylor, P. A., Lees, C. J., and Blazer, B. R. (2002). The infection of ex vivo activated and expanded CD4$^+$CD25$^+$ immune regulatory cells inhibits graft-vs-host disease (GVHD) lethality. *Blood* **99,** 4601–4609.

Thangada, S., Khanna, K. M., Blaho, V. A., Oo, M. L., Im, D. S., Guo, C. Y., Lefrancois, L., and Hla, T. (2010). Cell surface residence of sphingosine 1-phosphate receptor 1 on lymphocytes determines lymphocyte egress kinetics. *J. Exp. Med.* **207,** 1475–1483.

Thornton, A. M., and Shevach, E. M. (1998). $CD4^+CD25^+$ immunoregulatory T cells suppress polyclonal T cell activation in vitro by inhibiting interleukin-2 production. *J. Exp. Med.* **188,** 287–296.

Tsuji, M., Komatsu, N., Suzuki, K., Kanagawa, O., Honjo, T., Hori, S., and Fagarasan, S. (2009). Preferential generation of follicular B helper T cells from $Foxp3^+$ T cells in gut Peyer's patches. *Science* **323,** 1488–1492.

Velldhoen, M., Moncrieffe, H., Hocking, R. J., Atkins, C. J., and Stockinger, B. (2006). Modulation of dendritic cell function by naïve and regulatory $CD4^+$ T cells. *J. Immunol.* **176,** 6202–6210.

Vendetti, S., Riccomi, A., Negri, D. R., Veglia, F., Sciaraffia, E., and de Magistris, M. T. (2009). Development of antigen-specific T cells in mediastinal lymph nodes after intranasal immunization. *Methods* **49,** 334–339.

Vendetti, S., Davidson, T. S., Veglia, F., Riccomi, A., Negri, D. R. M., Lindstedt, R., Pasquali, P., Shevach, E. M., and De Magistris, M. T. (2010). Polyclonal Treg cells enhance the activity of a mucosal adjuvant. *Immunol. Cell Biol.* **88,** 698–706.

Walker, L. S. K., Chodos, A., Eggena, M., Dooms, H., and Abbas, A. K. (2003). Antigen-dependent proliferation of $CD4^+CD25^+$ regulatory T cells in vivo. *J. Exp. Med.* **198,** 249–258.

Weber, S. E., Harbertson, J., Godebu, E., Mros, G. A., Padrick, R. C., Carson, B. D., Ziegler, S. F., and Bradley, L. M. (2006). Adaptive islet-specific regulatory CD4 T cells control autoimmune diabetes and mediate the disappearance of pathogenic Th1 cells in vivo. *J. Immunol.* **176,** 4730–4739.

Xu, L. L., Kitani, A., Fuss, I., and Strober, W. (2007). Cutting edge: Regulatory T cells induce $CD4^+CD25^-Foxp3^-$ T cells or are self-induced to become Th17 cells in the absence of exogenous TGF-β. *J. Immunol.* **178,** 6725–6729.

Zhang, H., Podojil, J. R., Chang, J., Luo, X., and Miller, S. D. (2010). TGF-β-induced myelin peptide-specific regulatory T cells mediate antigen-specific suppression of induction of experimental autoimmune encephalomyelitis. *J. Immunol.* **184,** 6629–6636.

Zhao, D. M., Thornton, A. M., DiPaolo, R. J., and Shevach, E. M. (2006). Activated $CD4+CD25^+$ T cells selectively kill B lymphocytes. *Blood* **107,** 3925–3932.

CHAPTER **5**

Extrathymic Generation of Regulatory T Cells—Chances and Challenges for Prevention of Autoimmune Disease

Carolin Daniel[*,†] and **Harald von Boehmer**[*,‡]

Contents

[*] Department of Cancer Immunology and AIDS, Laboratory of Lymphocyte Biology, Dana Farber Cancer Institute, Boston, Massachusetts, USA
[†] Department of Pathology, Harvard Medical School, Boston, Massachusetts, USA
[‡] Harvard Faculty of Arts and Sciences, Harvard University, Cambridge, Massachusetts, USA

Advances in Immunology, Volume 112
ISSN 0065-2776, DOI: 10.1016/B978-0-12-387827-4.00005-X

Abstract Fopx3$^+$ expressing regulatory T cells (Tregs) function as an indispensable cellular constituent of the immune system by establishing and maintaining immunological self-tolerance. T cell receptor (TCR) ligands of high agonist activity, when applied *in vivo* under subimmunogenic conditions, convert naive but not activated T cells into stable Tregs expressing Foxp3. Tolerogenic vaccination with strong-agonist mimetopes of self-antigens may function as a safe and highly specific instrument in the prevention of autoimmune disease by promoting self-antigen-specific tolerance. In this review, we address the requirements for generation of dominant tolerance exerted by Foxp3$^+$ Tregs in autoimmune disease with special focus on type 1 diabetes (T1D). Further understanding of differentiation of T cells into Tregs at the cellular and molecular level will facilitate development of additional tolerogenic vaccination strategies that can be used in prevention as well as therapeutically to combat unwanted immunity.

1. INTRODUCTION

1.1. Historical perspective of dominant tolerance executed by Foxp3$^+$ regulatory T cells

More than a century ago, Paul Ehrlich proposed the concept of "horror autotoxicus," indicating that generally the immune system avoids the generation of immune responses against self. He showed that goats were able to generate antibodies against the blood of other goats, but not against their own blood, and thereby provided plausible evidence for the existence of immunological self-tolerance (Ehrlich, 1906).

Self and non-self discrimination may be acquired during development and then maintained throughout life. The idea that tolerance to self could result from deletion of immature self-reactive lymphocytes was first proposed in the clonal selection theory by Burnet (1959) and Lederberg (1959). Both Burnet and Lederberg proposed that immature lymphocytes with receptors for "self" are deleted when encountering "self." Their idea of "negative selection" implied that consequences of antigen receptor engagement, such as activation and inactivation of genes, differed in immature versus mature lymphocytes, resulting in apoptosis in the former and activation, expansion and differentiation into effector cells in the latter. "Negative selection," also called central or "recessive" tolerance, was studied by analyzing the fate of developing thymocytes in response to stimulation by conventional peptide–major histocompatibility complex (MHC) complexes in TCR-transgenic mice (Kisielow *et al.*, 1988), or in response to superantigens in wild-type mice (Kappler *et al.*, 1987). It became clear from these studies that there was a period during the ontogeny of lymphocytes when binding of receptors to antigen resulted in cell death or anergy (Ramsdell and Fowlkes, 1990) rather than proliferation and development into effector cells. Deletion of self-reactive lymphocytes was shown to be enhanced by promiscuous expression of ectopic antigens in the thymus (Derbinski *et al.*, 2001; Jolicoeur *et al.*, 1994) as well as by immigration of dendritic cells (DCs) presenting antigen from the periphery (Bonasio *et al.*, 2006). It became evident, however, that even the presence of additional "recessive" tolerance mechanisms in secondary lymphoid organs such as deletion and/or anergy (Rocha and von Boehmer, 1991; Rocha *et al.*, 1993) were insufficient to account for self–non-self-discrimination and that "dominant" immunoregulation was of essential relevance.

Strong evidence for the existence of dominant tolerance was presented by seminal studies from Ohki *et al.* (1987), in which tolerance was induced by thymic epithelial grafts in birds. It became clear that xenogeneic transplants in the form of quail wing buds grafted into chickens at embryonic day 4 were subject to rejection at the first or second week after birth. This rejection could be prevented when the epithelial thymic rudiment from the donor was co-implanted, indicating that thymic epithelial cells are able to present self-antigens to the differentiating T cells in a tolerogenic fashion and to induce tolerance to histocompatibility antigens including those that were specifically expressed by wing tissue (Ohki *et al.*, 1987). When the amount of implanted quail thymic epithelial rudiment was monitored, it became evident that only a part of the peripheral T cells had to differentiate in the context of the quail epithelial cells to generate tolerance to quail tissues. This observation supported the notion of intrathymic (i.t.) regulatory T cell generation. These regulatory T cells (Tregs) coresided with specific antiquail immunoreactive T cells and inhibited them from reacting against the quail wing antigenic determinants, thereby qualifying this mechanism of tolerance as "dominant." These studies were supported and extended by experiments from Salaun *et al.* (1990), analyzing the role of thymic epithelium in the establishment of tissue tolerance using a murine chimeric system. In this experimental setting, embryonic thymic rudiments that contained no hematopoietic cells from C3H (H-2^k) donors were grafted to newborn athymic BALB/c (H-2^d) mice. Chimeras harboring normal T cell numbers and function rejected third-party skin grafts but permanently accepted transplants syngeneic to the thymic epithelium, indicating that thymic epithelium is able to induce tolerance to histocompatibility antigens (Salaun *et al.*, 1990). These seminal studies by the group of Le Douarin and colleagues (Ohki *et al.*, 1987; Salaun *et al.*, 1990) therefore made critical contributions to the establishment of "dominant" tolerance executed by a committed lineage of Tregs. These experiments provided convincing evidence for the notion that suppressive properties of Tregs are "imprinted" and maintained even in different environments in which the original inductive stimuli are not present. In addition, it became clear from thymectomy studies in mice, performed between day 2 and 4 of life, that these maneuvers generated T cell-mediated lesions, which could be alleviated by the transfer of thymocytes or splenocytes from adult euthymic mice (Asano *et al.*, 1996; Bonomo *et al.*, 1995; Nishizuka and Sakakura, 1969; Sakaguchi *et al.*, 1982a,b). It was concluded from these findings that a specific cell population generated in the thymus after 3 days of life is able to execute dominant tolerance.

Initially, one of the greatest impediments to the study of regulatory cell phenomena was the lack of reliable surface markers to identify Tregs (Fowell and Mason, 1993; Powrie and Mason, 1990; Powrie *et al.*, 1993; Sakaguchi *et al.*, 1985). Work by Sakaguchi in 1995 (Sakaguchi *et al.*, 1995) provided convincing evidence that the high-affinity IL-2 receptor α chain (CD25) was expressed on a subset of T cells with regulatory characteristics and could serve as an appropriate surface marker for Tregs. Further attention to the essential role of dominant tolerance in preventing autoimmunity came from studies showing that autoimmune disorders that appeared in mice following thymectomy in the neonatal period could be prevented by transfer of $CD4^+CD25^+$ cells from normal donors (Sakaguchi *et al.*, 1995). In addition, these data provided evidence that transfer of $CD25^-$ T cells from normal BALB/c mice into syngeneic athymic nu/nu animals resulted in severe autoimmunity. Cotransfer of $CD4^+CD25^+$ T cells prevented the development of autoimmune disease. Moreover, it became apparent from a genetic study in mice with a transgenic TCR specific for myelin basic protein, that a few T cells expressing endogenously rearranged TCRs were capable of preventing the spontaneous development of experimental autoimmune encephalomyelitis (EAE) (Lafaille *et al.*, 1994).

These studies culminated in the identification of the transcription factor Foxp3 as the master regulator in Tregs and thus led to the identification of Foxp3 expressing Tregs that suppress the functions of other cells involved in adaptive and innate immunity (Fontenot *et al.*, 2003; Khattri *et al.*, 2003). Moreover, ectopic expression of Foxp3 in conventional $CD4^+$ T cells was suited to confer a Treg phenotype identical to that of $CD4^+CD25^+$ Tregs (Fontenot *et al.*, 2003; Hori *et al.*, 2003a). It was known from previous genetic analyses that mutations in the Foxp3 gene were associated with severe autoimmunity in humans and mice (Brunkow *et al.*, 2001; Chatila *et al.*, 2000). Acute depletion of $Foxp3^+$ cells in $Foxp3^{DTR}$ knockin and $Foxp3^{DTR}$ BAC transgenic mice harboring an ''ablatable'' population of Tregs which express human diphtheria toxin receptor showed that Treg-mediated suppression is indispensable for the prevention of fatal autoimmunity throughout the life of normal mice (Kim *et al.*, 2007; Lahl *et al.*, 2007). Scurfy mice lacking a functional $Foxp3^+$ protein because of a natural mutation suffer from various spontaneous and early-onset organ-specific autoimmune diseases due to hyperactivation of $CD4^+$ T cells (Brunkow *et al.*, 2001).

Based on evidence that adoptive transfer of Tregs can rescue the fatal disease in scurfy mice, it was proposed that Foxp3$^+$ Tregs are of critical relevance in the prevention of autoimmunity in mice. It became clear that a phenotypically similar syndrome known as immune dysregulation, polyendocrinopathy, enteropathy, X-linked (IPEX) develops in humans harboring mutations of the Foxp3 gene. With regard to clinical symptoms, it became apparent that the disease can widely vary but commonly comprises T1D, eczema, and enteropathy (Ochs *et al.*, 2005) with the symptoms appearing soon after birth. IPEX patients were shown to harbor a variety of mutations which critically impact the disease phenotype. The variability results from the fact that IPEX patients may harbor either a null or other mutations, the latter resulting in a fully translated, but functionally impaired, Foxp3 protein. Most severe forms of IPEX occurred in patients who have a null mutation or mutations in the DNA-binding forkhead domain. In contrast, mutations outside of the forkhead domain often cause mild or late-onset forms of disease (Ochs *et al.*, 2005).

Currently, there are over 80 diseases classified as autoimmune, affecting approximately 5% of the human population. In addition, autoimmune diseases are among the main 10 leading causes of death and affect at least 10 million individuals in the USA (Vang *et al.*, 2008). The incidence of autoimmune T1D has risen steadily in developed countries from the 1950s to today, with the current prediction that it will double in children under the age of 5 years by 2020 (Patterson *et al.*, 2009). Therefore, it appears of crucial importance to further improve our understanding of immunological self-tolerance at the cellular and mechanistic level in order to specifically interfere with autoimmune disease. Advances in this field should then enable investigators to generate specific tolerogenic vaccination protocols, thereby replacing general immunosuppression which often has fatal consequences. In this review, we discuss the requirements for the extrathymic (e.t.) generation of antigen-specific Tregs *in vivo* and their role in the prevention of autoimmunity such as T1D.

2. INTRA- AND EXTRATHYMIC INDUCTION OF FOXP3$^+$ TREGS

2.1. Requirements for i.t. differentiation of Foxp3$^+$ Tregs

Following the initial description of CD4$^+$CD25$^+$ Tregs, it was proposed that thymic selection processes play a critical role in their generation. All early observations concerning CD4$^+$CD25$^+$ Tregs were made with polyclonal populations of cells expressing diverse $\alpha\beta$ TCRs, and in fact, initially, very little was known about the origin of these cells besides the notion that they could be found in i.t. and e.t. lymphoid tissue. Based on findings from various TCR transgenic systems in which $CD4^+CD25^+$ Tregs with known specificity for neo-self-antigens were generated i.t., an essential role of thymic epithelium for Treg development was strongly suggested. Studies by the group of Caton *et al.* using the hemagglutinin (HA) TCR transgenic system first provided evidence that (HA)-specific $CD4^+$ T cells were selected into the Treg lineage when the agonist ligand was coexpressed (Jordan *et al.*, 2000). Such mice clearly harbored increased numbers of $CD4^+CD25^+$ Tregs that expressed only the transgenic TCR. In fact, such mice contained both $CD4^+CD25^+$ and $CD4^+CD25^-$ cells expressing the transgenic TCR, and the former could suppress the proliferation of the latter when cocultured, indicating that the $CD4^+$ as a whole was nonresponsive while the isolated $CD4^+CD25^-$ but not the $CD4^+CD25^+$ cells proliferated upon antigenic stimulation *in vitro* (Jordan *et al.*, 2001). Similar observations were subsequently obtained by Apostolou *et al.* (2002) in thymus transplantation experiments using the same TCR transgenic system or in the ovalbumin-specific DO11.10 TCR (Walker *et al.*, 2003). These studies revealed that expression of agonist ligands by radioresistant thymic tissue represents a potent means of i.t. Treg generation. The critical role of the thymic epithelium for induction of Tregs was underlined by the fact that transplantation of thymi from antigen-transgenic mice into otherwise antigen-deficient hosts was sufficient to instruct the selection of a large fraction of antigen-specific $CD4^+$ T cells into the Treg lineage (Apostolou *et al.*, 2002). Work by van Santen *et al.* (van Santen *et al.*, 2004) challenged this conclusion. By the use of a transgenic mouse system with controlled expression of a T cell epitope, they draw the hypothesis that diversion of differentiating thymocytes into the Treg lineage by encounter of agonist ligands was nonexistent and instead was the result of reduced sensitivity of the $CD4^+CD25^+$ subset to agonist-induced clonal deletion. In contrast, Tai *et al.* (2005) again showed the requirement of agonist ligands for i.t. Treg generation in the same transgenic system as used by van Santen *et al.* (2004), thereby underlining that recognition of agonist TCR ligands functions not only as a key event for negative selection but also for the initiation of the developmental program leading to Treg generation (Tai *et al.*, 2005). It was demonstrated that cross-presentation of such ligands by hematopoietic cells is not required but can be sufficient (Aschenbrenner *et al.*, 2007) and that Treg induction required a lower ligand density than negative selection (Hinterberger *et al.*, 2010).

Foxp3 induction leading to Treg differentiation was shown to occur relatively late during thymic differentiation. The Treg lineage differentiation process can start with $CD4^+CD8^-$ single-positive cells in the thymic medulla, but a contribution of $Foxp3^+$-expressing double-positive cells in the thymic cortex is not entirely excluded (Liston *et al.*, 2008). It became clear that Foxp3 is not required for the initial Treg fate determination of developing thymocytes, but more likely necessary to complete Treg differentiation once the initial Treg fate is determined (Gavin *et al.*, 2007; Lin *et al.*, 2007). By the use of mice in which the Foxp3-encoding region was replaced by a green fluorescent protein (GFP) reporter, Gavin *et al.* (2007) and Lin *et al.* (2007) could demonstrate that GFP^+ thymocytes, which do not express the Foxp3 protein, share some expression patterns of genes and some characteristics with Tregs, including *in vitro* anergy. However, these Foxp3-less "wannabes" did not exhibit suppressive properties *in vitro*. These studies underlined the conclusion that Foxp3 is important for stabilizing and maintaining the Treg phenotype as well as for their suppressive function (Gavin *et al.*, 2007; Lin *et al.*, 2007). In line with these findings, it was demonstrated that the expression of a hypomorphic Foxp3 allele caused a distinct reduction in the suppressive capacities of Tregs (Wan and Flavell, 2005).

Since engagement of TCR signaling was shown to instruct Treg development, the hypothesis was proposed that a TCR from a Treg should instruct Treg development when expressed as a transgene (Bautista *et al.*, 2009; Leung *et al.*, 2009). Unexpectedly, some observations revealed that this was not the case, perhaps due to saturable niches for the generation of Tregs in the thymus. When the number of precursors with the specific transgenic Treg-derived TCR was reduced, an increase of Tregs expressing that particular TCR could be observed (Bautista *et al.*, 2009; Leung *et al.*, 2009). The antigenic determinants responsible for thymic selection and differentiation are not fully understood; it is clear, however, that instruction of Treg generation results from TCR signals induced by i.t. (Tai *et al.*, 2005) and/or e.t. (Apostolou and von Boehmer, 2004; Kretschmer *et al.*, 2005) ligands. I.t. expression of peripheral antigens can be regulated by AIRE (autoimmune regulator) (Mathis and Benoist, 2007). Given the fact that AIRE-controlled i.t. ligands are able to instruct Treg development, one might assume that such ligands include peripheral antigens. Although Tregs possess a polyclonal TCR repertoire (Hori *et al.*, 2002, 2003b), there is evidence that a large proportion recognize self-antigens and that their primary role is the regulation and suppression of autoreactive T cells (Hsieh and Rudensky, 2005). It became clear, however, that TCR repertoires of Tregs and conventional $CD4^+$ $\alpha\beta$ T cells are partially overlapping (Hsieh *et al.*, 2004; Pacholczyk *et al.*, 2006) which might, at least in part, result from saturable thymic niches for selection of Tregs in that not all cells with an appropriate TCR can be selected into the Treg lineage (Bautista *et al.*, 2009; Leung *et al.*, 2009). To further analyze the TCR specificities of Tregs, studies in Rudensky's group (Hsieh *et al.*, 2004) isolated and sequenced TCRα genes from $CD25^+$ or $CD25^-$ $CD4^+$ TCli TCRβ transgenic T cells (Mombaerts *et al.*, 1992). TCli-$\alpha\beta$ TCR $RAG1^{-/-}$ mice were used as the source of conventional $CD4^+CD25^-$ T cells. These T cells are monospecific for human CLIP and do not express endogenous TCRα chains (Wong *et al.*, 2000). Upon retroviral transduction of these cells with Treg- or naive $CD4^+$ T cell-TCRα libraries followed by the analysis of reactivity of Treg- versus non-Treg-derived TCR, it became clear that Treg-derived TCR exhibited an increase in self-reactivity (Hsieh *et al.*, 2004). In addition, it was demonstrated that the presence or absence of the Foxp3 gene itself did not impact the efficiency or sensitivity of negative selection of thymocytes by a high-affinity TCR ligand (Hsieh *et al.*, 2006) and that activated T cells in Foxp3-deficient animals displayed TCRs utilized by Tregs in Foxp3-sufficient mice (Hsieh *et al.*, 2006). These observations supported the notion that in the absence of Foxp3, T cell precursors with self-reactive TCRs—supporting induction of Foxp3 and differentiation into Tregs—are not subject to deletion but, in contrast, following maturation are activated and contributed to the autoimmune pathology observed in Foxp3-deficient mice.

Recent work from Kristin Hogquist's group provided a useful tool to further address the role of TCR signal strength and T cell activation during Treg development *in vivo*: a transgenic mouse expressing GFP from the immediate early gene *Nr4a1* (Nur77) locus was created (Moran *et al.*, 2011). GFP expression was demonstrated to be upregulated in lymphocytes by antigen receptor stimulation but not by inflammatory stimuli. In T cells, GFP induction could be seen during positive selection. Moreover, maintenance of GFP expression required MHC interaction and GFP expression levels directly correlated with the strength of the TCR signal. Using the $Nur77^{GFP}$ mouse as a reporter of TCR signal strength, it became clear that $CD4^+$ $Foxp3^+$ Tregs harbored higher levels of GFP, thereby underlining the concept that $CD4^+Foxp3^+$ Tregs receive stronger TCR stimulation upon selection than cells undergoing conventional positive selection (Moran *et al.*, 2011). It became also apparent from these studies that there was a distinct overlap in the levels of GFP expression between Tregs and non-Tregs.

It became evident that in addition to TCR signals, the initiation of Treg generation is regulated by environmental cosignals which can either support or prevent this cell fate decision. CD28 costimulatory signals were proposed to have an essential cell-intrinsic role in the i.t. differentiation of Tregs. Genetic ablation of either CD28 or its ligands B7-1 and B7-2 caused severely reduced proportions and numbers of thymic Treg (Salomon *et al.*, 2000; Tai *et al.*, 2005; Tang *et al.*, 2003). This scenario was the result of a T cell-intrinsic function of CD28 signaling in Treg differentiation, as bystander wild-type T cells did not restore Treg development of CD28-deficient thymocytes in mixed bone marrow chimeras (Tai *et al.*, 2005). A lack of costimulation could result in quantitatively altered signal strength of the underlying TCR stimulus itself. Thus, the Treg-differentiation process could be hampered in the absence of CD28 costimulation.

Additionally, it was suggested that the Lck-binding domain of the CD28 cytoplasmic tail possesses a critical role for the induction of Foxp3 (Tai *et al.*, 2005), thereby suggesting that binding of Lck to the CD28 cytosolic tail results in an enhancement of the intensity and duration of the antigen-specific TCR signal (Holdorf *et al.*, 1999, 2002).

Deficiency of CTLA4, a negative costimulatory molecule, results in a lethal lymphoproliferative autoimmune syndrome that is due to the failure of both cell-intrinsic and non-cell-autonomous functions of CTLA4 in the maintenance of tolerance (Bachmann *et al.*, 1999; Tivol *et al.*, 1995; Waterhouse *et al.*, 1995). The potential role of CTLA4 in i.t. differentiation of Tregs is not well understood. It was shown, however, that CTLA4 inhibited the early events of cell cycle progression (Brunner *et al.*, 1999; Krummel and Allison, 1996), which could affect Treg differentiation, as proliferation of thymocytes following encountering of antigen appears to counteract Treg differentiation (Wirnsberger *et al.*, 2009).

It is well accepted that IL-2 signaling is essential for Treg homeostasis *in vivo*, whereby IL-2 needs to be acquired in a paracrine fashion, which has been extensively studied in both IL-2- and IL-2R-deficient animals (Almeida *et al.*, 2002; Malek *et al.*, 2002). It was proposed that during i.t. Treg differentiation, the IL-2 signaling pathway may cooperate with the instructive TCR stimulus. Ablation of either IL-2 or of IL-2 receptor subunits (CD25 or CD122) was shown to result in severe impairment of dominant tolerance accompanied by massive autoimmunity (Almeida *et al.*, 2002; Kramer *et al.*, 1995; Schorle *et al.*, 1991; Suzuki *et al.*, 1995, 1999). As IL- 2 signaling is essentially required for the survival of Tregs (Bayer *et al.*, 2005; Setoguchi *et al.*, 2005), the role of IL-2 as a true differentiation factor has been debated. STAT5, activated downstream of the IL-2R and other common γ-chain cytokine receptors, was proposed to represent a transcription factor involved in direct regulation of Foxp3 expression (Burchill *et al.*, 2008). In line with this notion, STAT5 was shown to directly bind the Foxp3 promoter and Foxp3 conserved noncoding sequence (CNS) element. Further, studies by Burchill (Burchill *et al.*, 2007) and Yao (Yao *et al.*, 2007) demonstrated that, upon induced ablation of a conditional *Stat5* allele in double positive (DP) thymocytes, a substantial reduction in $CD4^+$SP thymocytes expressing Foxp3 could be seen. Moreover, the expression of a constitutively active STAT5 was able to expand thymic Tregs as well as to rescue thymic Treg numbers in the absence of IL-2 (Burchill *et al.*, 2003, 2008). However, the question whether STAT5 precisely executes a nonredundant function in driving Foxp3 expression remains unresolved, as the expression of a Bcl-2 transgene was suited to rescue the differentiation of STAT5-deficient Tregs, thereby suggesting that induction of Foxp3 can be achieved in the absence of STAT5 (Josefowicz and Rudensky, 2009). In addition, the presence of Foxp3 expressing thymic $CD4^+$ T cells in the context of IL-2 or IL-2Rα deficiency underlines the conclusion that IL-2 is not absolutely required for i.t. Treg generation (D'Cruz and Klein, 2005; Fontenot *et al.*, 2005).

Besides its well-established role for *in vitro* induction of Tregs (Chen *et al.*, 2003; Zheng *et al.*, 2002), the requirement of TGFβ for the i.t. generation of Foxp3 Treg has been subject to discussion. Initially, it was proposed that T cell-specific ablation of the TGFβ receptor type II did not significantly impinge on the i.t. development of Tregs (Li *et al.*, 2006; Marie *et al.*, 2005, 2006). Subsequently, it was demonstrated that T cell-specific deletion of TGFβRI (Liu *et al.*, 2008) or TGFβRII (Ouyang *et al.*, 2010a) resulted in a reduction of the initial wave of thymically induced Tregs. In support of this notion, ablation of TGFβRI subunit in DP thymocytes caused a substantial but transient impairment in $Foxp3^+$ Treg generation during the first week of life followed by the recovery of Foxp3 thymocytes numbers to those seen in wild-type animals (Liu *et al.*, 2008). Another proposed scenario is that the requirement of TGFβ for i.t. Tregs induction was masked by IL-2, supporting survival and expansion of residual Tregs. The absence of both TGFβ-signaling pathways caused a complete deficiency of thymic Tregs (Liu *et al.*, 2008); however, it is not clear whether this is a cell autonomous effect.

2.2. Requirements for e.t. generation of $Foxp3^+$ Tregs

$Foxp3^+$ Tregs can also be generated e.t from peripheral naive T cells. Currently, it is not clear whether i.t. versus e.t. differentiation of Tregs serves different biological needs, whether these cells have different or redundant functions, and whether the requirements for i.t. versus e.t. Treg induction are distinct. In addition, the contributions of i.t.- versus e.t.-generated $Foxp3^+$ Tregs to the total Treg pool under steady state or in the course of immune challenge are not well defined. There is, however, accumulating evidence that e.t. Tregs are critical players in organs such as the gut, where T cells are continuously stimulated by microbes in the presence of locally generated TGFβ and retinoic acid derived from DCs (Izcue *et al.*, 2009).

E.t.-induced Tregs are being studied, as they can be easily generated and bear therapeutic potential. It became clear that i.t.- versus e.t.-generated Tregs display close similarity, with regard to their regulatory function, surface phenotype, stability as well as their global gene-expression pattern. Transcriptional profiling of $Foxp3^+$-expressing Tregs in comparison with naive or activated T cells demonstrated a distinct subset of differentially expressed genes comprising some genes normally upregulated in activated T cells, such as IL2ra (CD25), Ctla4 (CTLA4), and Tnfrsf18 (glucocorticoid-induced TNF receptor), representing signature target genes for Foxp3 (Chen *et al.*, 2006; Feuerer *et al.*, 2010; Marson *et al.*, 2007; Zheng *et al.*, 2007).

E.t.-induced Treg can be generated in peripheral lymphoid organs; they can be induced by subimmunogenic delivery of strong-agonist ligands under conditions that avoid the functional activation of antigen-presenting cells (APCs) (Apostolou and von Boehmer, 2004; Daniel *et al.*, 2010, 2011a; Kretschmer *et al.*, 2005; Verginis *et al.*, 2008). These experiments confirmed anecdotal evidence of generating "dominant" tolerance by subimmunogenic antigen delivery (Dietrich and Weigle, 1963). The absence of Foxp3$^+$ Treg in Rag$^{-/-}$ TCR-transgenic animals expressing only one particular TCR without co-expression of a TCR agonist ligand supported the concept that Tregs were in fact due to *de novo* generation rather than expansion of already committed Tregs. Best conversion into Foxp3$^+$ Tregs can be achieved in T cells that undergo only limited proliferation, whereas higher doses of TCR agonists result in more extensive proliferation and diminished conversion. Such high doses of TCR ligands were proposed to lead to the activation of the PI3K/Akt/mTOR pathway (Merkenschlager and von Boehmer, 2010) that could interfere with e.t. Foxp3 induction (Sauer *et al.*, 2008). Robust T cell activation was shown to negatively impact the e.t. generation of Treg based on a cell cycle-dependent maintenance of a silenced state of the Foxp3 locus (Josefowicz *et al.*, 2009). The role of TGFβ was addressed in the e.t. Treg conversion *in vivo*. When congenic BALB/c mice were injected with CFSE-labeled Thy1.2$^+$ HA-TCR transgenic naive T cells from dominant negative TGFβ-receptor II (dnTGFβRII) or wild-type mice followed by the delivery of antigen in the form of DEC205 fusion antibody, it became clear that the percentages of Foxp3$^+$ Tregs from dnTGFβRII donors were substantially reduced (Kretschmer *et al.*, 2005). However, the fact that T cells from dnTGFβRII animals proliferated vigorously could explain that the absolute number of converted Foxp3$^+$ Tregs at the end of the 2-week conversion period was significantly lowered (Kretschmer *et al.*, 2005), as T cell proliferation is counterproductive to Treg conversion.

With regard to lifespan and stability, it became apparent that e.t. CD4$^+$CD25$^+$ Tregs generated from naive T cells through confrontation with their agonist TCR ligand could survive for long periods in an intermitotic state in peripheral lymphoid tissue without losing their commitment to suppressive activity (Klein *et al.*, 2003; Polansky *et al.*, 2008). This feature is of particular importance, as it permits the prospective induction of such cells to suppress unwanted immune responses. The e.t. Foxp3$^+$ Tregs generated by subimmunogenic antigen delivery are stable (Polansky *et al.*, 2008), correlating with demethylation of the Foxp3 locus, and maintain their function even in an immunogenic context (Rubtsov *et al.*, 2010), as reencounter of antigen under immunogenic conditions does cause neither loss of Tregs nor loss of their activity (Klein *et al.*, 2003) but results in expansion of Tregs (Kretschmer *et al.*, 2005). Lessons learned from these artificial transgenic systems have been confirmed in

WT mice, where antigen-specific T cells are rather infrequent and where the *de novo* induction of $Foxp3^+$ Treg in WT mice has been visualized by antigen-specific tetramer staining (Verginis *et al.*, 2008).

2.3. Signaling pathways and regulatory DNA elements controlling Foxp3 expression

Several transcription factors downstream of TCR signaling and CD28 costimulation as NFAT, NFκB, and AP-1 have been implicated in the Treg-differentiation process. It became clear from *in vitro* studies that TCR signaling in the presence of TGFβ (Chen *et al.*, 2003), the downstream signaling transducers Smad2 and Smad3 as well as retinoic acid (Benson *et al.*, 2007; Coombes *et al.*, 2007; Mucida *et al.*, 2009; Sun *et al.*, 2007) function as important inducers of $Foxp3^+$ expression in peripheral naive T cells *in vitro*. It should be noted, however, that *in vitro* generation of $Foxp3^+$ cells in short-term cultures yields unstable Tregs that do not resemble Tregs generated *in vivo* (Polansky *et al.*, 2008). Importantly, recent studies indicated that induction and maintenance of Foxp3 expression are two separate processes being controlled by distinct *cis*-regulatory elements within the Foxp3 gene, while maintenance appears to be achieved by "epigenetic" mechanisms (Floess *et al.*, 2007). The expression of Foxp3 is tightly regulated by multiple transcription factors, chromatin configuration, and *cis*-regulatory elements. Following TCR activation, the binding of transcription factors such as AP-1, NFAT, CREB, and ATF to the Foxp3 promoter and enhancer elements is induced (Kim and Leonard, 2007; Tone *et al.*, 2008). In addition to the promoter, at least three CNS elements have been identified and contributed to regulation of the Foxp3 locus as well as to the maintenance of Foxp3 expression (Huehn *et al.*, 2009; Kim and Leonard, 2007; Tone *et al.*, 2008; Zheng *et al.*, 2010). The CNS being relevant for the maintenance of Foxp3 expression contains a region rich in CpG dinucleotides, which are targets for DNA methylation (Baron *et al.*, 2007; Floess *et al.*, 2007; Polansky *et al.*, 2008). While this region is methylated in conventional T cells, it is completely demethylated in Tregs, induced i.t. or generated e.t. *in vivo* (Polansky *et al.*, 2008). In contrast, it was noted that these elements remain at least partially methylated in Tregs generated *in vitro* that do not stably express Foxp3 (Baron *et al.*, 2007; Floess *et al.*, 2007; Polansky *et al.*, 2008). One possible consequence of continuous methylation of these CpG-containing elements in *in vitro* generated Tregs is that transcription factors like the methyl-sensitive CREB are not able to bind stably to this region. Accordingly, demethylation of this particular CNS appears to facilitate binding of transcription factors that mediate stable heritable maintenance of Foxp3 expression. In fact, a demethylated state of this region was shown to correlate with a stable Treg phenotype (Huehn *et al.*, 2009; Polansky *et al.*, 2008). Recent work from Rudensky's group revealed that loss of DNA methylation at this region specifically promotes the binding of CREB, ATF, c-rel, and Foxp3 itself, which is supportive of the notion that Foxp3 could form part of a positive feedback loop, maintaining its own expression (Zheng *et al.*, 2010).

It was proposed that TGFβ cooperates with TCR signals to induce Foxp3 in part by antagonizing cell cycle-dependent recruitment of maintenance DNA methyltransferase I (DNMT1) to the Foxp3 locus leading to its inactivation (Josefowicz *et al.*, 2009). Not mutually exclusive is the scenario that signals capable of inducing Foxp3 initiate chromatin remodeling and establish a poised state of the Foxp3 locus. Robust proliferation accompanied by propagation of CpG methylation by DNMT1 may prevent establishment of this permissive state (Josefowicz and Rudensky, 2009). TGFβ may mediate Treg induction through both direct and indirect mechanisms, with Smad3/2 binding to the Foxp3-CNS (Tone *et al.*, 2008).

2.4. PI3K–mTOR–Akt signaling pathway—Role in e.t. Treg generation and function

Recent studies providing genetic and pharmacological evidence demonstrated that the PI3K–Akt–mTOR signaling pathway interferes with Foxp3 induction *in vitro* and *in vivo* (Merkenschlager and von Boehmer, 2010). Rapamycin was originally identified as an antifungal compound derived from *Streptomyces hygroscopicus*, which can be found in soil samples collected from Easter Island (locally called Rapa nui) (Dennis *et al.*, 1999). It became clear that rapamycin had poor antibiotic properties; however, closer inspection revealed its potent immunosuppressive characteristics (Schmelzle and Hall, 2000). Initially, it was assumed that the immunosuppressive properties result from its ability to inhibit cell division (Schmelzle and Hall, 2000). Attempts to identify the mechanism by which rapamycin meditated its effects resulted in the discovery of the target of rapamycin 1 (TOR1) and TOR2 in yeast and later on of the mammalian homolog, mTOR (Schmelzle and Hall, 2000). The kinase mTOR, an evolutionary conserved member of the phosphatidylinositol-3-OH kinase (PI3-related kinase family), is now known to possess critical roles in the regulation of proliferation and survival as well as metabolism, protein synthesis, and energy balance. It became evident that environmental cues such as amino acids, growth factors, and insulin are integrated by this kinase complex in order to regulate cellular metabolism (Guertin and Sabatini, 2007). mTOR is encoded as a single gene whose protein product signals via two distinct complexes (mTORC1 and mTORC2). Activation of mTORC1 results in phosphorylation of p70 ribosomal S6 kinase 1 or its downstream substrate S6 as well as in phosphorylation of the translational inhibitor eukaryotic initiation factor

4E-binding protein 1 (Beugnet *et al.*, 2003). mTORC1 was shown to function as a critical regulator of protein translation. Additionally, activation of mTORC1 was demonstrated to inhibit autophagy and promote lipid metabolism as well as mitochondrial biogenesis (Laplante and Sabatini, 2009). Collectively, these activities targeted by mTORC1 result in the promotion of cell growth and differentiation. Signaling by the kinases PI3K, PDK1, and Akt was shown to activate mTORC1. In contrast, the detailed mechanisms of upstream mTORC2 activators are not yet fully elucidated. Activation of mTORC2 promotes phosphorylation of the hydrophobic motif of Akt at Ser473 and other kinases (SGK1 and PKC-α) (Guertin *et al.*, 2006). Recent work of Jonathan Powell's group supported the concept that mTORC1 signaling results in Th1 and Th17 differentiation, while signaling through mTORC2 promotes Th2 differentiation (Delgoffe *et al.*, 2011). In the absence of mTOR, naive $CD4^+$ T cells differentiate into Tregs (Delgoffe *et al.*, 2009). It became apparent in addition that inhibition of both signaling pathways mTORC1 and mTORC2 results in efficient promotion of Treg differentiation. That is, T cell activation in the absence of mTOR leads to a default Treg pathway. Until recently, the mechanisms that link PI3K–Akt–mTOR signaling to Foxp3 expression have been unknown. Merkenschlager and colleagues showed that early withdrawal of TCR signaling through pharmacological inhibition of PI3K–Akt–mTOR signaling promotes robust Foxp3 induction (Sauer *et al.*, 2008). Efficient induction of Foxp3 expression requires proteins that limit the activity of PI3K such as Cbl-b (Wohlfert *et al.*, 2006) or phosphatase and tensin homolog (PTEN) phosphatase (Sauer *et al.*, 2008), which antagonizes PI3K function, consequently inhibiting downstream signaling through Akt. In line with these observations, a constitutive activation of Akt was shown to interfere with stable induction of Foxp3 (Haxhinasto *et al.*, 2008). mTORC2 activates Akt by phosphorylating it on Ser473 (Delgoffe *et al.*, 2009). It became clear from recent studies that forkhead-box O (Foxo) transcription factors link PI3K–Akt–mTOR signaling to Foxp3 (Harada *et al.*, 2010; Ouyang *et al.*, 2010b). The Foxo family of transcription factors are mammalian orthologs of *Caenorhabditis elegans* Daf16 and *Drosophila melanogaster* dFoxo with critical roles in the control of diverse cellular functions. Three of the four Foxo family proteins (Foxo1, Foxo3, and Foxo4) were shown to be regulated by the evolutionary conserved PI3K–Akt pathway (Hedrick, 2009). It became clear from earlier studies that Foxo transcription factors possess a critical role in the homeostasis of the immune system (Hedrick, 2009). Recent work demonstrated that mice with a T cell-specific deletion of Foxo1 and Foxo3 succumbed to a lethal inflammatory disease accompanied by massive proliferation, activation, and differentiation of $CD4^+$ T cells (Ouyang *et al.*, 2010b). Activity of Foxo factors was shown to be regulated at multiple levels, most strikingly by Akt-mediated phosphorylation.

Phosphorylation by Akt was demonstrated to inactivate Foxo proteins by excluding them from the nucleus. Additionally, mTORC2-dependent phosphorylation of Foxo1 and Foxo3a leads to their inactivation by similarly promoting their sequestration into the cytoplasm (Merkenschlager and von Boehmer, 2010). Both of these transcription factors play a role in Foxp3 induction. Therefore, in the presence of mTORC2 activation, Foxo1- and Foxo3a-mediated Foxp3 expression is diminished. Foxo-binding motifs in the Fopx3 promoter region were identified (Harada *et al.*, 2010) in accordance with several conserved Foxo consensus sequences in the Foxp3 locus (Ouyang *et al.*, 2010b). Reporter gene assays focused on the importance of a site close to the Foxp3 transcription start site (Harada *et al.*, 2010; Ouyang *et al.*, 2010b), suggesting that Foxp3 is a direct Foxo target gene.

Interestingly, it was shown that Foxo proteins and NFκB are inversely regulated following T cell activation. NFκB is maintained in the cytoplasm through interaction with inhibitory IκB proteins. Upon TCR stimulation and costimulatory signals by CD28 activation of the IκB complex occurs, which triggers phosphorylation and degradation of IκB, thereby initiating nuclear translocation of NFκB (Hayden and Ghosh, 2008). In contrast, in quiescent T cells, Foxo3 factors are located in the nucleus as discussed above. Following T cell activation and induction of PI3K–Akt signaling, they are inactivated and thereby recruited to the cytoplasm (Hedrick, 2009). Further studies are required to identify the cellular and molecular mechanisms by which Foxo factors cross talk with NFκB and other transcription factors in order to control expression of Foxp3. Currently, the proposed scenario is that the PI3K–Akt–mTOR signaling pathway limits the induction of Foxp3 expression but does not impact the maintenance of Foxp3^{+} Tregs (Merkenschlager and von Boehmer, 2010). The former signals are, at least partially, integrated by Foxo transcription factors.

3. ENHANCEMENT OF E.T. TREG GENERATION

It became clear from various studies that treatment with the mTOR-inhibitor rapamycin and its analog everolimus is suited to specifically promote the generation of Foxp3^{+} Treg *in vitro* (Cobbold *et al.*, 2009; Daniel *et al.*, 2010) and *in vivo* (Daniel *et al.*, 2010, 2011a). We could show that the rapamycin-mediated increase in Foxp3 Treg generation results from its interference with activation of the PI3K–mTOR signaling pathway, from its ability to inhibit cellular proliferation and thereby from the lowered activation of DNMT1 as well as from dampening of T cell activation mediated by the ATP-gated P2X7 receptor controlling Ca^{2+} influx (Daniel *et al.*, 2010). Rapamycin-induced Treg generation was shown to be associated with histone H3K4me2 and 3 methylation near the Foxp3 transcriptional start site (Sauer *et al.*, 2008). Earlier studies hypothesized that rapamycin specifically reduced activity of mTORC1 without affecting mTORC2. It was therefore concluded that the ability of rapamycin to promote Foxp3 Treg induction resulted from its interference with mTORC1. It became, however, evident from more recent studies that mTOR inhibitors as rapamycin or everolimus do inhibit mTORC2 as well, thereby resulting in a reduction of Akt activity *in vitro* and *in vivo* (Mancini *et al.*, 2010; Powell and Delgoffe, 2010; Sarbassov *et al.*, 2006; Zeng *et al.*, 2007). This fact appears to be in line with recent studies on the role of Foxo factors in regulating Foxp3 expression. As discussed above, inactivation of Foxo proteins occurs by Akt-mediated phosphorylation which promotes their exclusion from the nucleus. Rapamycin- or everolimus-mediated inhibition of mTORC2 and thereby Akt activity may support the maintenance of active Foxo factors within the nucleus in order to promote induction of Foxp3 expression. mTOR inhibitors as rapamycin or everolimus were shown not only to increase the conversion rate into Treg but also to stabilize Foxp3 expression (Daniel *et al.*, 2010). The stabilizing effect of everolimus on Foxp3 expression became evident when *in vitro* converted Foxp3$^+$ Tregs were subsequently cultured in the absence of TGFβ. While Tregs that had been converted before in the presence of TGFβ alone almost completely lost Foxp3 expression after 4 days of culture, Tregs generated before in the presence of TGFβ and everolimus maintained higher levels of Foxp3$^+$ expression (Daniel *et al.*, 2010). The stabilizing effect of everolimus on Foxp3 expression could also be observed when the drug was added to restimulation cultures that had been converted previously in the presence of TGFβ only. Other compounds such as inhibitors of DNMT1 as 5-Aza-deoxycytidine are often limited in use due to their high toxicity. These drugs were shown to promote Foxp3$^+$ generation less efficiently. A recent pharmacological approach with compounds that simultaneously inhibited mTORC1 and mTORC2 revealed a potent induction of Fopx3$^+$ Tregs indicating that compounds in this class could act as potent agents inducing tolerance and immunosuppression (Delgoffe *et al.*, 2011).

From previous studies by Jonathan Sprent's group, it had become clear that the injection of specific IL-2 monoclonal antibodies resulted in a substantially increased biological activity of preexisting IL-2 (Boyman *et al.*, 2006). From these observations, it was concluded that the increase in proliferation occurred through the formation of immune complexes. Extending these results, we and others designed an *in vivo* approach which induced rapid and specific expansion of Foxp3 expressing cells by injecting mice with IL-2 mixed with a particular IL-2 monoclonal antibody. Injection of these IL-2/IL-2ab complexes for a short period of time (3 days) resulted in a substantial increase in Treg numbers (Daniel *et al.*, 2010; Webster *et al.*, 2009). The achieved expansion of Treg was, however, transient, and expanded Tregs were demonstrated to survive for only 1–2 weeks. Such Tregs are highly activated and harbor superior suppressive function. Characterization of $CD25^{+}Foxp3^{-}$ cells that were also increased after application of IL-2/IL-2ab complexes showed that these cells do not represent cytokine producing effector cells but anergic cells. Some of these cells were shown to produce IL-10 as well. Therefore, the combined application of rapamycin and everolimus followed by administration of IL-2/IL-2ab complexes at the end of the 2-week conversion period makes it feasible to achieve highly efficient Treg conversion and their subsequent expansion (Daniel *et al.*, 2010). Further advances in the mechanistic understanding of signals that impact induction of Foxp3 will facilitate the development of novel approaches and drugs to enhance $Foxp3^{+}$ Treg generation for preventive and therapeutic interventions. As discussed above, Foxo transcription factors, especially the kinases that regulate their activity, may represent promising new drug targets.

4. TOLEROGENIC VACCINATION BY STRONG-AGONIST MIMETOPES OF SELF-ANTIGENS IN ORDER TO PREVENT AUTOIMMUNE DISEASE

4.1. Characteristics of autoreactive T cell epitopes and generation of $Foxp3^{+}$ Tregs

In the past, we carefully explored the optimal conditions for e.t. $Foxp3^{+}$ Treg conversion *in vivo* in order to induce tolerance to antigens, including transplantation antigens (Apostolou and von Boehmer, 2004; Daniel *et al.*, 2010, 2011a; Kretschmer *et al.*, 2005; Verginis *et al.*, 2008). In those systems unrelated to autoimmunity, we successfully generated tolerogenic vaccination protocols and used them, for instance, to induce prospective tolerance in female mice to a variety of male tissues, including skin and hematopoetic grafts (Verginis *et al.*, 2008). However, our initial attempts to prospectively generate insulin-specific Tregs in the non-obese diabetic (NOD) mouse model of autoimmune T1D by subimmunogenic delivery of the natural insulin B chain epitope failed. The poor efficacy of e.t. Treg generation could be caused by several distinct scenarios of which we will address one:

There is compelling evidence that autoantigens are presented and recognized by autoreactive T cell receptors in a unique manner (Fairchild *et al.*, 1993; Garcia *et al.*, 1999; Hahn *et al.*, 2005; He *et al.*, 2002; Liu *et al.*, 1995; Sethi *et al.*, 2011; Stadinski *et al.*, 2010b). Insulin was revealed to represent an essential autoantigen in T1D that needs to be recognized by the immune system in NOD mice in order to initiate the disease (Jaeckel *et al.*, 2004; Nakayama *et al.*, 2005). When Wegmann *et al.*

(1994) and Daniel *et al.* (1995) initially cloned T cells directly from islets of prediabetic NOD mice, it became evident that the majority of isolated CD4 clones reacted with insulin. More than 90% of such insulin-reactive clones were demonstrated to respond specifically to insulin B:9–23. Besides this well-established knowledge, there has been considerable difficulty in identifying the relevant epitope(s) for the T cells within the insulin B:9–23 peptide, supporting the notion that this peptide may bind to the MHC class II molecule I-A^{g7} and be recognized by T cells in different binding registers. This hypothesis was recently solved by work from John Kapplers group which showed that the relevant insulin B chain epitope is presented by MHCII I-A^{g7} molecules in an unfavored binding register, which is associated with weak agonist activity of the peptide–MHC complex (Stadinski *et al.*, 2010b). The poor binding is due to an incompatibility between the p9 amino acid of the insulin epitope and the unique I-A^{g7} p9 pocket polymorphism that is strongly associated with disease susceptibility (Stadinski *et al.*, 2010b). In this unfavored binding register, an arginine residue confronts another arginine in the positively charged p9 pocket of I-A^{g7}.

Elucidation of the crystal structures of human T1D susceptibility alleles HLA-DQ8 and DQ2 and the homologous molecule in the NOD mouse, I-A^{g7}, provided evidence for striking structural similarities in the peptide-binding pockets (Lee *et al.*, 2001). The one unique feature that confers disease susceptibility was identified in the peptide-binding groove of HLA-DQ8 and I-A^{g7}: most human HLA-DQ and mouse I-A alleles harbor an aspartic acid residue at position 57 in the β chain that points into the peptide-binding groove (Acha-Orbea and McDevitt, 1987; Todd *et al.*, 1987). In most HLA-DQ and I-A alleles, this aspartic acid forms an electrostatic salt bridge with arginine at position 76 in the α chain of the molecule. It was then shown that both HLA-DQ8 and I-A^{g7} lack aspartic acid at position 57 in the β chain. This variation was shown to disrupt the electrostatic interaction and leave the positive charge of the arginine unopposed (Corper *et al.*, 2000; Lee *et al.*, 2001). This then results in a positively charged surface of the peptide-binding groove. The aforementioned polymorphisms therefore strongly implicate a critical relevance of peptide presentation in autoimmune susceptibility.

Interestingly, recent work by Stadinski *et al.* (2010a) identified the peptide WE14 from chromogranin A as the antigen for the BDC2.5 T cell clone. Earlier studies had indicated that the secretory granules of islet β cells are the source for the antigen for those clones (Bergman and Haskins, 1994). It became now clear from peptide truncation and extension analysis that WE14 binds to MHCII I-A^{g7} in an atypical manner, in that the peptide occupies only the carboxy-terminal half of the MHC-binding groove (Stadinski *et al.*, 2010a). These findings further support the notion that in autoimmunity, such as T1D, autoreactive T cells respond to unusually presented self-ligands (Stadinski *et al.*, 2010a,b).

In line with these observations, atypical antigen presentation was shown by Garcia's group in the murine experimental allergic encephalomyelitis EAE model: the immunodominant epitope of myelin basic protein MBP1-11 was demonstrated to bind poorly to I-A^u (He *et al.*, 2002). Elucidated crystal structures of the I-A^u/MBP1-11 complex indicated that the MBP1-11 peptide only partially filled the peptide-binding groove. Peptide binding and structural work indicated in addition that the fourth residue of MBP1-11 was placed in a structurally incompatible p6 pocket of I-A^u, which also accounts for the short half-life of the peptide–MHC complex (He *et al.*, 2002). It became evident from structures of the trimolecular complex including TCRs recognizing the MBP1-11 peptide that partial filling of the peptide-binding groove also caused impairment of TCR interactions with this self-epitope. The observed scenario was that the partial occupancy of the peptide-binding groove not only impacted the stability of the peptide–MHC complex but also resulted in diminished contacts with the TCRs.

It has therefore been suggested that the unfavorable manner in which self-epitopes interact with the disease-promoting MHC molecule constitutes a scenario in which autoreactive T cells are able to escape negative selection in the thymus and to generate a pathogenic T cell response in the periphery. This hypothesis is well in line with a number of different studies indicating that T cells with specificity for weakly agonistic autoantigens evade negative selection in the thymus and mount an autoimmune attack in peripheral tissue (Fairchild *et al.*, 1993; Garcia *et al.*, 1999; Hahn *et al.*, 2005; Liu *et al.*, 1995; Stadinski *et al.*, 2010b; Wucherpfennig and Sethi, 2011).

We propose that, in autoimmunity, weak agonistic ligands do not only promote evasion from "recessive" tolerance—that is, negative selection—in the thymus but that such ligands also fail to efficiently induce "dominant" tolerance, that is, $Foxp3^+$ Tregs that are able to control the immune response in the periphery. This hypothesis is based on previous data showing that strong-agonist TCR ligands when applied under subimmunogenic conditions convert naive T cells into stable $Foxp3^+$ Tregs. It is well accepted that, in T1D, insulin represents an essential autoantigen that must be recognized by the immune system in order to initiate the disease (Jaeckel *et al.*, 2004; Nakayama *et al.*, 2005). When in the latter experiments insulin 1 and insulin 2 gene knockout NOD mice combined with a mutated proinsulin transgene that abrogated stimulation of autoreactive T cell clones were generated, it became clear that these mice did not develop T1D (Nakayama *et al.*, 2005). In contrast, mice that harbored at least one copy of the native insulin gene rapidly developed disease (Nakayama *et al.*, 2005). We therefore focused on insulin B:9–23 with the aim to generate insulin-specific $Foxp3^+$ Tregs that could interfere with the development of

T1D. We showed that, in autoimmune T1D, the intrinsic properties of self-reactive epitopes such as insulin critically impinge on the efficacy of e.t. Treg generation: We failed to induce dominant tolerance to natural insulin epitopes. In order to efficiently induce insulin-specific Foxp3$^+$ Treg that could interfere with the development of T1D in NOD mice, we generated a strongly agonistic variant of the weakly agonist natural insulin epitope by replacing the arginine residue (R22) in the natural insulin epitope that in the third MHC-binding register (Stadinski *et al.*, 2010b) confronts another arginine in the positively charged p9 pocket of I-A^{g7}, with glutamic acid. We demonstrated that the subimmunogenic delivery of this strongly agonistic insulin mimetope at a young age was suited to convert naive T cells into Foxp3$^+$ Treg with high efficacy (Fig. 5.1) and thereby to prevent T1D. It became clear that subimmunogenic doses of the insulin mimetope protected NOD mice for 40 weeks and longer from development of disease (Daniel *et al.*, 2011a,b). When the development of diabetes was analyzed as a function of insulin autoantibody (IAA) levels at a young age, prior to vaccination, it became clear that there was indeed a correlation of autoantibody indices and development of disease in that mice with higher IAA indices developed diabetes earlier. We found that the natural insulin epitope afforded only marginal protection in preventing disease; in contrast, the insulin mimetope prevented the onset of T1D in the mice with moderate levels of IAA. Only in mice with high IAA levels already at a young age, vaccination with the insulin mimetope did not achieve complete prevention of disease development. This observation can probably be best

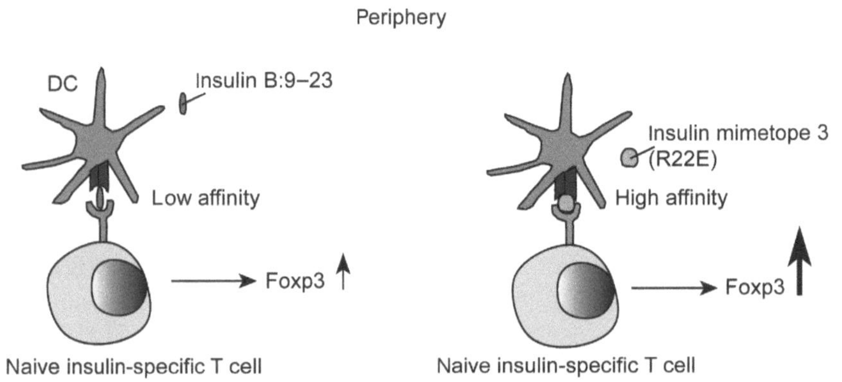

FIGURE 5.1 The extrathymic conversion of naive T cells into Foxp3$^+$ regulatory T cells can be accomplished *in vivo* by delivering strong-agonist ligands under subimmunogenic conditions. In the case of T1D, poor-agonistic insulin epitopes (insulin B:9–23; Stadinski *et al.*, 2010a,b) fail to efficiently induce extrathymic Foxp3$^+$ Tregs. A strongly agonistic insulin mimetope (R22E) was suited to convert naive T cells into Foxp3$^+$ Tregs with high efficacy and thereby to prevent T1D (Daniel *et al.*, 2011a,b). (For color version of this figure, the reader is referred to the Web version of this chapter.)

explained by the assumption that mice with very high indices of IAA already contain too many activated T cells that are resistant to Foxp3$^+$ Treg conversion. Mice vaccinated with subimmunogenic doses of the insulin mimetope exhibited significantly lower insulitis scores in comparison to untreated mice or mice that had received the natural insulin epitope. When pancreatic cryosections from NOD mice of various treatment groups were analyzed for Foxp3 expression by immunehistochemistry, it became clear that in contrast to the natural insulin epitope, tolerogenic vaccination with the strong-agonist mimetope resulted in elevated numbers of Foxp3$^+$ cells. Also in NOD Foxp3 GFP reporter mice, subimmunogenic vaccination with the strong-agonistic insulin mimetope resulted in increased percentages and absolute numbers of GFP$^+$ cells in pancreatic lymph nodes, while treatment with the natural insulin epitope resulted only in a marginal increase of Foxp3 expression. The concept of induced dominant tolerance was further supported by the fact that insulin-specific TCR transgenic effector T cells proliferated least extensively when transferred together with antigen into mimetope-vaccinated NOD mice, which became equally apparent at 8 or 30 weeks of age. Taken together, we described a strongly agonistic insulin mimetope that, when applied under subimmunogenic conditions, efficiently generates Foxp3$^+$ Tregs which suppress the proliferation of IFNγ and IL-17 cytokine-secreting (Lu *et al.*, 2010) insulin-specific effector cells (Daniel *et al.*, 2011a,b). In contrast, natural insulin epitopes were found to be ineffective in preventing T1D irrespective of the applied dose.

In line with our findings and done in short-term studies in experiments unrelated to autoimmune disease, it became evident that TCR ligand affinity also critically impacts peripheral induction of Foxp3 *in vivo* (Gottschalk *et al.*, 2010). By kinetic analyzes, these authors could show that Foxp3$^+$ Tregs generated by low-affinity ligands did not stay.

Thus the recent results on e.t. Treg induction show that low doses of strong-agonist ligands are suited to generate stable Foxp3$^+$ Tregs with high efficacy (Daniel *et al.*, 2010; Gottschalk *et al.*, 2010; Kretschmer *et al.*, 2005; Polansky *et al.*, 2008), whereas even high doses of poorly agonistic ligands fail to induce stable Foxp3$^+$ Treg (Daniel *et al.*, 2011a,b; Gottschalk *et al.*, 2010). These findings support the notion that TCR ligand density cannot compensate for diminished agonistic activity in determining the efficacy and persistence of Foxp3$^+$ Tregs. It was therefore proposed that qualitative differences in TCR stimulation are relevant for stable generation of Foxp3$^+$ Tregs (Gottschalk *et al.*, 2010). The proposed scenario could be that a strong TCR–peptide–MHC interaction specifically triggers survival signals, as opposed to proapoptotic signals which are induced by triggering of many complexes with weaker interactions (Gottschalk *et al.*, 2010). As a means to explain how the TCR can distinguish between ligands with weak or strong-agonistic activity, studies by Stefanova *et al.* (2003) provided evidence for competing feedback loops in T cells that permit functional discrimination between self- and foreign ligands. It became clear that recruitment of the tyrosine phosphatase 1 (SHP-1) following engagement of a weak-agonistic ligand triggered a negative response circuit leading to Lck inactivation by SHP-1-mediated dephosphorylation, which resulted in cessation of the TCR signal. In contrast, it was shown that strong-agonistic ligands activated a positive feedback loop involving modification of Lck by ERK, which interferes with the recruitment of SHP-1 and inactivation of Lck, thereby sustaining the TCR signal (Stefanova *et al.*, 2003). Ronald Schwartz's group (Singh and Schwartz, 2003) provided additional evidence for an intrinsic ''calibration'' process of TCR sensitivity to antigenic stimulation. Adoptive transfer experiments of monoclonal naive TCR transgenic T cells into recipient mice presenting different constant levels of the cognate antigen showed that peripheral $CD4^{+}$ T cells undergo adaptation in response to persistent antigenic stimulation. Their finding that T cells residing in a host with a lower level of antigen presentation show a greater responsiveness supported the notion of a cell-intrinsic tuning of activation thresholds (Singh and Schwartz, 2003).

In addition, it became apparent that cell-extrinsic mechanisms could account for the discrimination between ''high-potency'' and ''high-density'' ligands. It was suggested that prolonged stable interactions between T cells and APCs were required for effective induction of proliferation and effector functions (Celli *et al.*, 2008; von Andrian and Mempel, 2003). In addition, the maturation state of APCs was revealed to be of high importance for the T cell fate under tolerizing conditions (Steinman *et al.*, 2003). Moreover, ligand dose and potency were indicated to impact the kinetics and type of effector T cell APC contact formation (Henrickson *et al.*, 2008; Skokos *et al.*, 2007), and thereby also to impinge on the threshold for *in vivo* induction of T cell proliferation versus tolerance. Accordingly, it was assumed that low doses of peptides could lead to short interactions, which, given a strong-agonist ligand, could be sufficient to provide the subimmunogenic stimulation necessary for efficient and stable Foxp3 induction in the absence of strong cell division *in vivo* (Daniel *et al.*, 2011a,b; Gottschalk *et al.*, 2010).

Since the discovery of the TCR, studies of TCR-induced changes resulted in extension of our understanding of the biochemical pathways that are triggered downstream of the TCR (Fooksman *et al.*, 2010; Smith-Garvin *et al.*, 2009; van der Merwe and Dushek, 2011). Nevertheless, links between many signaling pathways and cellular outcomes remain poorly defined (Smith-Garvin *et al.*, 2009). Future studies will be necessary to fill gaps in our knowledge by providing answers how TCR–peptide–MHC binding is communicated from the extracellular portions of the TCR to signaling pathways across the T cell membrane (van der Merwe and Dushek, 2011). It is currently thought that quantitative and qualitative differences in the strength of TCR–peptide–MHC interactions critically influence signaling events downstream of TCR recognition and thereby the type of cell fate decision instructed in peripheral T cells (van der Merwe and Dushek, 2011).

Three main types of TCR triggering mechanism have been proposed, which involve aggregation, conformational change, and segregation. Recently published studies have provided evidence for each type of mechanism, thereby concluding that all three could be of relevance (van der Merwe and Dushek, 2011). This mechanistic complexity may result from the uniquely demanding nature of TCR-mediated antigen recognition which requires high sensitivity combined with precise discrimination (van der Merwe and Dushek, 2011). Therefore, in terms of molecular aspects impacting T cell activation, it is of interest that a strong-agonist TCR ligand is required for conversion of naive T cells into Treg even though it must be delivered under subimmunogenic conditions and by avoiding activation of the PI3K/Akt/mTOR pathway (Apostolou and von Boehmer, 2004; Gottschalk *et al.*, 2010; Kretschmer *et al.*, 2005). Considering popular models of T cell activation such as TCR cross-linking and costimulation at a first glance, this might appear somewhat counterintuitive (Acuto and Cantrell, 2000). However, conformational changes induced in single receptors by TCR ligands were likewise found as an important mode of TCR triggering (Irvine *et al.*, 2002; van der Merwe and Dushek, 2011). Perhaps this mode of T cell activation leads to conversion of T cells into Tregs rather than to induction of effector function such as cytokine secretion. Further progress in our understanding of TCR triggering will need more detailed information about the structure of the full TCR–CD3 complex and how this structure changes on peptide–MHC binding.

5. CONCLUDING REMARKS

In the setting of autoimmune T1D, it has become clear that subimmunogenic application of a strong-agonist insulin mimetope converts naive insulin-specific T cells into stable Foxp3$^+$ Tregs with high efficacy resulting in the most efficient diabetes prevention in mice recorded to date. In contrast, it became clear that immunogenic high doses of antigens promoted T cell proliferation and activation of the PI3K/Akt/mTOR signaling pathway (Merkenschlager and von Boehmer, 2010) that could interfere with the e.t. generation of Foxp3 Tregs (Kretschmer *et al.*, 2005; Merkenschlager and von Boehmer, 2010; Sauer *et al.*, 2008). Currently, the described tolerogenic vaccination approach is limited to prevention rather than treatment of disease, because it converts naive and not already activated T cells. It is proposed that treatment of recent onset disease may become feasible when compounds are developed that are suited to specifically interfere with the silencing of the Foxp3 locus in activated T cells.

Since the early 1990s, mucosal exposure of insulin and many of its immunogenic epitopes were used in animal models to achieve prevention of T1D. These attempts did not have the specific goal to induce Foxp3$^+$ Tregs. Oral insulin at a dose of 1 mg twice a week for 5 weeks followed by weekly treatment was able to delay diabetes onset and reduce diabetes incidence in NOD mice (Zhang *et al.*, 1991). Adjuvants such as cholera toxin B subunit or incomplete Freund's adjuvant were likewise used for insulin immunization studies (Bergerot *et al.*, 1997). Similar to oral treatment, intranasal aerosol insulin treatment of NOD mice delayed diabetes incidence in NOD mice (Aspord and Thivolet, 2002; Harrison *et al.*, 1996). In addition to whole insulin, insulin-derived peptides, such as the insulin B:9–23-mutated B24–C33, and proinsulin II, were also shown to reduce diabetes incidence (Chen *et al.*, 2001; Daniel and Wegmann, 1996; Martinez *et al.*, 2003). These approaches all significantly reduced pathology and incidence of diabetes; however, they were not able to completely prevent disease development presumably because they did not explore the possibility of specifically converting T cells into Foxp3$^+$ Tregs. From our recent work, it became evident that the subimmunogenic application of a strong-agonist insulin mimetope resulted in a distinct reduction of IAA levels (Daniel *et al.*, 2011a,b). In contrast, immunization trials with insulin in the presence of IFA were shown to increase titers of IAA (Orban *et al.*, 2010). These results therefore suggested that different mechanisms of tolerization were involved and that perhaps a Th1–Th2 shift contributed to the previously published results (Alleva *et al.*, 2002).

In NOD mice, insulin was shown to represent an essential target of the autoimmune attack (Jaeckel *et al.*, 2004; Nakayama *et al.*, 2005). The understanding of the scenario in humans is less advanced even though in humans' i.t. expression levels of insulin control T1D (Pugliese *et al.*, 1997). Several studies showed that a variable nucleotide tandem repeat found upstream of the insulin gene was associated with T1D susceptibility and resistance in humans (Onengut-Gumuscu and Concannon, 2006). Moreover, recent studies suggest that prepro-insulin functions as the most important autoantigen in childhood-onset T1D (Patterson *et al.*, 2009). Thus insulin has a major influence on human disease as well.

It became apparent through recent fine mapping of the extended MHC region, 8 Mb of chromosome 6p21, which was realized by the development of high throughput and genom-wide single nucleotide polymorphism genotyping, that major susceptibility to T1D does indeed map to the HLA class II region (Howson *et al.*, 2009). Additionally, the genome-wide association study (Concannon *et al.*, 2009) of human T1D confirmed that from the 26 identified loci, at least 6 loci are shared between the NOD mouse model and humans at risk for T1D and 19 were shown to be associated with immune regulation (Concannon *et al.*, 2009; Turley *et al.*, 2003).

The hallmark characteristics of T1D present in most patients are at least one HLA class II haplotype associated with susceptibility and the presence of islet autoantibodies such as IAAs (Ziegler and Nepom, 2010). It became clear that among children with a family history of T1D, there is a first peak incidence of islet autoantibodies as early as at 1–2 years of age (Bonifacio *et al.*, 2008; Hummel *et al.*, 2004; Naserke *et al.*, 1999). These studies propose that the events leading to destruction of pancreatic β cells are initiated with increased frequency in early infancy. It appears important to note that among these autoantibodies, antibodies against insulin or proinsulin are often among the first to appear (Hummel *et al.*, 2004). In addition, it became evident that the IAAs are of high-affinity IgG1 class, which could suggest rapid affinity maturation (Achenbach *et al.*, 2004; Castano *et al.*, 1993). The discussed findings therefore support the view that reactivity against insulin is frequently associated with the initiation of the disease process in juvenile T1D.

Currently, our proposed tolerogenic vaccination strategy is focused on prevention of disease. Therefore, a possible clinical protocol could aim at inducing insulin-specific Foxp3$^+$ Tregs in individuals at genetic risk of developing disease and in the presence of moderate levels of insulin autoantibodies. Data from the Diabetes Prevention Trial (DPT-Trial) and the European Nicotinamide Diabetes Intervention Trial (ENDIT-Trial) supported the notion that in human T1D, a chronic autoimmune process gradually results in destruction of pancreatic β-cells (Bluestone *et al.*, 2010). It became clear that the risk of diabetes is as high as 90% in first-degree relatives of patients who are positive for at least two autoantibodies. In contrast, the risk of diabetes development was shown to be less than 20% in relatives who are positive for only one autoantibody (Bluestone *et al.*, 2010). Based on this knowledge, the timing of antigen-specific tolerance strategies seems to be of critical relevance. It seems therefore reasonable to predict that tolerogenic vaccination could be most successful in the earliest stages of disease. Tolerogenic vaccination was shown to require only a single epitope because the induced Tregs suppresses immune responses of T cells to antigen from the same source through bystander suppression (Verginis *et al.*, 2008), whereby the vicinity to antigen-presenting cells but not the specificity of the suppressed cells is important (Fig. 5.2). It is the distinct advantage of this strategy that Foxp3$^+$ Tregs of one specificity can suppress effector cells with specificity for any antigen from the same source.

With the advances in our understanding of Foxp3$^+$ Treg generation and function, there is significant promise that the development of tolerogenic vaccination strategies can be achieved in the future in order to safely and specifically interfere with unwanted immune responses such as autoimmunity, allergy, and transplantation, thereby replacing indiscriminate immunosuppression.

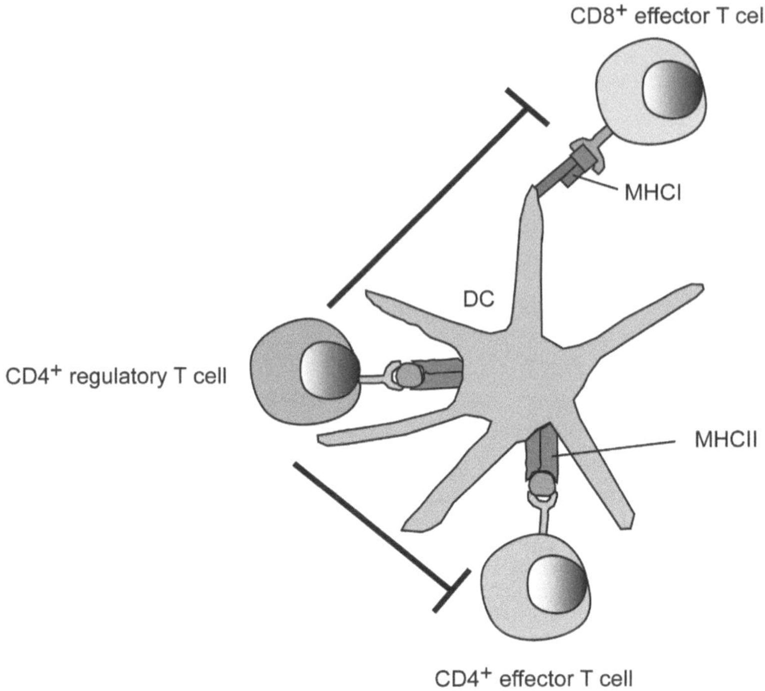

FIGURE 5.2 Tolerogenic vaccination requires only a single epitope. Extrathymically induced Tregs suppress immune reactions of neighboring effector T cells to antigen from the same source through bystander suppression. In order for Tregs to exert their suppressive functions, the vicinity to antigen-presenting cells but not the specificity of the suppressed cells is relevant (Verginis *et al.*, 2008). (For color version of this figure, the reader is referred to the Web version of this chapter.)

ACKNOWLEDGMENTS

The studies were supported by National Institutes of Health Grant NIH-AI-53102 (to H. v. B.). C. D. was supported by a Leopoldina research fellowship (BMBF-LPD 9901/8-184) and by the LOEWE (LipidSignaling Forschungszentrum Frankfurt) program of the Federal State of Hessen, Germany.

REFERENCES

Acha-Orbea, H., and McDevitt, H. O. (1987). The first external domain of the nonobese diabetic mouse class II I-A beta chain is unique. *Proc. Natl. Acad. Sci. USA* **84,** 2435–2439.

Achenbach, P., Koczwara, K., Knopff, A., Naserke, H., Ziegler, A. G., and Bonifacio, E. (2004). Mature high-affinity immune responses to (pro)insulin anticipate the autoimmune cascade that leads to type 1 diabetes. *J. Clin. Invest.* **114,** 589–597.

Acuto, O., and Cantrell, D. (2000). T cell activation and the cytoskeleton. *Annu. Rev. Immunol.* **18,** 165–184.

Alleva, D. G., Gaur, A., Jin, L., Wegmann, D., Gottlieb, P. A., Pahuja, A., Johnson, E. B., Motheral, T., Putnam, A., Crowe, P. D., Ling, N., Boehme, S. A., *et al.* (2002). Immunological characterization and therapeutic activity of an altered-peptide ligand, NBI-6024, based on the immunodominant type 1 diabetes autoantigen insulin B-chain (9–23) peptide. *Diabetes* **51,** 2126–2134.

Almeida, A. R., Legrand, N., Papiernik, M., and Freitas, A. A. (2002). Homeostasis of peripheral CD4+ T cells: IL-2R alpha and IL-2 shape a population of regulatory cells that controls CD4+ T cell numbers. *J. Immunol.* **169,** 4850–4860.

Apostolou, I., and von Boehmer, H. (2004). In vivo instruction of suppressor commitment in naive T cells. *J. Exp. Med.* **199,** 1401–1408.

Apostolou, I., Sarukhan, A., Klein, L., and von Boehmer, H. (2002). Origin of regulatory T cells with known specificity for antigen. *Nat. Immunol.* **3,** 756–763.

Asano, M., Toda, M., Sakaguchi, N., and Sakaguchi, S. (1996). Autoimmune disease as a consequence of developmental abnormality of a T cell subpopulation. *J. Exp. Med.* **184,** 387–396.

Aschenbrenner, K., D'Cruz, L. M., Vollmann, E. H., Hinterberger, M., Emmerich, J., Swee, L. K., Rolink, A., and Klein, L. (2007). Selection of Foxp3+ regulatory T cells specific for self antigen expressed and presented by Aire+ medullary thymic epithelial cells. *Nat. Immunol.* **8,** 351–358.

Aspord, C., and Thivolet, C. (2002). Nasal administration of CTB-insulin induces active tolerance against autoimmune diabetes in non-obese diabetic (NOD) mice. *Clin. Exp. Immunol.* **130,** 204–211.

Bachmann, M. F., Kohler, G., Ecabert, B., Mak, T. W., and Kopf, M. (1999). Cutting edge: Lymphoproliferative disease in the absence of CTLA-4 is not T cell autonomous. *J. Immunol.* **163,** 1128–1131.

Baron, U., Floess, S., Wieczorek, G., Baumann, K., Grutzkau, A., Dong, J., Thiel, A., Boeld, T. J., Hoffmann, P., Edinger, M., Turbachova, I., Hamann, A., *et al.* (2007). DNA demethylation in the human FOXP3 locus discriminates regulatory T cells from activated FOXP3(+) conventional T cells. *Eur. J. Immunol.* **37,** 2378–2389.

Bautista, J. L., Lio, C. W., Lathrop, S. K., Forbush, K., Liang, Y., Luo, J., Rudensky, A. Y., and Hsieh, C. S. (2009). Intraclonal competition limits the fate determination of regulatory T cells in the thymus. *Nat. Immunol.* **10,** 610–617.

Bayer, A. L., Yu, A., Adeegbe, D., and Malek, T. R. (2005). Essential role for interleukin-2 for CD4(+)CD25(+) T regulatory cell development during the neonatal period. *J. Exp. Med.* **201,** 769–777.

Benson, M. J., Pino-Lagos, K., Rosemblatt, M., and Noelle, R. J. (2007). All-trans retinoic acid mediates enhanced T reg cell growth, differentiation, and gut homing in the face of high levels of co-stimulation. *J. Exp. Med.* **204,** 1765–1774.

Bergerot, I., Ploix, C., Petersen, J., Moulin, V., Rask, C., Fabien, N., Lindblad, M., Mayer, A., Czerkinsky, C., Holmgren, J., and Thivolet, C. (1997). A cholera toxoid-insulin conjugate as an oral vaccine against spontaneous autoimmune diabetes. *Proc. Natl. Acad. Sci. USA* **94,** 4610–4614.

Bergman, B., and Haskins, K. (1994). Islet-specific T-cell clones from the NOD mouse respond to beta-granule antigen. *Diabetes* **43,** 197–203.

Beugnet, A., Tee, A. R., Taylor, P. M., and Proud, C. G. (2003). Regulation of targets of mTOR (mammalian target of rapamycin) signalling by intracellular amino acid availability. *Biochem. J.* **372,** 555–566.

Bluestone, J. A., Herold, K., and Eisenbarth, G. (2010). Genetics, pathogenesis and clinical interventions in type 1 diabetes. *Nature* **464,** 1293–1300.

Bonasio, R., Scimone, M. L., Schaerli, P., Grabie, N., Lichtman, A. H., and von Andrian, U. H. (2006). Clonal deletion of thymocytes by circulating dendritic cells homing to the thymus. *Nat. Immunol.* **7,** 1092–1100.

Bonifacio, E., Pfluger, M., Marienfeld, S., Winkler, C., Hummel, M., and Ziegler, A. G. (2008). Maternal type 1 diabetes reduces the risk of islet autoantibodies: Relationships with birthweight and maternal HbA(1c). *Diabetologia* **51,** 1245–1252.

Bonomo, A., Kehn, P. J., Payer, E., Rizzo, L., Cheever, A. W., and Shevach, E. M. (1995). Pathogenesis of post-thymectomy autoimmunity. Role of syngeneic MLR-reactive T cells. *J. Immunol.* **154,** 6602–6611.

Boyman, O., Kovar, M., Rubinstein, M. P., Surh, C. D., and Sprent, J. (2006). Selective stimulation of T cell subsets with antibody-cytokine immune complexes. *Science* **311,** 1924–1927.

Brunkow, M. E., Jeffery, E. W., Hjerrild, K. A., Paeper, B., Clark, L. B., Yasayko, S. A., Wilkinson, J. E., Galas, D., Ziegler, S. F., and Ramsdell, F. (2001). Disruption of a new forkhead/winged-helix protein, scurfin, results in the fatal lymphoproliferative disorder of the scurfy mouse. *Nat. Genet.* **27,** 68–73.

Brunner, M. C., Chambers, C. A., Chan, F. K., Hanke, J., Winoto, A., and Allison, J. P. (1999). CTLA-4-mediated inhibition of early events of T cell proliferation. *J. Immunol.* **162,** 5813–5820.

Burchill, M. A., Goetz, C. A., Prlic, M., O'Neil, J. J., Harmon, I. R., Bensinger, S. J., Turka, L. A., Brennan, P., Jameson, S. C., and Farrar, M. A. (2003). Distinct effects of STAT5 activation on CD4+ and CD8+ T cell homeostasis: Development of CD4+CD25+ regulatory T cells versus CD8+ memory T cells. *J. Immunol.* **171,** 5853–5864.

Burchill, M. A., Yang, J., Vogtenhuber, C., Blazar, B. R., and Farrar, M. A. (2007). IL-2 receptor beta-dependent STAT5 activation is required for the development of Foxp3+ regulatory T cells. *J. Immunol.* **178,** 280–290.

Burchill, M. A., Yang, J., Vang, K. B., Moon, J. J., Chu, H. H., Lio, C. W., Vegoe, A. L., Hsieh, C. S., Jenkins, M. K., and Farrar, M. A. (2008). Linked T cell receptor and cytokine signaling govern the development of the regulatory T cell repertoire. *Immunity* **28,** 112–121.

Burnet, F. M. (1959). The Clonal Selection Theory. Cambridge University Press, Cambridge.

Castano, L., Ziegler, A. G., Ziegler, R., Shoelson, S., and Eisenbarth, G. S. (1993). Characterization of insulin autoantibodies in relatives of patients with type I diabetes. *Diabetes* **42,** 1202–1209.

Celli, S., Garcia, Z., Beuneu, H., and Bousso, P. (2008). Decoding the dynamics of T cell-dendritic cell interactions in vivo. *Immunol. Rev.* **221,** 182–187.

Chatila, T. A., Blaeser, F., Ho, N., Lederman, H. M., Voulgaropoulos, C., Helms, C., and Bowcock, A. M. (2000). JM2, encoding a fork head-related protein, is mutated in X-linked autoimmunity-allergic disregulation syndrome. *J. Clin. Invest.* **106,** R75–R81.

Chen, W., Bergerot, I., Elliott, J. F., Harrison, L. C., Abiru, N., Eisenbarth, G. S., and Delovitch, T. L. (2001). Evidence that a peptide spanning the B-C junction of proinsulin is an early autoantigen epitope in the pathogenesis of type 1 diabetes. *J. Immunol.* **167,** 4926–4935.

Chen, W., Jin, W., Hardegen, N., Lei, K. J., Li, L., Marinos, N., McGrady, G., and Wahl, S. M. (2003). Conversion of peripheral CD4+. *J. Exp. Med.* **198,** 1875–1886.

Chen, C., Rowell, E. A., Thomas, R. M., Hancock, W. W., and Wells, A. D. (2006). Transcriptional regulation by Foxp3 is associated with direct promoter occupancy and modulation of histone acetylation. *J. Biol. Chem.* **281,** 36828–36834.

Cobbold, S. P., Adams, E., Farquhar, C. A., Nolan, K. F., Howie, D., Lui, K. O., Fairchild, P. J., Mellor, A. L., Ron, D., and Waldmann, H. (2009). Infectious tolerance via the consumption of essential amino acids and mTOR signaling. *Proc. Natl. Acad. Sci. USA* **106,** 12055–12060.

Concannon, P., Chen, W. M., Julier, C., Morahan, G., Akolkar, B., Erlich, H. A., Hilner, J. E., Nerup, J., Nierras, C., Pociot, F., Todd, J. A., and Rich, S. S. (2009). Genome-wide scan for linkage to type 1 diabetes in 2,496 multiplex families from the Type 1 Diabetes Genetics Consortium. *Diabetes* **58,** 1018–1022.

Coombes, J. L., Siddiqui, K. R., Arancibia-Carcamo, C. V., Hall, J., Sun, C. M., Belkaid, Y., and Powrie, F. (2007). A functionally specialized population of mucosal CD103+ DCs induces Foxp3+ regulatory T cells via a TGF-beta and retinoic acid-dependent mechanism. *J. Exp. Med.* **204,** 1757–1764.

Corper, A. L., Stratmann, T., Apostolopoulos, V., Scott, C. A., Garcia, K. C., Kang, A. S., Wilson, I. A., and Teyton, L. (2000). A structural framework for deciphering the link between I-Ag7 and autoimmune diabetes. *Science* **288,** 505–511.

Daniel, D., and Wegmann, D. R. (1996). Intranasal administration of insulin peptide B: 9–23 protects NOD mice from diabetes. *Ann. N. Y. Acad. Sci.* **778,** 371–372.

Daniel, D., Gill, R. G., Schloot, N., and Wegmann, D. (1995). Epitope specificity, cytokine production profile and diabetogenic activity of insulin-specific T cell clones isolated from NOD mice. *Eur. J. Immunol.* **25,** 1056–1062.

Daniel, C., Wennhold, K., Kim, H. J., and von Boehmer, H. (2010). Enhancement of antigen-specific Treg vaccination in vivo. *Proc. Natl. Acad. Sci. USA* **107,** 16246–16251.

Daniel, C., Weigmann, B., and von Boehmer, H. (2011a). Prevention of type 1 diabetes in mice by tolerogenic mice with a strong agonist insulin mimetope. *J. Exp. Med.* **208,** 1501–1510.

Daniel, C., Ploegh, H., and von Boehmer, H. (2011b). Antigen-specific induction of regulatory T cells in vivo and in vitro. *Methods Mol. Biol.* **707,** 173–185.

D'Cruz, L. M., and Klein, L. (2005). Development and function of agonist-induced CD25+Foxp3+ regulatory T cells in the absence of interleukin 2 signaling. *Nat. Immunol.* **6,** 1152–1159.

Delgoffe, G. M., Kole, T. P., Zheng, Y., Zarek, P. E., Matthews, K. L., Xiao, B., Worley, P. F., Kozma, S. C., and Powell, J. D. (2009). The mTOR kinase differentially regulates effector and regulatory T cell lineage commitment. *Immunity* **30,** 832–844.

Delgoffe, G. M., Pollizzi, K. N., Waickman, A. T., Heikamp, E., Meyers, D. J., Horton, M. R., Xiao, B., Worley, P. F., and Powell, J. D. (2011). The kinase mTOR regulates the differentiation of helper T cells through the selective activation of signaling by mTORC1 and mTORC2. *Nat. Immunol.* **12,** 295–303.

Dennis, P. B., Fumagalli, S., and Thomas, G. (1999). Target of rapamycin (TOR): Balancing the opposing forces of protein synthesis and degradation. *Curr. Opin. Genet. Dev.* **9,** 49–54.

Derbinski, J., Schulte, A., Kyewski, B., and Klein, L. (2001). Promiscuous gene expression in medullary thymic epithelial cells mirrors the peripheral self. *Nat. Immunol.* **2,** 1032–1039.

Dietrich, F. M., and Weigle, W. O. (1963). Induction of tolerance to heterologous proteins and their catabolism in C57BL/6 mice. *J. Exp. Med.* **117,** 621–631.

Ehrlich, P. (1906). Collected Studies on Immunity. Wiley, New York.

Fairchild, P. J., Wildgoose, R., Atherton, E., Webb, S., and Wraith, D. C. (1993). An autoantigenic T cell epitope forms unstable complexes with class II MHC: A novel route for escape from tolerance induction. *Int. Immunol.* **5,** 1151–1158.

Feuerer, M., Hill, J. A., Kretschmer, K., von Boehmer, H., Mathis, D., and Benoist, C. (2010). Genomic definition of multiple ex vivo regulatory T cell subphenotypes. *Proc. Natl. Acad. Sci. USA* **107,** 5919–5924.

Floess, S., Freyer, J., Siewert, C., Baron, U., Olek, S., Polansky, J., Schlawe, K., Chang, H. D., Bopp, T., Schmitt, E., Klein-Hessling, S., Serfling, E., *et al.* (2007). Epigenetic control of the foxp3 locus in regulatory T cells. *PLoS Biol.* **5,** e38.

Fontenot, J. D., Gavin, M. A., and Rudensky, A. Y. (2003). Foxp3 programs the development and function of CD4+CD25+ regulatory T cells. *Nat. Immunol.* **4,** 330–336.

Fontenot, J. D., Rasmussen, J. P., Gavin, M. A., and Rudensky, A. Y. (2005). A function for interleukin 2 in Foxp3-expressing regulatory T cells. *Nat. Immunol.* **6,** 1142–1151.

Fooksman, D. R., Vardhana, S., Vasiliver-Shamis, G., Liese, J., Blair, D. A., Waite, J., Sacristan, C., Victora, G. D., Zanin-Zhorov, A., and Dustin, M. L. (2010). Functional anatomy of T cell activation and synapse formation. *Annu. Rev. Immunol.* **28,** 79–105.

Fowell, D., and Mason, D. (1993). Evidence that the T cell repertoire of normal rats contains cells with the potential to cause diabetes. Characterization of the CD4+ T cell subset that inhibits this autoimmune potential. *J. Exp. Med.* **177,** 627–636.

Garcia, K. C., Teyton, L., and Wilson, I. A. (1999). Structural basis of T cell recognition. *Annu. Rev. Immunol.* **17,** 369–397.

Gavin, M. A., Rasmussen, J. P., Fontenot, J. D., Vasta, V., Manganiello, V. C., Beavo, J. A., and Rudensky, A. Y. (2007). Foxp3-dependent programme of regulatory T-cell differentiation. *Nature* **445,** 771–775.

Gottschalk, R. A., Corse, E., and Allison, J. P. (2010). TCR ligand density and affinity determine peripheral induction of Foxp3 in vivo. *J. Exp. Med.* **207,** 1701–1711.

Guertin, D. A., and Sabatini, D. M. (2007). Defining the role of mTOR in cancer. *Cancer Cell* **12,** 9–22.

Guertin, D. A., Stevens, D. M., Thoreen, C. C., Burds, A. A., Kalaany, N. Y., Moffat, J., Brown, M., Fitzgerald, K. J., and Sabatini, D. M. (2006). Ablation in mice of the mTORC components raptor, rictor, or mLST8 reveals that mTORC2 is required for signaling to Akt-FOXO and PKCalpha, but not S6K1. *Dev. Cell* **11,** 859–871.

Hahn, M., Nicholson, M. J., Pyrdol, J., and Wucherpfennig, K. W. (2005). Unconventional topology of self peptide-major histocompatibility complex binding by a human autoimmune T cell receptor. *Nat. Immunol.* **6,** 490–496.

Harada, Y., Harada, Y., Elly, C., Ying, G., Paik, J. H., DePinho, R. A., and Liu, Y. C. (2010). Transcription factors Foxo3a and Foxo1 couple the E3 ligase Cbl-b to the induction of Foxp3 expression in induced regulatory T cells. *J. Exp. Med.* **207,** 1381–1391.

Harrison, L. C., Dempsey-Collier, M., Kramer, D. R., and Takahashi, K. (1996). Aerosol insulin induces regulatory CD8 gamma delta T cells that prevent murine insulin-dependent diabetes. *J. Exp. Med.* **184,** 2167–2174.

Haxhinasto, S., Mathis, D., and Benoist, C. (2008). The AKT-mTOR axis regulates de novo differentiation of CD4+Foxp3+ cells. *J. Exp. Med.* **205,** 565–574.

Hayden, M. S., and Ghosh, S. (2008). Shared principles in NF-kappaB signaling. *Cell* **132,** 344–362.

He, X. L., Radu, C., Sidney, J., Sette, A., Ward, E. S., and Garcia, K. C. (2002). Structural snapshot of aberrant antigen presentation linked to autoimmunity: The immunodominant epitope of MBP complexed with I-Au. *Immunity* **17,** 83–94.

Hedrick, S. M. (2009). The cunning little vixen: Foxo and the cycle of life and death. *Nat. Immunol.* **10,** 1057–1063.

Henrickson, S. E., Mempel, T. R., Mazo, I. B., Liu, B., Artyomov, M. N., Zheng, H., Peixoto, A., Flynn, M. P., Senman, B., Junt, T., Wong, H. C., Chakraborty, A. K., *et al.* (2008). T cell sensing of antigen dose governs interactive behavior with dendritic cells and sets a threshold for T cell activation. *Nat. Immunol.* **9,** 282–291.

Hinterberger, M., Aichinger, M., da Costa, O. P., Voehringer, D., Hoffmann, R., and Klein, L. (2010). Autonomous role of medullary thymic epithelial cells in central CD4(+) T cell tolerance. *Nat. Immunol.* **11,** 512–519.

Holdorf, A. D., Green, J. M., Levin, S. D., Denny, M. F., Straus, D. B., Link, V., Changelian, P. S., Allen, P. M., and Shaw, A. S. (1999). Proline residues in CD28 and the Src homology (SH)3 domain of Lck are required for T cell costimulation. *J. Exp. Med.* **190,** 375–384.

Holdorf, A. D., Lee, K. H., Burack, W. R., Allen, P. M., and Shaw, A. S. (2002). Regulation of Lck activity by CD4 and CD28 in the immunological synapse. *Nat. Immunol.* **3,** 259–264.

Hori, S., Collette, A., Demengeot, J., and Stewart, J. (2002). A new statistical method for quantitative analyses: Application to the precise quantification of T cell receptor repertoires. *J. Immunol. Methods* **268,** 159–170.

Hori, S., Nomura, T., and Sakaguchi, S. (2003a). Control of regulatory T cell development by the transcription factor Foxp3. *Science* **299,** 1057–1061.

Hori, S., Takahashi, T., and Sakaguchi, S. (2003b). Control of autoimmunity by naturally arising regulatory CD4+ T cells. *Adv. Immunol.* **81,** 331–371.

Howson, J. M., Walker, N. M., Clayton, D., and Todd, J. A. (2009). Confirmation of HLA class II independent type 1 diabetes associations in the major histocompatibility complex including HLA-B and HLA-A. *Diabetes Obes. Metab.* **11**(Suppl. 1), 31–45.

Hsieh, C. S., and Rudensky, A. Y. (2005). The role of TCR specificity in naturally arising CD25+ CD4+ regulatory T cell biology. *Curr. Top. Microbiol. Immunol.* **293,** 25–42.

Hsieh, C. S., Liang, Y., Tyznik, A. J., Self, S. G., Liggitt, D., and Rudensky, A. Y. (2004). Recognition of the peripheral self by naturally arising CD25+ CD4+ T cell receptors. *Immunity* **21,** 267–277.

Hsieh, C. S., Zheng, Y., Liang, Y., Fontenot, J. D., and Rudensky, A. Y. (2006). An intersection between the self-reactive regulatory and nonregulatory T cell receptor repertoires. *Nat. Immunol.* **7,** 401–410.

Huehn, J., Polansky, J. K., and Hamann, A. (2009). Epigenetic control of FOXP3 expression: The key to a stable regulatory T-cell lineage? *Nat. Rev. Immunol.* **9,** 83–89.

Hummel, M., Bonifacio, E., Schmid, S., Walter, M., Knopff, A., and Ziegler, A. G. (2004). Brief communication: Early appearance of islet autoantibodies predicts childhood type 1 diabetes in offspring of diabetic parents. *Ann. Intern. Med.* **140,** 882–886.

Irvine, D. J., Purbhoo, M. A., Krogsgaard, M., and Davis, M. M. (2002). Direct observation of ligand recognition by T cells. *Nature* **419,** 845–849.

Izcue, A., Coombes, J. L., and Powrie, F. (2009). Regulatory lymphocytes and intestinal inflammation. *Annu. Rev. Immunol.* **27,** 313–338.

Jaeckel, E., Lipes, M. A., and von Boehmer, H. (2004). Recessive tolerance to preproinsulin 2 reduces but does not abolish type 1 diabetes. *Nat. Immunol.* **5,** 1028–1035.

Jolicoeur, C., Hanahan, D., and Smith, K. M. (1994). T-cell tolerance toward a transgenic beta-cell antigen and transcription of endogenous pancreatic genes in thymus. *Proc. Natl. Acad. Sci. USA* **91,** 6707–6711.

Jordan, M. S., Riley, M. P., von Boehmer, H., and Caton, A. J. (2000). Anergy and suppression regulate CD4(+) T cell responses to a self peptide. *Eur. J. Immunol.* **30,** 136–144.

Jordan, M. S., Boesteanu, A., Reed, A. J., Petrone, A. L., Holenbeck, A. E., Lerman, M. A., Naji, A., and Caton, A. J. (2001). Thymic selection of CD4+CD25+ regulatory T cells induced by an agonist self-peptide. *Nat. Immunol.* **2,** 301–306.

Josefowicz, S. Z., and Rudensky, A. (2009). Control of regulatory T cell lineage commitment and maintenance. *Immunity* **30,** 616–625.

Josefowicz, S. Z., Wilson, C. B., and Rudensky, A. Y. (2009). Cutting edge: TCR stimulation is sufficient for induction of Foxp3 expression in the absence of DNA methyltransferase 1. *J. Immunol.* **182,** 6648–6652.

Kappler, J. W., Roehm, N., and Marrack, P. (1987). T cell tolerance by clonal elimination in the thymus. *Cell* **49,** 273–280.

Khattri, R., Cox, T., Yasayko, S. A., and Ramsdell, F. (2003). An essential role for Scurfin in CD4+CD25+ T regulatory cells. *Nat. Immunol.* **4,** 337–342.

Kim, H. P., and Leonard, W. J. (2007). CREB/ATF-dependent T cell receptor-induced FoxP3 gene expression: A role for DNA methylation. *J. Exp. Med.* **204,** 1543–1551.

Kim, J. M., Rasmussen, J. P., and Rudensky, A. Y. (2007). Regulatory T cells prevent catastrophic autoimmunity throughout the lifespan of mice. *Nat. Immunol.* **8,** 191–197.

Kisielow, P., Bluthmann, H., Staerz, U. D., Steinmetz, M., and von Boehmer, H. (1988). Tolerance in T-cell-receptor transgenic mice involves deletion of nonmature CD4+8+ thymocytes. *Nature* **333,** 742–746.

Klein, L., Khazaie, K., and von Boehmer, H. (2003). In vivo dynamics of antigen-specific regulatory T cells not predicted from behavior in vitro. *Proc. Natl. Acad. Sci. USA* **100,** 8886–8891.

Kramer, S., Schimpl, A., and Hunig, T. (1995). Immunopathology of interleukin (IL) 2-deficient mice: Thymus dependence and suppression by thymus-dependent cells with an intact IL-2 gene. *J. Exp. Med.* **182,** 1769–1776.

Kretschmer, K., Apostolou, I., Hawiger, D., Khazaie, K., Nussenzweig, M. C., and von Boehmer, H. (2005). Inducing and expanding regulatory T cell populations by foreign antigen. *Nat. Immunol.* **6,** 1219–1227.

Krummel, M. F., and Allison, J. P. (1996). CTLA-4 engagement inhibits IL-2 accumulation and cell cycle progression upon activation of resting T cells. *J. Exp. Med.* **183,** 2533–2540.

Lafaille, J. J., Nagashima, K., Katsuki, M., and Tonegawa, S. (1994). High incidence of spontaneous autoimmune encephalomyelitis in immunodeficient anti-myelin basic protein T cell receptor transgenic mice. *Cell* **78,** 399–408.

Lahl, K., Loddenkemper, C., Drouin, C., Freyer, J., Arnason, J., Eberl, G., Hamann, A., Wagner, H., Huehn, J., and Sparwasser, T. (2007). Selective depletion of Foxp3+ regulatory T cells induces a scurfy-like disease. *J. Exp. Med.* **204,** 57–63.

Laplante, M., and Sabatini, D. M. (2009). An emerging role of mTOR in lipid biosynthesis. *Curr. Biol.* **19,** R1046–R1052.

Lederberg, J. (1959). Genes and antibodies. *Science* **129,** 1649–1653.

Lee, K. H., Wucherpfennig, K. W., and Wiley, D. C. (2001). Structure of a human insulin peptide-HLA-DQ8 complex and susceptibility to type 1 diabetes. *Nat. Immunol.* **2,** 501–507.

Leung, M. W., Shen, S., and Lafaille, J. J. (2009). TCR-dependent differentiation of thymic Foxp3+ cells is limited to small clonal sizes. *J. Exp. Med.* **206,** 2121–2130.

Li, M. O., Sanjabi, S., and Flavell, R. A. (2006). Transforming growth factor-beta controls development, homeostasis, and tolerance of T cells by regulatory T cell-dependent and -independent mechanisms. *Immunity* **25,** 455–471.

Lin, W., Haribhai, D., Relland, L. M., Truong, N., Carlson, M. R., Williams, C. B., and Chatila, T. A. (2007). Regulatory T cell development in the absence of functional Foxp3. *Nat. Immunol.* **8,** 359–368.

Liston, A., Nutsch, K. M., Farr, A. G., Lund, J. M., Rasmussen, J. P., Koni, P. A., and Rudensky, A. Y. (2008). Differentiation of regulatory Foxp3+ T cells in the thymic cortex. *Proc. Natl. Acad. Sci. USA* **105,** 11903–11908.

Liu, G. Y., Fairchild, P. J., Smith, R. M., Prowle, J. R., Kioussis, D., and Wraith, D. C. (1995). Low avidity recognition of self-antigen by T cells permits escape from central tolerance. *Immunity* **3,** 407–415.

Liu, Y., Zhang, P., Li, J., Kulkarni, A. B., Perruche, S., and Chen, W. (2008). A critical function for TGF-beta signaling in the development of natural CD4+CD25+Foxp3+ regulatory T cells. *Nat. Immunol.* **9,** 632–640.

Lu, L. F., Boldin, M. P., Chaudhry, A., Lin, L. L., Taganov, K. D., Hanada, T., Yoshimura, A., Baltimore, D., and Rudensky, A. Y. (2010). Function of miR-146a in controlling Treg cell-mediated regulation of Th1 responses. *Cell* **142,** 914–929.

Malek, T. R., Yu, A., Vincek, V., Scibelli, P., and Kong, L. (2002). CD4 regulatory T cells prevent lethal autoimmunity in IL-2Rbeta-deficient mice. Implications for the nonredundant function of IL-2. *Immunity* **17,** 167–178.

Mancini, M., Petta, S., Martinelli, G., Barbieri, E., and Santucci, M. A. (2010). RAD 001 (everolimus) prevents mTOR and Akt late re-activation in response to imatinib in chronic myeloid leukemia. *J. Cell. Biochem.* **109,** 320–328.

Marie, J. C., Letterio, J. J., Gavin, M., and Rudensky, A. Y. (2005). TGF-beta1 maintains suppressor function and Foxp3 expression in CD4+CD25+ regulatory T cells. *J. Exp. Med.* **201,** 1061–1067.

Marie, J. C., Liggitt, D., and Rudensky, A. Y. (2006). Cellular mechanisms of fatal early-onset autoimmunity in mice with the T cell-specific targeting of transforming growth factor-beta receptor. *Immunity* **25,** 441–454.

Marson, A., Kretschmer, K., Frampton, G. M., Jacobsen, E. S., Polansky, J. K., MacIsaac, K. D., Levine, S. S., Fraenkel, E., von Boehmer, H., and Young, R. A. (2007). Foxp3 occupancy and regulation of key target genes during T-cell stimulation. *Nature* **445,** 931–935.

Martinez, N. R., Augstein, P., Moustakas, A. K., Papadopoulos, G. K., Gregori, S., Adorini, L., Jackson, D. C., and Harrison, L. C. (2003). Disabling an integral CTL epitope allows suppression of autoimmune diabetes by intranasal proinsulin peptide. *J. Clin. Invest.* **111,** 1365–1371.

Mathis, D., and Benoist, C. (2007). A decade of AIRE. *Nat. Rev. Immunol.* **7,** 645–650.

Merkenschlager, M., and von Boehmer, H. (2010). PI3 kinase signalling blocks Foxp3 expression by sequestering Foxo factors. *J. Exp. Med.* **207,** 1347–1350.

Mombaerts, P., Clarke, A. R., Rudnicki, M. A., Iacomini, J., Itohara, S., Lafaille, J. J., Wang, L., Ichikawa, Y., Jaenisch, R., Hooper, M. L., *et al.* (1992). Mutations in T-cell antigen receptor genes alpha and beta block thymocyte development at different stages. *Nature* **360,** 225–231.

Moran, A. E., Holzapfel, K. L., Xing, Y., Cunningham, N. R., Maltzman, J. S., Punt, J., and Hogquist, K. A. (2011). T cell receptor signal strength in Treg and iNKT cell development demonstrated by a novel fluorescent reporter mouse. *J. Exp. Med.* **208,** 1279–1289.

Mucida, D., Pino-Lagos, K., Kim, G., Nowak, E., Benson, M. J., Kronenberg, M., Noelle, R. J., and Cheroutre, H. (2009). Retinoic acid can directly promote TGF-beta-mediated Foxp3 (+) Treg cell conversion of naive T cells. *Immunity* **30,** 471–472.

Nakayama, M., Abiru, N., Moriyama, H., Babaya, N., Liu, E., Miao, D., Yu, L., Wegmann, D. R., Hutton, J. C., Elliott, J. F., and Eisenbarth, G. S. (2005). Prime role for an insulin epitope in the development of type 1 diabetes in NOD mice. *Nature* **435,** 220–223.

Naserke, H. E., Bonifacio, E., and Ziegler, A. G. (1999). Immunoglobulin G insulin autoantibodies in BABYDIAB offspring appear postnatally: Sensitive early detection using a protein A/G-based radiobinding assay. *J. Clin. Endocrinol. Metab.* **84,** 1239–1243.

Nishizuka, Y., and Sakakura, T. (1969). Thymus and reproduction: Sex-linked dysgenesia of the gonad after neonatal thymectomy in mice. *Science* **166,** 753–755.

Ochs, H. D., Ziegler, S. F., and Torgerson, T. R. (2005). FOXP3 acts as a rheostat of the immune response. *Immunol. Rev.* **203,** 156–164.

Ohki, H., Martin, C., Corbel, C., Coltey, M., and Le Douarin, N. M. (1987). Tolerance induced by thymic epithelial grafts in birds. *Science* **237,** 1032–1035.

Onengut-Gumuscu, S., and Concannon, P. (2006). Recent advances in the immunogenetics of human type 1 diabetes. *Curr. Opin. Immunol.* **18,** 634–638.

Orban, T., Farkas, K., Jalahej, H., Kis, J., Treszl, A., Falk, B., Reijonen, H., Wolfsdorf, J., Ricker, A., Matthews, J. B., Tchao, N., Sayre, P., *et al.* (2010). Autoantigen-specific regulatory T cells induced in patients with type 1 diabetes mellitus by insulin B-chain immunotherapy. *J. Autoimmun.* **34,** 408–415.

Ouyang, W., Beckett, O., Ma, Q., and Li, M. O. (2010a). Transforming growth factor-beta signaling curbs thymic negative selection promoting regulatory T cell development. *Immunity* **32,** 642–653.

Ouyang, W., Beckett, O., Ma, Q., Paik, J. H., DePinho, R. A., and Li, M. O. (2010b). Foxo proteins cooperatively control the differentiation of Foxp3+ regulatory T cells. *Nat. Immunol.* **11,** 618–627.

Pacholczyk, R., Ignatowicz, H., Kraj, P., and Ignatowicz, L. (2006). Origin and T cell receptor diversity of Foxp3+CD4+CD25+ T cells. *Immunity* **25,** 249–259.

Patterson, C. C., Dahlquist, G. G., Gyurus, E., Green, A., and Soltesz, G. (2009). Incidence trends for childhood type 1 diabetes in Europe during 1989–2003 and predicted new cases 2005–20: A multicentre prospective registration study. *Lancet* **373,** 2027–2033.

Polansky, J. K., Kretschmer, K., Freyer, J., Floess, S., Garbe, A., Baron, U., Olek, S., Hamann, A., von Boehmer, H., and Huehn, J. (2008). DNA methylation controls Foxp3 gene expression. *Eur. J. Immunol.* **38,** 1654–1663.

Powell, J. D., and Delgoffe, G. M. (2010). The mammalian target of rapamycin: Linking T cell differentiation, function, and metabolism. *Immunity* **33,** 301–311.

Powrie, F., and Mason, D. (1990). OX-22high CD4+ T cells induce wasting disease with multiple organ pathology: Prevention by the OX-22low subset. *J. Exp. Med.* **172,** 1701–1708.

Powrie, F., Leach, M. W., Mauze, S., Caddle, L. B., and Coffman, R. L. (1993). Phenotypically distinct subsets of CD4+ T cells induce or protect from chronic intestinal inflammation in C. B-17 scid mice. *Int. Immunol.* **5,** 1461–1471.

Pugliese, A., Zeller, M., Fernandez, A., Jr., Zalcberg, L. J., Bartlett, R. J., Ricordi, C., Pietropaolo, M., Eisenbarth, G. S., Bennett, S. T., and Patel, D. D. (1997). The insulin gene is transcribed in the human thymus and transcription levels correlated with allelic variation at the INS VNTR-IDDM2 susceptibility locus for type 1 diabetes. *Nat. Genet.* **15,** 293–297.

Ramsdell, F., and Fowlkes, B. J. (1990). Clonal deletion versus clonal anergy: The role of the thymus in inducing self tolerance. *Science* **248,** 1342–1348.

Rocha, B., and von Boehmer, H. (1991). Peripheral selection of the T cell repertoire. *Science* **251,** 1225–1228.

Rocha, B., Tanchot, C., and von Boehmer, H. (1993). Clonal anergy blocks in vivo growth of mature T cells and can be reversed in the absence of antigen. *J. Exp. Med.* **177,** 1517–1521.

Rubtsov, Y. P., Niec, R. E., Josefowicz, S., Li, L., Darce, J., Mathis, D., Benoist, C., and Rudensky, A. Y. (2010). Stability of the regulatory T cell lineage in vivo. *Science* **329,** 1667–1671.

Sakaguchi, S., Takahashi, T., and Nishizuka, Y. (1982a). Study on cellular events in post-thymectomy autoimmune oophoritis in mice. II. Requirement of Lyt-1 cells in normal female mice for the prevention of oophoritis. *J. Exp. Med.* **156,** 1577–1586.

Sakaguchi, S., Takahashi, T., and Nishizuka, Y. (1982b). Study on cellular events in post-thymectomy autoimmune oophoritis in mice. I. Requirement of Lyt-1 effector cells for oocytes damage after adoptive transfer. *J. Exp. Med.* **156,** 1565–1576.

Sakaguchi, S., Fukuma, K., Kuribayashi, K., and Masuda, T. (1985). Organ-specific autoimmune diseases induced in mice by elimination of T cell subset. I. Evidence for the active participation of T cells in natural self-tolerance; deficit of a T cell subset as a possible cause of autoimmune disease. *J. Exp. Med.* **161,** 72–87.

Sakaguchi, S., Sakaguchi, N., Asano, M., Itoh, M., and Toda, M. (1995). Immunologic self-tolerance maintained by activated T cells expressing IL-2 receptor alpha-chains (CD25). Breakdown of a single mechanism of self-tolerance causes various autoimmune diseases. *J. Immunol.* **155,** 1151–1164.

Salaun, J., Bandeira, A., Khazaal, I., Calman, F., Coltey, M., Coutinho, A., and Le Douarin, N. M. (1990). Thymic epithelium tolerizes for histocompatibility antigens. *Science* **247,** 1471–1474.

Salomon, B., Lenschow, D. J., Rhee, L., Ashourian, N., Singh, B., Sharpe, A., and Bluestone, J. A. (2000). B7/CD28 costimulation is essential for the homeostasis of the CD4+CD25+ immunoregulatory T cells that control autoimmune diabetes. *Immunity* **12,** 431–440.

Sarbassov, D. D., Ali, S. M., Sengupta, S., Sheen, J. H., Hsu, P. P., Bagley, A. F., Markhard, A. L., and Sabatini, D. M. (2006). Prolonged rapamycin treatment inhibits mTORC2 assembly and Akt/PKB. *Mol. Cell* **22,** 159–168.

Sauer, S., Bruno, L., Hertweck, A., Finlay, D., Leleu, M., Spivakov, M., Knight, Z. A., Cobb, B. S., Cantrell, D., O'Connor, E., Shokat, K. M., Fisher, A. G., *et al.* (2008). T cell receptor signaling controls Foxp3 expression via PI3K, Akt, and mTOR. *Proc. Natl. Acad. Sci. USA* **105,** 7797–7802.

Schmelzle, T., and Hall, M. N. (2000). TOR, a central controller of cell growth. *Cell* **103,** 253–262.

Schorle, H., Holtschke, T., Hunig, T., Schimpl, A., and Horak, I. (1991). Development and function of T cells in mice rendered interleukin-2 deficient by gene targeting. *Nature* **352,** 621–624.

Sethi, D. K., Schubert, D. A., Anders, A. K., Heroux, A., Bonsor, D. A., Thomas, C. P., Sundberg, E. J., Pyrdol, J., and Wucherpfennig, K. W. (2011). A highly tilted binding mode by a self-reactive T cell receptor results in altered engagement of peptide and MHC. *J. Exp. Med.* **208,** 91–102.

Setoguchi, R., Hori, S., Takahashi, T., and Sakaguchi, S. (2005). Homeostatic maintenance of natural Foxp3(+) CD25(+) CD4(+) regulatory T cells by interleukin (IL)-2 and induction of autoimmune disease by IL-2 neutralization. *J. Exp. Med.* **201,** 723–735.

Singh, N. J., and Schwartz, R. H. (2003). The strength of persistent antigenic stimulation modulates adaptive tolerance in peripheral CD4+ T cells. *J. Exp. Med.* **198,** 1107–1117.

Skokos, D., Shakhar, G., Varma, R., Waite, J. C., Cameron, T. O., Lindquist, R. L., Schwickert, T., Nussenzweig, M. C., and Dustin, M. L. (2007). Peptide-MHC potency governs dynamic interactions between T cells and dendritic cells in lymph nodes. *Nat. Immunol.* **8,** 835–844.

Smith-Garvin, J. E., Koretzky, G. A., and Jordan, M. S. (2009). T cell activation. *Annu. Rev. Immunol.* **27,** 591–619.

Stadinski, B. D., Delong, T., Reisdorph, N., Reisdorph, R., Powell, R. L., Armstrong, M., Piganelli, J. D., Barbour, G., Bradley, B., Crawford, F., Marrack, P., Mahata, S. K., *et al.* (2010a). Chromogranin A is an autoantigen in type 1 diabetes. *Nat. Immunol.* **11,** 225–231.

Stadinski, B. D., Zhang, L., Crawford, F., Marrack, P., Eisenbarth, G. S., and Kappler, J. W. (2010b). Diabetogenic T cells recognize insulin bound to IAg7 in an unexpected, weakly binding register. *Proc. Natl. Acad. Sci. USA* **107,** 10978–10983.

Stefanova, I., Hemmer, B., Vergelli, M., Martin, R., Biddison, W. E., and Germain, R. N. (2003). TCR ligand discrimination is enforced by competing ERK positive and SHP-1 negative feedback pathways. *Nat. Immunol.* **4,** 248–254.

Steinman, R. M., Hawiger, D., and Nussenzweig, M. C. (2003). Tolerogenic dendritic cells. *Annu. Rev. Immunol.* **21,** 685–711.

Sun, C. M., Hall, J. A., Blank, R. B., Bouladoux, N., Oukka, M., Mora, J. R., and Belkaid, Y. (2007). Small intestine lamina propria dendritic cells promote de novo generation of Foxp3 T reg cells via retinoic acid. *J. Exp. Med.* **204,** 1775–1785.

Suzuki, H., Kundig, T. M., Furlonger, C., Wakeham, A., Timms, E., Matsuyama, T., Schmits, R., Simard, J. J., Ohashi, P. S., Griesser, H., *et al.* (1995). Deregulated T cell activation and autoimmunity in mice lacking interleukin-2 receptor beta. *Science* **268,** 1472–1476.

Suzuki, H., Zhou, Y. W., Kato, M., Mak, T. W., and Nakashima, I. (1999). Normal regulatory alpha/beta T cells effectively eliminate abnormally activated T cells lacking the interleukin 2 receptor beta in vivo. *J. Exp. Med.* **190,** 1561–1572.

Tai, X., Cowan, M., Feigenbaum, L., and Singer, A. (2005). CD28 costimulation of developing thymocytes induces Foxp3 expression and regulatory T cell differentiation independently of interleukin 2. *Nat. Immunol.* **6,** 152–162.

Tang, Q., Henriksen, K. J., Boden, E. K., Tooley, A. J., Ye, J., Subudhi, S. K., Zheng, X. X., Strom, T. B., and Bluestone, J. A. (2003). Cutting edge: CD28 controls peripheral homeostasis of CD4+CD25+ regulatory T cells. *J. Immunol.* **171,** 3348–3352.

Tivol, E. A., Borriello, F., Schweitzer, A. N., Lynch, W. P., Bluestone, J. A., and Sharpe, A. H. (1995). Loss of CTLA-4 leads to massive lymphoproliferation and fatal multiorgan tissue destruction, revealing a critical negative regulatory role of CTLA-4. *Immunity* **3,** 541–547.

Todd, J. A., Bell, J. I., and McDevitt, H. O. (1987). HLA-DQ beta gene contributes to susceptibility and resistance to insulin-dependent diabetes mellitus. *Nature* **329,** 599–604.

Tone, Y., Furuuchi, K., Kojima, Y., Tykocinski, M. L., Greene, M. I., and Tone, M. (2008). Smad3 and NFAT cooperate to induce Foxp3 expression through its enhancer. *Nat. Immunol.* **9,** 194–202.

Turley, S., Poirot, L., Hattori, M., Benoist, C., and Mathis, D. (2003). Physiological beta cell death triggers priming of self-reactive T cells by dendritic cells in a type-1 diabetes model. *J. Exp. Med.* **198,** 1527–1537.

van der Merwe, P. A., and Dushek, O. (2011). Mechanisms for T cell receptor triggering. *Nat. Rev. Immunol.* **11,** 47–55.

van Santen, H. M., Benoist, C., and Mathis, D. (2004). Number of T reg cells that differentiate does not increase upon encounter of agonist ligand on thymic epithelial cells. *J. Exp. Med.* **200,** 1221–1230.

Vang, T., Miletic, A. V., Arimura, Y., Tautz, L., Rickert, R. C., and Mustelin, T. (2008). Protein tyrosine phosphatases in autoimmunity. *Annu. Rev. Immunol.* **26,** 29–55.

Verginis, P., McLaughlin, K. A., Wucherpfennig, K. W., von Boehmer, H., and Apostolou, I. (2008). Induction of antigen-specific regulatory T cells in wild-type mice: Visualization and targets of suppression. *Proc. Natl. Acad. Sci. USA* **105,** 3479–3484.

von Andrian, U. H., and Mempel, T. R. (2003). Homing and cellular traffic in lymph nodes. *Nat. Rev. Immunol.* **3,** 867–878.

Walker, L. S., Chodos, A., Eggena, M., Dooms, H., and Abbas, A. K. (2003). Antigen-dependent proliferation of CD4+ CD25+ regulatory T cells in vivo. *J. Exp. Med.* **198,** 249–258.

Wan, Y. Y., and Flavell, R. A. (2005). Identifying Foxp3-expressing suppressor T cells with a bicistronic reporter. *Proc. Natl. Acad. Sci. USA* **102,** 5126–5131.

Waterhouse, P., Penninger, J. M., Timms, E., Wakeham, A., Shahinian, A., Lee, K. P., Thompson, C. B., Griesser, H., and Mak, T. W. (1995). Lymphoproliferative disorders with early lethality in mice deficient in Ctla-4. *Science* **270,** 985–988.

Webster, K. E., Walters, S., Kohler, R. E., Mrkvan, T., Boyman, O., Surh, C. D., Grey, S. T., and Sprent, J. (2009). In vivo expansion of T reg cells with IL-2-mAb complexes: Induction of resistance to EAE and long-term acceptance of islet allografts without immunosuppression. *J. Exp. Med.* **206,** 751–760.

Wegmann, D. R., Norbury-Glaser, M., and Daniel, D. (1994). Insulin-specific T cells are a predominant component of islet infiltrates in pre-diabetic NOD mice. *Eur. J. Immunol.* **24,** 1853–1857.

Wirnsberger, G., Mair, F., and Klein, L. (2009). Regulatory T cell differentiation of thymocytes does not require a dedicated antigen-presenting cell but is under T cell-intrinsic developmental control. *Proc. Natl. Acad. Sci. USA* **106,** 10278–10283.

Wohlfert, E. A., Gorelik, L., Mittler, R., Flavell, R. A., and Clark, R. B. (2006). Cutting edge: Deficiency in the E3 ubiquitin ligase Cbl-b results in a multifunctional defect in T cell TGF-beta sensitivity in vitro and in vivo. *J. Immunol.* **176,** 1316–1320.

Wong, P., Goldrath, A. W., and Rudensky, A. Y. (2000). Competition for specific intrathymic ligands limits positive selection in a TCR transgenic model of CD4+ T cell development. *J. Immunol.* **164,** 6252–6259.

Wucherpfennig, K. W., and Sethi, D. (2011). T cell receptor recognition of self and foreign antigens in the induction of autoimmunity. *Semin. Immunol.* **23**(2), 84–91.

Yao, Z., Kanno, Y., Kerenyi, M., Stephens, G., Durant, L., Watford, W. T., Laurence, A., Robinson, G. W., Shevach, E. M., Moriggl, R., Hennighausen, L., Wu, C., *et al.* (2007). Nonredundant roles for Stat5a/b in directly regulating Foxp3. *Blood* **109,** 4368–4375.

Zeng, Z., Sarbassov, D. D., Samudio, I. J., Yee, K. W., Munsell, M. F., Ellen, J. C., Giles, F. J., Sabatini, D. M., Andreeff, M., and Konopleva, M. (2007). Rapamycin derivatives reduce mTORC2 signaling and inhibit AKT activation in AML. *Blood* **109,** 3509–3512.

Zhang, Z. J., Davidson, L., Eisenbarth, G., and Weiner, H. L. (1991). Suppression of diabetes in nonobese diabetic mice by oral administration of porcine insulin. *Proc. Natl. Acad. Sci. USA* **88,** 10252–10256.

Zheng, S. G., Gray, J. D., Ohtsuka, K., Yamagiwa, S., and Horwitz, D. A. (2002). Generation ex vivo of TGF-beta-producing regulatory T cells from CD4+CD25- precursors. *J. Immunol.* **169,** 4183–4189.

Zheng, Y., Josefowicz, S. Z., Kas, A., Chu, T. T., Gavin, M. A., and Rudensky, A. Y. (2007). Genome-wide analysis of Foxp3 target genes in developing and mature regulatory T cells. *Nature* **445,** 936–940.

Zheng, Y., Josefowicz, S., Chaudhry, A., Peng, X. P., Forbush, K., and Rudensky, A. Y. (2010). Role of conserved non-coding DNA elements in the Foxp3 gene in regulatory T-cell fate. *Nature* **463,** 808–812.

Ziegler, A. G., and Nepom, G. T. (2010). Prediction and pathogenesis in type 1 diabetes. *Immunity* **32,** 468–478.

INDEX

CONTENTS OF RECENT VOLUMES

Volume 88

Volume 89

Volume 90

Volume 91

Volume 92

Volume 93

Volume 94

Volume 95

Volume 96

Volume 97

Volume 98

Volume 99

Volume 100

Volume 101

Volume 102

Volume 103

Volume 104

Volume 105

Volume 106

Volume 107

Volume 108

Volume 109

Volume 110

Volume 111

FSC
www.fsc.org
MIX
Papier | Fördert
gute Waldnutzung
FSC® C083411

Kolibri 360 GmbH
Ferdinand-Jühlke-Straße 7
99095 Erfurt, Deutschland
produktsicherheit@kolibri360.de